A History of Global Health

A History *of* Global Health

Interventions into the Lives of Other Peoples

Randall M. Packard
Institute for the History of Medicine
Johns Hopkins University

Johns Hopkins University Press | *Baltimore*

Printed in the United States of America on acid-free paper
9 8 7 6 5 4 3 2

Johns Hopkins University Press
2715 North Charles Street
Baltimore, Maryland 21218-4363
www.press.jhu.edu

Library of Congress Cataloging-in-Publication Data
Names: Packard, Randall M., 1945– , author.
Title: A history of global health : interventions into the lives of other peoples / Randall M. Packard.
Description: Baltimore : Johns Hopkins University Press, 2016. | Includes bibliographical references and index. | Description based on print version record and CIP data provided by publisher; resource not viewed.
Identifiers: LCCN 2015042993 (print) | LCCN 2015041793 (ebook) | ISBN 9781421420349 (electronic) | ISBN 1421420341 (electronic) | ISBN 9781421420325 (hardcover : alk. paper) | ISBN 1421420325 (hardcover : alk. paper) | ISBN 9781421420332 (pbk. : alk. paper) | ISBN 1421420333 (pbk. : alk. paper)
Subjects: | MESH: Global Health—history. | History, 19th Century. | History, 20th Century. | History, 21st Century.
Classification: LCC RA441 (print) | LCC RA441 (ebook) | NLM WA 11.1 | DDC 362.1—dc23
LC record available at http://lccn.loc.gov/2015042993

A catalog record for this book is available from the British Library.

For my grandsons, Max and Dom, who I hope will live in a world in which "health for all" is more than a slogan

CONTENTS

ABBREVIATIONS

ARV	antiretroviral drug
BCG	bacille Calmette-Guérin (vaccine)
CCM	country coordinating mechanism
CDC	Centers for Disease Control and Prevention
CRHP	Comprehensive Rural Health Project (Jamkhed)
CVI	Children's Vaccine Initiative
DALY	disability-adjusted life-year
DDT	dichlorodiphenyltrichloroethane
DEVTA	deworming and enhanced vitamin A
DSP	Departamento do Salud Pública/Department of Public Health (Mexico)
ECLA	Economic Commission for Latin America
EPI	Expanded Programme on Immunization
EPTA	Expanded Programme for Technical Assistance
FAO	Food and Agriculture Organization
FSA	Farm Security Administration
GATT	General Agreement on Tariffs and Trade
GAVI	Global Alliance for Vaccines and Immunizations
GFF	Global Financing Facility (World Bank)
GOBI	growth monitoring, oral rehydration, breastfeeding, and immunization
GPA	Global Program on AIDS
GPEI	Global Polio Eradication Initiative
HAART	highly active antiretroviral therapy
HSS	health-systems strengthening
IC	Interim Commission
ICA	International Cooperation Administration
IHB	International Health Board (Rockefeller Foundation)
IHD	International Health Division (Rockefeller Foundation)
IIAA	Institute of Inter-American Affairs, Health Division
ILO	International Labour Organization

IMF	International Monetary Fund
IPPF	International Planned Parenthood Federation
IPV	inactivated polio vaccine
IRB	institutional review board
ISA	Institute for Social Anthropology (Smithsonian Institution)
ITC	International Tuberculosis Campaign
ITN	insecticide-treated bed net
KAP	knowledge, attitudes, and practices
LFA	local fund agent
LNHO	League of Nations Health Organization
MEP	Malaria Eradication Programme
MSF	Médicins Sans Frontières/Doctors Without Borders
NGO	nongovernmental organization
NPC	National Planning Committee (India)
OFRRO	Office of Foreign Relief and Rehabilitation Operations
OPR	Office of Population Research (Princeton University)
OPV	oral polio vaccine
PAHO	Pan American Health Organization
PASB	Pan American Sanitary Bureau
PEPFAR	President's Emergency Plan for AIDS Relief
PHC	primary health care
PIH	Partners In Health
PrEP	pre-exposure prophylactic (drug therapy)
RBM	Roll Back Malaria Partnership
RCT	randomized controlled trial
SAP	structural adjustment policy
SEP	Smallpox Eradication Programme
SPHC	selective primary health care
TPC	Technical Preparatory Committee
UNDP	United Nations Development Programme
UNESCO	United Nations Educational, Scientific and Cultural Organization
UNFPA	United Nations Fund for Population Activities
UNICEF	United Nations International Children's Emergency Fund
UNRRA	United Nations Relief and Rehabilitation Administration
USAID	United States Agency for International Development
VAC	vitamin A capsule
WFB	World Food Board
WHA	World Health Assembly
WHO	World Health Organization
WTO	World Trade Organization

ILLUSTRATIONS AND TABLES

Illustrations

Tables

A History of Global Health

INTRODUCTION

Ebola

In early December 2013, a two-year-old boy named Emile Ouomuono became very ill. He had a fever and black stools and was vomiting. Emile lived in the village of Meliandou in southeast Guinea, a few miles from the country's borders with Liberia and Sierra Leone. Four days after developing these symptoms, he died. Emile was buried in his village. Soon afterward, on December 13, his mother developed similar symptoms and also died. The boy's three-year-old sister died with the same symptoms on December 29. His grandmother, in whose house Emile had stayed when he became ill, also developed symptoms of the disease. She sought the help of a male nurse who lived in the town of Guéckédou, a bustling trading hub where people converged from Liberia and Sierra Leone. The nurse tried to offer assistance but possessed neither the knowledge nor the medical supplies needed to treat her. The grandmother returned to Meliandou, where she died on January 1. In early February, the nurse who had treated Emile's grandmother also developed a fever. When the nurse's condition deteriorated, he sought help from a doctor in Macenta, in the next prefecture. The nurse stayed just one night in Macenta—sleeping in the doctor's house—and died the next day. The doctor died several days later.[1]

Whatever was killing the residents of this remote part of Guinea was on the move. But word of the outbreak did not reach health officials in Conakry until January 24, when a doctor in the town of Tekolo called a superior to report that something strange was happening in a village under his jurisdiction. It took another six weeks before a team of physicians from the humanitarian relief organization Médecins Sans Frontières (Doctors Without Borders) was able to reach Guéckédou and collect samples, which were subsequently tested in a laboratory in Lyon, France. On March 20, the lab identified the condition as Ebola. By then the disease had reached Conakry and was spreading farther, into Liberia and Sierra Leone. On March 25, the US Centers for Disease Control and Prevention (CDC) confirmed that an outbreak of Ebola fever was occurring in Liberia, Guinea, and Sierra Leone.[2]

The disease continued to spread. By December 2014, there were 12,000 laboratory-confirmed cases in the three affected countries, and additional cases had occurred in Mali, Senegal, and Nigeria. More than 7,000 people had died from the disease. The outbreak was the largest since the disease first appeared in Zaire in the 1970s, and it evoked a massive emergency response from global-health organizations trying desperately to contain the outbreak before it spread across the globe.

Ebola is a terrifying disease. In its most overt form it produces horrific symptoms that include high fever, vomiting, diarrhea, and massive hemorrhaging. The majority of infected people, if not properly treated, die from the disease within days of developing symptoms. Yet Ebola often presents itself with less clear symptoms, which makes it difficult to distinguish from other diseases, including malaria and cholera. Without a laboratory test, it is impossible to make a definitive diagnosis.

Ebola does not spread easily. A person infected with the virus can remain symptomless for 10–21 days. During this time, he or she is not infectious to others. The disease only spreads through direct contact with an infected person's bodily fluids once symptoms have begun to appear. But because Ebola is a severely debilitating disease, which kills quickly, a person who develops symptoms normally has limited opportunities to infect others. Despite these constraints, Ebola spread rapidly from Guinea to neighboring countries. Why?

The answer to this question is complicated, involving a number of converging factors. At the beginning of the epidemic, much of the media's and public-health groups' attention focused on the local customs and behaviors of the population affected by the epidemic. We were told about the local consumption of "bushmeat," which may have infected the first person who contracted the disease; local aversion to using Western medical services; and burial practices that brought family and friends into close contact with diseased bodies. All of these behaviors occurred and no doubt played a role in the epidemic. But they were not why Ebola burned rapidly through these three countries. Moreover, focusing on these behaviors deflected attention from other, more fundamental causes of the epidemic and represented examples of cultural modeling and victim blaming, which have been part of how those living in the global north have often made sense of the health problems of those living in the global south.[3]

The epidemic was also fueled by the poverty both of those it affected and of the countries in which it occurred. The forest area in southeastern Guinea, where the epidemic first appeared, was one of the poorest in the world. There was no industry or electricity, and roads were nearly impassable at certain times of year. Families existed on a combination of agriculture, foraging, and hunting in the forest. Average annual earnings were less than US$1 a day, and 45 percent of the children suffered chronic malnutrition. Local resources were further depleted by the influx of thousands of refugees from the civil wars that wracked Liberia and Sierra Leone. The recent expansion of agribusiness and mining had contributed to deforestation and forced residents to forage and hunt deeper in the forest, which may have increased their exposure to the animal-borne viruses. Finally, the need for local residents to seek economic opportunities outside of this impoverished forest zone produced patterns of labor migration that contributed to the spread of the epidemic.

But the most immediate cause for the rapid spread of Ebola was the absence of functioning health systems. Ebola was able to go undiagnosed for three months and to spread rapidly throughout the region because the three countries in which it initially occurred lacked basic health-care services. The medical workers who first encountered the disease possessed neither the supplies nor, in some instances, the medical knowledge needed to manage Ebola cases. There is no cure for Ebola, but basic palliative care—including, most importantly, intravenous fluids for rehydrating patients—can dramatically reduce mortality; 90 percent of the patients who were transported to hospitals in Europe and the United States survived. Clinics and even hospitals in the three countries most affected by Ebola lacked these basic medical supplies. Health workers were also unprepared to control the spread of infection, lacking a basic infrastructure and the necessary protective equipment. Patients were not isolated, and health workers took few precautions to protect themselves from being infected.

Not surprisingly, health workers were among the first to contract the disease and die from it, further weakening already-stretched health systems and contributing to the spread of the disease.[4] Moreover, some health workers, especially those who had not been paid for weeks, abandoned their posts for fear of dying from the disease. As Ebola spread, clinics and hospitals became places where people with the disease went to die. The association of Ebola deaths with health services caused family caregivers to avoid these services and treat their loved ones at home. This decision contributed to the spread of the disease among family caretakers. Health services, far from containing the outbreak, often amplified it.

The epidemic spread as well because the health services that existed were not integrated into the communities they served. The early spread of the disease was linked to local burial practices. Burial activities, like burial practices everywhere, are deeply imbedded in local cultural systems. But health workers might have intervened in ways that would have limited the spread of infection from the diseased bodies of Ebola victims to mourning relatives, had the health workers been viewed as part of the community. Only after the epidemic was well under way were local efforts made to work with communities and modify their burial practices.

In addition to lacking fully prepared health services, Guinea, Liberia, and Sierra Leone did not have laboratory services to test for Ebola or epidemiological services to keep track of infected persons. The absence of a network of laboratories that could have tested for Ebola, and the difficulties of conveying samples to distant medical centers, delayed both recognition of what was killing people and efforts to mount an effective response to the disease. Lack of knowledge of the nature and seriousness of the epidemic also meant that people who were infected with Ebola but had not yet developed symptoms were able to travel to neighboring regions and, eventually, to large metropolitan areas like Conakry and Monrovia, where they succumbed to the disease, becoming foci for its farther spread. Only after it was recognized that the region faced an epidemic of Ebola were efforts begun to track patient contacts and contain the spread of the disease.

The CDC summarized the state of health service unpreparedness in Liberia in October 2014:

> Ebola emergency preparedness plans at the county and hospital level were lacking. . . . In all counties, there was insufficient personal protective equipment to care for patients with Ebola. Health care providers had not received training on the donning and removal of personal protective equipment. No training on case investigation, case management, contact tracing, or safe burial practices had been provided at either the county or hospital level. No Ebola surveillance systems were in place.[5]

As Anthony Fauci of the National Institutes of Health noted in December 2014, "if the West African countries stricken by the current Ebola outbreak had a reasonable health-care infrastructure, the outbreak would not have gotten out of control."[6]

The health systems in the rural areas of Guinea, Liberia, and Sierra Leone were among the weakest in the world. Thousands of children died every year from malaria and respiratory and diarrheal diseases, due to a lack of available medical care. The three countries also had some of the highest infant mortality rates in the world. Yet their medical systems shared characteristics with rural health services in many countries in Africa and other parts of the global south in terms of their lack of trained medical personnel, supplies, and infrastructure. In India, for example, 66 percent of the rural population do not have access to preventive medicines and 33 percent have to travel over 30 kilometers to get needed medical treatments. Also, 3660 of that country's primary health centers lack either an operating theater or a lab, or both. The poor condition of state-run health services has led 80 percent of the patients to seek care from private-sector health providers. In the rural areas, however, medically trained physicians provide few of these private services.[7] A survey in rural Madhya Pradesh found that 67 percent of health-care providers had no medical qualifications at all.[8] The poor quality of rural health services has contributed to major health disparities between urban health and rural health. The infant mortality, neonatal mortality, and prenatal mortality rates in India's rural areas are nearly double those in urban areas.[9] In many places, urban health services, while possessing more facilities, are also plagued by limitations in supplies, infrastructure, and medical personnel.

The Ebola epidemic that occurred in West Africa in 2014 and 2015 was a symptom of a larger, global health-care crisis. At the best of times in many countries in Africa, Asia, and Latin America, clinical coverage is inadequate, forcing patients to travel long distances for treatment. Once they get to medical facilities, they find drug shortages or outages of basic medicines to treat common health problems. Basic equipment, such as disposable gloves and syringes, rehydration fluids, and bandages, is unavailable. Patients or their family caretakers are required to purchase what is needed outside the hospital. Staff are underpaid or not paid for months on end, encouraging them to set up parallel private services on the side instead of attending to public-sector patients. The ties between local health services and the communities they serve are often tenuous, at best. It is not surprising that local populations have little faith in these services. Health systems also lack the surveillance capabilities, including testing laboratories, necessary to track and report emerging epidemics in a timely manner. In short, health services in many parts of the world lack what Partners In Health cofounder Paul Farmer refers to as the four *S*'s: "Staff, Stuff, Space, and Systems."[10]

These conditions tend to be viewed by those in the global north as just the natural state of affairs in so-called underdeveloped parts of the world: "It's Africa or India, after all. What do you expect?" Yet these conditions have a history. They are the product of the unwillingness or inability of governments to fund health services, particularly in rural areas. Health systems have also been weakened by civil wars, such as those that occurred in Sierra Leone and Liberia before the Ebola outbreak, that continue to disrupt large areas of the globe. But they are also the product of a long history of neglect on the part of multinational and bilateral aid donors, including the United States. It needs to be pointed out that over the 10 years preceding the Ebola epidemic, Liberia, Guinea, and Sierra Leone received nearly US$1 billion—directed toward improving health conditions—from the President's Emergency Plan for AIDS Relief; the President's Malaria Initiative; the World Bank; and the Global Fund to Fight AIDS, Tuberculosis and Malaria. Yet very little of this aid was directed toward training health workers or building a health-care infrastructure.

Little attention has been given, as well, to the underlying determinants of ill health. Many rural regions may have access to clinics but lack basic sanitation or a clean water supply. Sewage, contaminated water, overcrowded housing, and unhealthy diets also haunt the populations of many low-income countries. These conditions continue to undermine the health of millions of people across the globe.

Investments in efforts to improve the health of peoples living in resource-poor countries (what we now call global health) by governments and organizations situated in the global north have focused primarily on the application of biomedical technologies—vaccines, antiretroviral drugs, insecticide-treated bed nets, vitamin A capsules—to eliminate specific health problems through vertically organized programs. Yet these programs are only loosely connected to the recipients' national health systems, which struggle to provide even the most basic forms of care.

Today, global health is a multibillion-dollar enterprise. It is driven by growing recognition of the interconnectedness of the world's populations; fears that deadly diseases like Ebola or avian flu can rapidly spread around the globe; the threat of bioterrorism; humanitarian concerns for the well-being of the world's poor; a desire to reduce global inequalities in health; efforts to promote economic development; and the political and economic interests of donor countries. It is funded by large multinational organizations, such as the World Bank, UNICEF, the World Health Organization, and the Global Fund to Fight AIDS, Tuberculosis and Malaria. It is also supported by bilateral organizations, including the United States' Agency for International Development, Britain's Department for International Development, and China's Department of Foreign Aid. Public/private partnerships, such as the Global Alliance for Vaccines and Immunizations, and private philanthropies, including the Bill & Melinda Gates Foundation, also fund global-health efforts.

These organizations employ an army of researchers, program developers, physicians, project officers, health educators, project evaluators, and health workers, scattered across the world. They also support a vast development industry, including thousands of nongovernmental organizations (NGOs) and development contractors, as well as pharmaceutical and chemical corporations that produce the biomedical technologies that have become the cornerstone of global-health interventions.

These organizations also support schools of public health that train students who will join the ranks of global-health workers and have faculty whose research depends on global-health funding. Global health has become a major focus of higher education. Across the United States and in many European countries, thousands of students take courses in global health every year and sign up to work in short-term health projects in Africa, Asia, and Latin America. Medical students, in particular, view short-term residencies in low-income countries as a kind of necessary rite of passage. These students, many of whom have been inspired by the work of Paul Farmer and Doctors Without Borders, hope to make a difference in the lives of others.

At its core, global health is about efforts to improve the health of peoples living in countries that used to be called underdeveloped, or Third World, and are now known as low-income countries. These efforts are not new. They have a history that stretches back to the early twentieth century. The current scale and complexity of global-health assistance is unprecedented, but the central motivations, organizing principles, and modes of operation that characterize it are not. It is the history of these efforts—what used to be called technical assistance or development assistance (under the banner of international health or world health) and, before that, tropical medicine or colonial medicine—that I explore in this book. The volume is an effort to explain why, despite the investment of billions of dollars in programs aimed at improving global health, basic health services, public-health infrastructure, and the underlying social and economic determinants of ill health have received so little attention.

I first became aware of the limits of international-health interventions while serving as a Peace Corps volunteer in a trachoma-eradication campaign in eastern Uganda in the late 1960s. The campaign, which I described briefly in an earlier book,[11] was well intentioned but poorly designed and implemented. It was sponsored by the Uganda Foundation for the Blind, a local nongovernmental organization that had no direct connection to the Ugandan Ministry of Health. It was, like many global-health programs today, a freestanding vertical program. It was staffed by teams of recent college graduates who had no formal public-health or medical training. There were no provisions for training Ugandans to work on the project or take it over after we left. The treatment required patients to attend our clinic once a week for 12 weeks. This was a challenge for many of those who lived miles from the clinic. In addition, the campaign was limited to one district, which was surrounded by other districts where the disease existed but in which there were no trachoma-control programs. It thus ignored the fact that people frequently moved back and forth between districts. Our patients would disappear for weeks at a time, interrupting their treatment. Overall, the program's dropout rate was high. Finally, many of the people in the district lacked the economic resources needed to practice the basic sanitary measures that could prevent infection. It was little wonder that, after two years of work, the estimated prevalence of trachoma in the district had actually risen slightly.

This experience shaped my thinking about global health and influenced the course I have taught since 1992 on the History of International Health and Development.[12] A central argument of the course and this book is that there have been remarkable continuities in how health interventions have been conceived and implemented over the past century. Many of the problems I experienced in Uganda have a history that stretches back to the early twentieth century and still persists today. These trends have worked against the development of effective basic-health systems and efforts to address the social determinants of health:

1. Health interventions have been largely developed outside the countries where the health problems exist, with few attempts to seriously incorporate local perspectives or community participation in the planning process. During the first half of the twentieth century, this planning occurred primarily in the metropolitan capitols of European and American colonial powers. After World War II, new centers of international-health planning and governance emerged in Geneva, Atlanta, and New York City, and at conference centers in Bellagio, Italy; Talloires, France; and Alma-Ata, in the former Soviet Union.
2. Health planning has privileged approaches based on the application of biomedical technologies that prevent or eliminate health problems one at a time.
3. Little attention has been given to supporting the development of basic health services.

4. The planning of health interventions has often occurred in a crisis environment, in which there was an imperative to act fast. This mindset has privileged interventions that are simple, easy to implement, and have the potential to quickly make a significant impact. On the other hand, it has discouraged longer-term interventions aimed at building health infrastructure, training personnel, and addressing the underlying determinants of ill health.
5. Global-health interventions have been empowered by faith in the superiority of Western medical knowledge and technology and a devaluing of the knowledge and abilities of the local populations.
6. Health has been linked to social and economic development, but this connection has focused primarily on how improvements in health can stimulate economic development, while ignoring the impact that social and economic developments have on health. What we now call the social determinants of health have received little attention.

There have been moments when these continuities have broken down, when alternative formulations and approaches to health have been proposed and, to some degree, implemented. Most importantly, the late 1920s and 1930s saw a growing interest in understanding and addressing the social and economic causes of ill health and in developing more-comprehensive approaches to health and health care. A second shift occurred in late 1970s, with the movement to promote "health for all" and primary health care. Yet each time, these movements toward more-comprehensive approaches to health failed to gain widespread support, and the world of global health returned to interventions that reflected the six trends described above. By placing the history of global-health strategies within a broader political and economic history of international development, this book attempts to explain why these continuities have persisted and why they have occasionally broken down, only to reemerge again. By doing so, I hope to provide an historical context for understanding the current global health-care crisis.

The book is based in part on earlier research I have conducted on various aspects of the history of international, or global, health and disease, yet it is largely synthetic, drawing on a wide array of published works in a number of disciplines. The recent outpouring of studies examining the history of international, or global, health has been impressive. Hundreds of books and articles have been published over the past two decades. Many of these studies have attempted to complicate earlier histories of major public-health institutions, such as the Rockefeller Foundation's International Health

Division, the League of Nations Health Organization, and the World Health Organization. They have explored the histories of a wider array of institutions involved in international-health activities; focused on the important role played by national, regional, and local actors in implementing health programs and shaping health strategies; drawn attention to the varieties of ways in which health interventions have impacted local societies; revealed greater levels of variation in how international health institutions operated in different settings; and reexamined earlier assumptions about the motivations, modes of operation, and impact of major health initiatives, such as the global smallpox-eradication campaign of the 1960s and 1970s. A great deal of recent attention has also been paid to the growing role of biomedical research in shaping health interventions.[13]

The production of historical and anthropological studies will continue apace. I have chosen this moment, however—when the limits of efforts to improve the health of peoples living in the global south have been so clearly revealed—to survey the changed landscape of global-health history and construct a historical narrative that links many of the studies that have been produced over the last two decades. I hope this overview will provide those interested in global health, be they policy makers, researchers, program managers, students, or fellow historians, with a better sense of why the field looks the way it does, while also suggesting alternatives to this vision. A number of the interventions described in this book are familiar to those engaged in global health. Yet the longer histories of these interventions and the complex sets of events that have shaped them often are not. By offering a more comprehensive historical narrative, this book will, I hope, encourage those in global health to think more critically about their field.

I recognize from the start that such a broadly synthetic history cannot be comprehensive. Space constraints have required me to choose which themes and examples to include and which to leave out. Not everyone will agree with these choices. For example, global health today encompasses much more than the kind of north–south interventions that I focus on here. There are other players involved in designing and implementing health interventions and clear examples of south–south and even south–north assistance. In addition, global health has come to include domestic health problems within donor countries, localized efforts to improve health within former and current recipient countries, and a much wider range of activities than was part of earlier conceptions of international, or world, health. These are not part of my story.

More importantly, global health is not just about the interventions of bilateral and international organizations. It is about the lives of people in countries across the globe and how they have been touched in one way or another by the efforts of strangers to improve their health. It is also about the roles that local public-health workers and laypeople have played in shaping these interventions and working to improve their own health. The stories of these individuals and their experiences, while occasionally highlighted here, regrettably play a smaller roll in this story than I would have liked. This is largely because there has been much less ground-level scholarship on the history of global health. Recent studies by anthropologists and a few historians have begun to explore these experiences, but a great deal more work needs to be done to recapture them.

Also largely missing from this story are women. It will become immediately apparent that this is a history that is dominated by men. This does not mean that women were not involved in global-health interventions. Women have always played a central role in health care and were a part of many international-health activities. Yet women hardly appear in the histories of public-health programs organized by international or bilateral organizations. With few exceptions, some of which are described in the following chapters, it was not until the 1970s, in the context of family-planning activities, that women began to take on larger leadership roles and not until the 1990s that they became prominent in global-health organizations. The photographs from international-health meetings prior to the 1990s contained few women. As late as 1984, a photograph of participants at the Bellagio children's vaccine meeting contained the images of only two women among the 33 conference participants. Clearly, the role of women in the early history of international health needs more attention.

These are important elements of the history of global health, yet at its core, this history remains predominantly about flows of goods, services, and strategies along well-trod, north–south pathways. This book hopes to provide a clearer understanding of global health in the early twenty-first century by exploring these historical paths. Why have particular ones been chosen while others have been, if not avoided, only occasionally explored? Why have certain sets of ideas, policies, and practices come to dominate global-health interventions at particular points in time? Why have other ideas, policies, and practices had less success in achieving and maintaining purchase? If we can answer these questions, we can begin to understand why people continue to die from preventable diseases and why epidemics like Ebola are able to find fertile ground to expand out to the world.

Finally, while I argue that health interventions have failed to address the basic health needs of people living in many parts of Africa, Asia, and Latin America over the course of the previous century, this book should not be read as an attack on the work that has been carried out by thousands of men and women who have dedicated their lives to improving the health of the world's populations. I may question the choices made in the past, and the political and economic interests that drove them, but I recognize that the selected paths often resulted in improvements in health. Deciding which avenues to follow is never simple. Arguments can be made for moving in different directions. On what basis does one choose between improving health by building sustainable systems of primary health care, or saving lives by distributing insecticide-treated bed nets or vitamin A capsules? In the end, the choices that were made were often shaped by complex sets of historical circumstances. We need to understand these forces and how they have defined and limited global-health interventions. We also need to acknowledge the limitations and consequences of the choices that have been made. It may be time to explore new pathways or, as Paul Farmer and his colleagues have recently suggested, to "reimagine global health."[14] It is my hope that this book will help us do this.

PART I

Colonial Entanglements

On November 15, 1932, representatives from several African colonial territories, British India, the League of Nations Health Organization, and the Rockefeller Foundation met in Cape Town, South Africa. The purpose of the meeting was to discuss questions relating to public-health administration and protection against epidemic diseases. Much of the conference focused on the problem of yellow fever. Growing concerns about the potential spread of yellow fever from its endemic locations in Latin America and west central Africa into other African colonial territories, and from there to South Asia, provided the background for these discussions.

A map labeled "African Air Routes, 1932" accompanied the conference report.[1] This map showed the routes of European airlines crisscrossing the continent and connecting Africa to the wider world. It was intended to illustrate that advances in air travel, which were bringing various parts of the world into closer contact with one another, were also creating pathways along which pathogens—specifically, yellow fever—could travel. The British, who had established colonies from Egypt to South Africa at the end of the nineteenth century, feared that yellow fever could be transported from West Africa to its colonies elsewhere in Africa, and from there to the jewel of its colonial empire in India, where the deadly disease had never been identified but the mosquitoes that transmitted it existed. The impact of this eastward spread of yellow fever would be devastating. The League of Nations Health Organization, which had been established after World War I as part of an effort to prevent future wars by ensuring the health and well-being of the world's populations, viewed the conference as an opportunity to expand its influence into the colonial world. The Rockefeller Foundation, which had established an International Health Board in 1913 to advance the science of hygiene across the globe, had begun a campaign to eradicate yellow fever in Latin America and West Africa after the war. In 1932, it was exploring the limits of the disease in East Africa. The conference thus highlighted the shared concerns of newly established international-health organizations and colonial health officials. It revealed the extent to which the interests of the two groups had become entangled.

Historians have tended to treat the histories of colonial medicine and early international-health organizations as separate topics. Yet recent studies by Warwick Anderson, Steven Palmer, Nancy Stepan, Deborah Neill, Sunil Amrith, and Helen Tilley have suggested that colonial medicine and international health shared a long and complex history, stretching from the end of the nineteenth century up through the 1950s.[2] Social networks linked colonial health authorities with international organizations. Colonial medical authorities and representatives of international-health organizations collaborated in efforts to control sleeping sickness, malaria, hookworm, and yellow fever. They traveled along the air routes represented in the 1932 map, engaging in international meetings like the 1932 Cape Town conference, and consulted with one another. They published articles in the same scientific journals and formed complex research networks that stretched across the globe. Colonial officials moved back and forth, serving at various times in colonial administrations and international-health organizations. Most importantly, the early interventions of international-health organizations were developed within colonial settings and, to a significant degree, were dependent on the coercive power of colonial rule. These interventions were also shaped by colonial ideas about "the pathology of native populations" and the inability of colonial peoples to improve their own health. All of these linkages contributed to the entanglement of colonial medicine and international health and shaped the subsequent development of international health. If we are to understand why disease-focused interventions were privileged over the development of basic health services and efforts to address the social determinants of health, we must start with these colonial beginnings.

Part I examines these interconnections, focusing primarily on the role played by America's colonial possessions in the Caribbean, Panama, and the Philippines in the formation of a set of practices, attitudes, and personnel that would come to inhabit international-health institutions and interventions over the first half of the twentieth century. The linkage between colonial medicine and international health was not limited to the United States, which was, in fact, a latecomer to the colonial division of the globe. European colonial physicians had developed their own ideas and practices to deal with the health of colonial populations. Moreover, similar connections tied European colonial physicians to emerging international organizations in Europe during the early twentieth century (part II). I have chosen to focus on early US involvement in overseas health because the United States, as the largest funder of international-health activities, came to play a dominant role in crafting international-health policies. US economic and political interests shaped many of the strategies that became part of global health, especially during its formative years.

In chapter 1, I describe US medical interventions within its colonial holdings in the Americas and the Philippines. I argue that it was in these colonial settings that a model of health interventions—the disease-elimination campaign—was perfected.[3] While alternative approaches were implemented and had a significant impact on improving health in Panama and Cuba, it was the disease-eradication campaign that became a central element of early international-health efforts.

In chapter 2, I trace the movement of US colonial medical officers into the Rockefeller Foundation's International Health Board (IHB), which grew to become the most influential international-health organization in the first decades of the twentieth century. I focus particularly on the role these former colonial medical officers played in directing the IHB's campaigns against hookworm and yellow fever. I also describe how colonial medical knowledge was incorporated into the training of a generation of international-health workers through schools of tropical medicine and hygiene that the Rockefeller Foundation established in the United States, Europe, Latin America, and Asia. Through these educational activities, the knowledge produced in colonial settings was transferred into the world of international health. Over time, the intersection of colonial medicine and international health transformed colonial ideas and practices into a kind of international expertise about the health problems of the "tropical" or "developing" world and the strategies needed to deal with them. Freed from their colonial moorings, these ideas became available for use in post- and noncolonial settings by international-health experts across the globe. They became naturalized as global-health science.

CHAPTER ONE

Colonial Training Grounds

William Gorgas and Yellow Fever

Yellow fever, which drew the attention of those attending the Pan African Health Conference in Cape Town in 1932, is a good place to begin to understand both the nature of colonial medicine and its entanglement with the emerging field of international health.[1] Yellow fever was a ghastly disease that, in its final stages, resembled Ebola, causing patients to vomit and defecate blood. Epidemics of yellow fever occurred regularly in port cities in the Caribbean, Latin America, and the southern United States during the eighteenth and early nineteenth centuries, creating heavy losses of life and disrupting commerce. The disease also occasionally reached northern port cities in Europe and America.

The causes of yellow fever were debated. Some believed it was a contagious disease that spread from place to place, often on board ships. These authorities called for the establishment of quarantines restricting the movements of ships, to prevent the spread of the disease. Others viewed yellow fever as a product of local environments and poor sanitation. These sanitationists viewed cleaning up environments as the best way to prevent the spread of yellow fever. By the end of the nineteenth century, both approaches were being employed, and nations in Europe and the Americas sought to limit the spread of yellow fever through a combination of international trade regulations, urban sanitation, and various forms of quarantine. In addition, new institutions and international conferences explored ways to prevent the spread of yellow fever. The Pan American Sanitary Bureau, which later became the Pan American Health Organization, was founded in 1902 as an instrument for coordinating hemisphere-wide efforts to prevent the spread of contagious diseases, the most important of which, at the time, was yellow fever. These efforts had only a limited impact on the occurrence of yellow fever.

It was not until the US military occupation of Cuba at the end of the nineteenth century and the discovery of the role of the *Aedes aegyptei* mosquito in the transmission of yellow fever that health authorities were able to effectively prevent and eliminate the disease from the port cities around the Atlantic where it had become a recurring threat. Yellow fever–eradication efforts in Cuba (and later in Panama) involved direct intervention in the lives of local populations by US health authorities and created a model for subsequent international-health interventions, as well as a training ground for men who would become international-health leaders. US efforts to control cholera in the Philippines, America's other colonial possession, reinforced this model and provided another colonial training ground.

Much of the history of yellow fever is familiar to historians and practitioners of public health. Carlos Finlay, Jesse Lazear, Walter Reed, and the US Yellow Fever Commission unraveled the mechanism by which yellow fever was transmitted by the female *Aedes aegyptei* mosquito; William Gorgas successfully applied this new knowledge to rid Havana and Panama of the disease; and Frederick Soper expanded Gorgas's methods to eradicate yellow fever from the Americas. All of these events are landmarks in a progressive narrative of international public health in which medical science overcomes age-old diseases and opens up tropical lands to economic development. The history of these events, however, is more complicated. Medical science played an important role. But multiple approaches to health were involved. In reexamining Gorgas's work, we need to ask why medical discovery and the disease-eradication campaign became the center of this history and formed a model for future international-health efforts. Why were alternative approaches forgotten or ignored?

William Crawford Gorgas was born October 3, 1854, in Toulminville, Alabama, near Mobile. He received his medical training at Bellevue Hospital Medical College in New York City, graduating in 1879. He then entered the US Army as an assistant surgeon. After he contracted and survived a bout of yellow fever that left him immune to the disease, the army assigned him to locations where the disease was prevalent. He spent several years combating yellow fever epidemics in Texas and Florida.[2]

In 1898, the United States declared war on Spain and invaded Cuba and the Philippines. Following several months of sea and land battles, the Spanish sued for peace and ceded control of Cuba, the Philippines, Puerto Rico, and Guam to the United States in the Treaty of Paris. The United States had overnight become an overseas colonial power. Its colonial possessions would become training grounds for a generation of US public-health workers who would later take up positions in emerging international-health organizations. While the 1898 invasion of Cuba was ostensibly intended to defend Cuban independence from Spain and stabilize trade relations in the region, Cuba had long been viewed by the United States as a hotbed of yellow fever and a source of the recurrent yellow fever epidemics that ravaged the American South. As Mariola Espinoza has convincingly argued, the invasion of Cuba was intended to put an end to this hemispheric threat of contagion.[3] Responsibility for the elimination of yellow fever in Cuba quickly fell to Colonel William Gorgas. The army sent Gorgas to command a hospital for US yellow fever patients in Cuba in 1898. The following year, he was appointed chief sanitary officer of Havana. Over the next two years he dedicated himself to cleaning up Havana and eliminating yellow fever.

Gorgas's initial approach to yellow fever in Havana followed the traditional strategy that has been employed by military and civil authorities in the United States: quarantine, combined with efforts to sanitize the city.[4] Havana, a major port city, was notorious for its lack of sanitation. In 1879, the US National Board of Health had sent a group of experts to examine sanitary conditions and the problem of yellow fever in Havana. Its report described unpaved, garbage-ridden streets; poor, overcrowded housing that lacked ventilation; and the absence of proper sanitation. Most households relied on privies to dispose of human waste, which contaminated the soil and the city's water supply. Referring to the pervasive use of privies, the report noted: "The effluvia therefrom pervades the houses, and the fluid contents saturate the soil and the soft porous coral rocks on which the city is built. Hence, all well water is ruined, and every ditch dug in the streets exhales an offensive odor. Thus Havana may be said to be built over a privy."[5]

Efforts were begun to sanitize Havana and the other large cities of the island as soon as US forces succeeded in pacifying Cuba. In Havana, Gorgas employed teams of sanitary officers to regularly scour the city, removing rubbish, dead animals, and untreated diseased persons from public places. Teams of street cleaners swept the streets and disposed of waste. Horse-drawn sprinklers regularly sprayed the streets with electrozone, a disinfectant manufactured through the electrolysis of seawater. His teams also inspected private residences and businesses. These activities had a dramatic impact on the health of the city. The overall death rate dropped from 98 per 1000 individuals in 1898 to 24.4 in 1900. But yellow fever continued.[6]

Recognizing that sanitation efforts had failed to control yellow fever, Gorgas adopted a new approach, based on the work of the Yellow Fever Commission. In February 1901, he began targeting the *Aedes aegyptei* mosquitoes. Gorgas directed his sanitation officers to attack the mosquitoes on two fronts: fumigating the houses of yellow fever patients and those of their neighbors, to kill any mosquitoes that may have been infected by the patient; and eliminating mosquito larvae by screening, covering, or oiling open water sources.[7] This campaign required meticulous reporting and surveillance. It also required the cooperation of local residents, which was not readily obtained, as they objected to the intrusive measures introduced by Gorgas. Fumigation was a messy business that resulted in the staining of fixtures and fabrics. Gorgas's inspectors also emptied or destroyed water containers and sprayed oil on the surface of water kept in cisterns. None of these activities were popular. Most Cubans did not view yellow fever as a serious health problem, since years of repeated yellow fever epidemics had left large portions of the Cuban population immune to the disease. Yellow fever was much more of a threat to newcomers, especially to the invading American military troops, most of whom were susceptible to the disease.

To achieve cooperation, Gorgas employed legal sanctions to enforce compliance and punish householders who refused to eliminate breeding sites.[8] The fact that the city was under US military occupation made the application of these sanctions possible. More-collaborative methods for developing community cooperation were never considered. The importance of military rule to Gorgas's success in eliminating yellow fever from Havana was emphasized by Gregorio M. Guitéras, who unsuccessfully tried to apply Gorgas's methods to eradicate yellow fever in the city of Laredo, Texas, in 1903. He concluded: "The Laredo epidemic has shown conclusively to my mind that results such as were obtained in Havana in the suppression of yellow fever during the American occupation cannot be obtained elsewhere, where the disease is widely spread, without the undisputed authority and the means that were at the command of the Government of Intervention in Cuba. These powers in reality amounted to martial law."[9]

Gorgas's campaign against *Aedes aegyptei* mosquitoes ended yellow fever transmission in Havana in less than a year. His victory was a major vindication of the mosquito theory of yellow fever, though not everyone was convinced. It was also a victory for the emerging specialty of tropical medicine. Moreover, it provided a strategy for attacking yellow fever in other parts of Latin America. In 1904, Gorgas was sent to direct sanitation efforts in Panama, where the United States had taken over the building of the Panama Canal from the French. The Isthmus Canal Convention

Colonel William C. Gorgas overseeing work on the Panama Canal. National Library of Medicine.

granted the United States control over a strip of land, mostly covered by jungle vegetation, that ran from the Pacific Ocean to the Atlantic and extended five miles on either side of the canal route. Excluded from US control were two major cities on either end of the route: Colón (located on the Atlantic Ocean) and Panama City (on the Pacific). Construction of the canal involved some 60,000 workers, about half of whom came from the West Indies. Control of the Canal Zone was directed by the Isthmus Canal Commission, which exercised authority over nearly every aspect of the lives of its employees.

Like Havana, the Canal Zone suffered greatly from yellow fever, as well as from malaria. The combined diseases had virtually halted French and American efforts to build the canal. Gorgas applied the lessons from Havana, launching a sanitation campaign aimed at eliminating *Aedes aegyptei* and *Anopheles* mosquitoes. This involved massive drainage operations and the installation of screens on houses. Gorgas calculated that his men drained 100 square miles of territory, constructing roughly 6.5 million feet of drainage ditches. They also applied 50,000 gallons of kerosene oil a month to breeding sites.[10] In the end, he succeeded in bringing both diseases under control, eradicating yellow fever.

In reviewing Gorgas's successful campaigns, I want to stress four points. First, in many respects Gorgas's methods resembled colonial medical campaigns that were conducted against cholera, plague, sleeping sickness, and malaria in Africa and South Asia by European colonial authorities. Specifically, in parallel with colonial medical campaigns elsewhere, the decision to attack yellow fever was based on external interests—in this case, US military and public-health needs—as much as on the interests of the colonized populations. The campaigns were imposed from above, with little concern for the ideas or the cooperation of local residents. Compliance was achieved through compulsion. These campaigns focused narrowly on a single disease, rather than on the development of broad-based health services. Gorgas insisted that wider sanitation efforts were unimportant for controlling yellow fever, as long as there was an effective system for controlling mosquitoes.[11] Colonial health services in general were designed to protect the health of Europeans and American colonial personnel and workers, who were essential to the colonial economy. Health services were clustered in urban areas and sites of economic production. The general health of colonial populations was left to European and American missionaries, who built mission hospitals and concentrated on improving maternal and child health, and to the occasional military-style disease campaigns, like Gorgas's efforts against yellow fever.

Second, the success of the disease-campaign approach to public health was dependent on the colonial environments that existed in places like Havana, Panama, and the Philippines.[12] Colonial contexts empowered public-health officials to act unilaterally, applying health strategies that served the narrow interests of colonial authorities. This empowerment was made possible by the ability of colonial officials to enforce sanitation laws. But it was also achieved through the cultural construct of subject populations being dependent and incapable of taking responsibility for their own health. Such attitudes ignored the existence of local bodies of medical knowledge and, in the case of Latin American countries in the early twentieth century, existing public-health infrastructures, such as those described by historian Steven Palmer.[13] Medical research into tropical diseases preceded the arrival of US colonial medical authorities.

Gorgas himself seldom used disparaging or racially charged language to describe the local populations of Panama and Havana. In fact, he paid little attention to the native populations at either location. His primary concern was the health of white Americans. As long as he had the power to fine those who resisted his regulations, he did not need to enlist their cooperation and, thus, did not have to be concerned about their character. His concern was mosquitoes and their elimination. He occasionally complained about the ignorance of his workers and their need for supervision.[14] But I found little evidence in his writings of the racialized discourse that characterized the language of many whites living in colonial settings at this time and was clearly articulated in the Philippines.[15]

Yet Gorgas shared a belief with many of his countryman that the development of the tropics depended on the success of white settlement. Indigenous populations were incapable of developing these areas. He felt that the greatest significance of his sanitary work in Cuba and Panama was that it demonstrated that white men could live in the tropics.[16] It is also true, as historians Alexandra Stern and Julie Greene have shown, that Panama was very much a colonial society in which racial ideas shaped social relations and economic opportunities. Segregation and racial discrimination were the rule.[17]

Third, the methods employed by Gorgas became a model for international-health efforts over the next century. They were a persistent feature of international, or global, health interventions up through the end of the twentieth century. Gorgas himself became an international celebrity within public-health circles. His successes and fame led him from the world of colonial medicine to that of international health, becoming one of a growing number of colonial medical authorities (including a number of physicians who had worked with Gorgas in Panama and Cuba) who traveled around the globe as technical advisors, working for foreign governments, attending international sanitary and medical conferences, and taking up positions in newly emerging international-health organizations. Gorgas became part of the community of authorities on tropical medicine, corresponding frequently with a number of prominent physicians and public-health experts in the United States and Europe and attending numerous medical and public-health conferences. Among his wide circle of correspondents was Sir Ronald Ross, whose work in India in the 1880s on the role of mosquitoes in transmitting malaria contributed to unraveling the source of yellow fever in Cuba. Ross visited Panama and forwarded Gorgas's annual reports on the situation in that country to the government of India, in an effort to encourage Indian health officials to more aggressively adopt vector-control strategies to combat malaria.[18] Gorgas also corresponded with the eminent American sanitationist Charles V. Chapin, with whom Gorgas exchanged ideas and debated the importance of various sanitation strategies. Malcolm Watson, who advanced knowledge of species sanitation through his work in Malaya and subsequently served as an advisor on malaria control in several British colonies in Africa, visited Panama and praised Gorgas's work as the "greatest sanitary achievement the world has seen."

Finally, despite the dominant roles of scientific discovery and the disease campaign in the history of efforts to improve health conditions in Cuba and Panama, these were not the only strategies deployed by Gorgas. It could be argued that other interventions played an equally important a role in the building of the Panama Canal. Yet they received little attention from health authorities at the time, or from later historians.

Alternative Approaches

Gorgas was primarily interested in yellow fever and malaria, which threatened the lives of American workers. Yet he was also concerned about pneumonia, which was taking the heaviest disease toll among the West Indian workers who made up the bulk of the canal labor force. Attacking mosquitoes would not prevent pneumonia deaths. Other methods had to be employed. Gorgas appointed a board to identify the conditions that were responsible for the pneumonia cases. The board studied a range of variables, including annual fluctuations in the incidence of the disease and differences in altitude, climate, months of employment, sleeping quarters, and clothing.[19] It concluded that the overcrowding of workers, some 84 to a room in dilapidated barracks inherited from the French, had contributed to the spread of infection and was primarily responsible for the high rates of pneumonia mortality.

Gorgas's solution to this problem was to recommend that the West Indian workers be given land on which to construct their own huts. He also encouraged them to bring their families with them. To assist in this endeavor he recommended that their wages be doubled, from US$0.11 to US$0.20 an hour. While this was considerably less than the salaries of white workers in the Canal Zone, Gorgas claimed that it was four times that of the salaries paid to workers in the countries surrounding the canal. By 1910, 30,000 of the total population of 37,000 West Indian workers lived in their own huts, many of whom settled there with their wives and children. Gorgas attributed the subsequent decline in pneumonia among this population to their resettlement and the elimination of overcrowding.[20] Had Gorgas not successfully addressed the pneumonia problem that plagued the West Indian workers, it is unlikely that the canal would have been completed.

While Gorgas is remembered as the father of the military-style disease campaign, which became a central weapon in later international-health efforts, he believed in the importance of addressing what we now call the social determinants of health. Commenting on the lessons learned from

Death rate from pneumonia among negroes

Year	*Death Rate (%)*
1906	18.74
1907	10.61
1908	2.60
1909	1.66
1910	1.66
1911	2.24
1912	1.30
1913*	0.42

Source: William Gorgas, "Recommendation as to Sanitation Concerning Employees of the Mines on the Rand, Made to the Transvaal Chamber of Mines," *Journal of the American Medical Association* 62, 24 (1914): 1857.

*First eight months

Panama, Gorgas noted in a 1915 lecture to the health officers of New York State:

> It is a health officer's duty to urge forward those measures in his community that will control individual diseases; but my long experience has taught me that it is still more his duty to take that broader view of life that goes to the root of bad hygiene, and do what he can to elevate the general social conditions of his community. This, my experience has taught me, can best be accomplished by increasing wages. Such measures tend at the same time to alleviate the poverty, misery, and suffering that are occurring among the poorest classes everywhere in modern communities.[21]

We should not be surprised by these remarks. Gorgas's career spanned the transition from a sanitationist approach to public health that was dominant in America during most of the nineteenth century to the so-called new public health that was ushered in by the development of bacteriology, as well as by the emergence of parasitology and the discovery of the role of mosquitoes in the transmission of malaria and yellow fever at the end of the century. Nineteenth-century sanitary approaches to public health focused on eliminating a broad range of environmental conditions that were viewed as causing disease. Much of this work was aimed at improving water supplies, housing, and nutrition. In the wake of the bacteriological discoveries of the 1880s, the new vision of public health aimed to improve health by identifying and attacking the germs and vectors that directly caused disease.

In place of searching for the causes of disease in the environment, the new public-health professionals sought them out under the microscope.[22] Powerful new methods of identifying disease overshadowed efforts to improve water supplies, clean the streets, and ameliorate the housing and living conditions of the poor.

It could be said that Gorgas came to Havana as a sanitationist, attempting to clean up the city, and left as an iconic representative of the new direction in public health, attacking mosquitoes. But his efforts to address pneumonia among West Indian workers showed that he retained a belief in the benefits of addressing the broader determinants of health. A closer look at the various public-health interventions he subsequently employed reveals his commitment to multiple approaches to public health.

Gorgas is often thought to have used the same type of campaign against yellow fever in Panama that he applied in Havana. Yet his work in Panama, while similar in many ways to that in Havana, was different in one fundamental respect. In Panama, Gorgas found that he was unable to enforce the covering and destruction of the open water sources that provided breeding grounds for *Aedes* mosquitoes in Panama City and Colón, which lay outside the Canal Zone. Panama itself was not under military rule, and local Panamanian authorities did not actively enforce sanitation regulations. He therefore decided to introduce a piped water supply, thereby making it easier to convince the populace to eliminate their dependence on cisterns, wells, and barrels. Gorgas's men introduced water and sewer lines, and paved the streets in Panama City and Colón. In short, in addition to attacking mosquitoes, he improved the overall sanitation and infrastructure of these cities.

Similarly, in 1913, the Transvaal Chamber of Mines in South Africa recruited Gorgas to review the gold industry's sanitary conditions and make recommendations aimed at reducing the extremely high rate of pneumonia cases and deaths among black mine workers. Gorgas's recommendations mirrored many of the measures he had put in place in Panama to reduce pneumonia among West Indian workers there. He proposed improvements in housing, including the creation of family housing. He also recommended improvements in diet, the provision of piped water and sewage removal, and the establishment of a centralized medical system. He made no recommendations related to wages, however.[23] His South African recommendations were only partially accepted, as they came with significant financial costs.[24] Yet they reflected a broader approach to occupational health than was reflected in much of his yellow fever work.

In that same year, however, the surgeon general of the US Army ordered Gorgas to visit the port city of Guayaquil, on the Pacific coast of Ecuador.

The city was experiencing repeated epidemics of yellow fever. The army was concerned that ships traveling through the soon-to-be-opened canal from that Ecuadorian city could reintroduce yellow fever into the Canal Zone, and from there it might spread across the Pacific. Gorgas visited the city and reported back to the surgeon general. He compared Guayaquil with Havana and described how yellow fever had been brought under control in the latter city. He recommended that the same methods be applied in Guayaquil. Yet he emphasized that the Havana campaign had been conducted under a military governor who was entirely in accord with Gorgas's measures. Gorgas therefore proposed that the president of Ecuador appoint a health officer for Guayaquil who would be empowered to levy fines on householders who refused to eliminate breeding sites, and that this individual would have the right to appeal to the president if ordinances were not enforced. In short, Gorgas recommended the imposition of a military-style regime similar to the one that had existed in Havana.[25]

Gorgas was not alone in valuing multiple approaches to public health. It was quite common during the first half of the twentieth century for public-health authorities to engage in very technical, targeted approaches to health and, at the same time, espouse ideas about the social basis of disease and the need for economic improvements, or what nineteenth-century German physician and social-medicine advocate Rudolf Virchow termed "medical economic welfare." For example, Angelo Celli, who directed Italy's campaign to control malaria by distributing quinine, argued in his 1901 textbook on malaria control that malaria was a social disease and that the eventual control of the disease required social reform, better wages, and agrarian development. Similarly, Marshall A. Barber, who developed the use of the larvicide Paris green (a highly toxic copper and arsenic compound) to attack mosquito larvae in the American South and helped develop mass-treatment methods for controlling hookworm in Malaya, claimed that "with only a moderate betterment of social conditions, malaria in the United States tends to disappear."[26] Even the strongest advocates of the new public-health direction were aware of the underlying social determinants of health. Both approaches were always present and were often in tension with one another. The social-determinants approach, however, only occasionally took center stage in the organization of international-health activities. Why this approach failed to gain wider support is one of the central questions explored in the rest of this book.

The immediate question, however, is why medical science and the disease-eradication campaign came to dominate narratives of Gorgas's success in improving health conditions in Cuba and Panama and be accepted as central to efforts to improve health conditions in other parts of the tropical world. Why were his other strategies ignored? Part of the reason was simply that malaria and yellow fever were diseases that plagued Americans in the tropics; pneumonia was not. Yet Gorgas's antimosquito campaigns also dominated, because they converged with three emerging narratives related to the power of tropical medicine. The first was deployed by early specialists in tropical medicine in the United States and Europe, who struggled to distinguish themselves, and the value of their work, from that of general physicians at the turn of the century.[27] These specialists recounted Gorgas's success in conquering yellow fever in order to confirm the power of tropical medicine. Second, the story of Gorgas's victory over yellow fever became part of popular accounts that explained the United States' ability to complete the Panama Canal in terms of America's scientific and technical expertise. US know-how had permitted the Americans to do what the French could not. Finally, Gorgas's success reinforced an emerging narrative of American imperialism in which colonial expansion was rationalized in terms of the benefits that America's scientific knowledge brought to the peoples of its colonial possessions.[28] This was a narrative that would become an important part of US foreign policy after World War II.

Yet there is a more specific reason why Gorgas's antimosquito campaigns became a central element in subsequent international-health activities (chapter 2). They were attractive to the heads of the newly created International Health Board (IHB) of the Rockefeller Foundation. The IHB was committed to a public-health approach that emphasized biomedical science and found in Gorgas's mosquito work an exemplar of this model of public health. Conversely, the IHB was averse to approaches involving social and economic changes. The IHB adopted Gorgas and his disease campaign and, in so doing, reified their central role in the subsequent history of global health.[29] Before examining the work of the IHB, however, we need to briefly visit America's other colonial training ground for future international-health leaders, the Philippines.

The Philippines: Sanitizing Colonial Bodies

The US occupation of Cuba lasted only a few years, thanks to the Teller Amendment, which stipulated that the annexation of Cuba would be temporary. America's occupation of the Philippines, on the other hand, lasted for decades and involved a much more intensive exercise of colonial power and intervention into the health of the Filipino population. Though the Spanish relinquished control of the Philippines in 1898, Filipino insurgents launched a guerilla war against the US occupation, and it took three years to subjugate the islands. As in Cuba, tropical diseases took a heavy toll among US troops, and the conquest of the archipelago involved a campaign against microbes and insects as much as against enemy troops. Outbreaks of plague and cholera during the insurrection were met with sometimes brutal campaigns involving forced quarantines, the razing and burning of infected villages, and the mass burial of disease victims. Understandably, the local population viewed such actions as weapons of war rather than of public health.[30]

Following the insurrection, military and sanitary operations continued to be linked. US authorities organized military-style health campaigns that were similar, in some respects, to those initiated by Gorgas in Havana and Panama. But the target of sanitation efforts in the Philippines during the early years of the American occupation was the Filipino population itself, rather than mosquitoes. Until the 1920s, relatively little was done to control malaria, other than initiate a segregation of American troops and issue quinine. Even less was done to control dengue fever, which was a persistent cause of debility among US troops serving in the Philippines up through World War II. The Filipino population came to be viewed as the primary source of disease and a threat to the health of American troops and administrators, due to their supposedly poor sanitary habits. As Warwick Anderson has documented, US health officials were obsessed by what they saw as the Filipinos' "indiscriminant defecation." American authorities also shared the common colonial view that local populations were incapable of improving health conditions without white tutelage. Henry du Rest Phelan, a US Army doctor who was assigned to Suriago in the Philippines in 1902, captured these attitudes when he observed that the ground beneath native houses was covered with "filth of all kinds, human excrement included." He viewed the Filipinos as lacking the capacity to improve their sanitary conditions without assistance: "They appear to me like so many children who need a strong hand to lead them in the path they are to follow."[31]

US authorities were not the first to introduce bacteriology and sanitation to the Philippines. By the 1880s, young Filipino researchers, some of whom had trained in Spain and France, were undertaking new research on polluted water, malaria, and cells. Influenced by the revolutionary new discoveries being made in bacteriology, they focused on the nature of pathogens and microbial pathogenesis in disease development and transmission. These researchers sought to refute long-standing Spanish colonialist allegations that the Philippine environment was intrinsically unhealthy, and that Filipino bodies were markedly diseased and pathological, attitudes that dominated

US views of the Filipino population.[32] In addition, many communities had garbage-collection services. Yet US colonial health officials generally ignored or disparaged the work of these Filipino scientists and public-health authorities.

To improve the health of the islands, the US occupational forces launched a series of campaigns aimed at educating and sanitizing the Filipino population. By contrast, relatively little was done to develop basic health services until the commonwealth period, under Filipino administration, after 1935. Responsibility for the sanitation campaigns fell to Victor G. Heiser, who was placed in charge of health in the Philippines from 1905 to 1915. Heiser was a member of the Public Health and Marine Hospital Services, which later became the US Public Health Service. He was assigned as chief quarantine officer in the Philippines in 1902, but before taking up his post, he was directed to attend the International Congress on Medicine in Cairo. There he met Gorgas, who had been dispatched to Egypt to attend the congress and learn about the sanitation measures the British had enacted in relation to the Suez Canal. Following the congress, Heiser made his way eastward through various British colonial possessions on his way to the Philippines: "All along the way I heard the same arguments: 'Orientals could not be sanitated [sic]; he always lived in filth and squalor; to persuade him to live in any other way was hopeless; . . . all efforts, therefore, should be concentrated on making living conditions safe for the European who was obliged to sojourn in his midst.' "[33]

Heiser shared these racial attitudes with regard to the lack of sanitary habits among tropical populations. He also had little respect for the Filipino health workers and scientists who gradually took over public-health functions in the islands during the 1910s and 1920s. He rejected, however, the idea that "our little brown brothers" could not be taught the sanitary standards of civilized peoples. To this end, he launched campaigns to "discipline their bodies" and clean up environmental nuisances. He enforced vaccination, household hygiene, and improvements in housing, water supplies, nutrition, and—above all—sewage disposal. Heiser employed a small army of inspectors to oversee his colonial charges, including 200 physicians. Heiser possessed almost complete military power and did not hesitate to invade the domestic spaces and businesses of the Filipino population. By 1903, his team had inspected nearly 2 million homes; 241,000 homes and 162,000 yards were cleaned and 11,200 cesspools and similar sewage systems emptied.[34]

A campaign directed by Allan McLaughlin, in response to a cholera outbreak in Manila in 1908, was an example of the military-style health campaigns launched to combat disease in the Philippines. McLaughlin organized 600 men into disinfecting squads that went about spraying carbolic over dwellings and "liming all closets and places where fecal matter existed or was likely to be deposited." The effort to disinfect the dejecta of the entire population necessitated the disinfection of entire districts. More than 150,000 pounds of lime and 700 gallons of carbolic acid were used. Anyone who tried to obstruct the disinfection operations was arrested and fined.[35] US colonial health officials also employed educational campaigns, the imposition of systems of pails for disposing of fecal waste, the cleansing of markets, and the application of building regulations requiring the construction of privies in all new buildings. Through these multiple strategies, they attempted to sanitize the Filipino body.

America's colonial possessions played an important role in fostering a set of public-health practices and attitudes regarding the control of diseases among dependent populations that would come to dominate international-health activities for much of the twentieth century. Following World War I, US medical personnel, who had served in Panama and the Philippines, were recruited by the Rockefeller Foundation's International Health Board to serve as country directors for hookworm- and yellow fever–elimination programs in various parts of the globe. They became part of a global network of international-health experts who shared a common vision of public health that shaped the early history of international-health activities. The incorporation of colonial physicians into the IHB led to the consolidation of medical science and the disease campaign as central elements of international health. The following chapter describes the building of this network and its impact on the early history of international health.

CHAPTER TWO

From Colonial to International Health

The International Health Commission and the British Empire

The Rockefeller Foundation created an International Health Commission in 1913, with the goal of extending the foundation's efforts to eradicate hookworm in the American South to other countries around the globe where hookworm was seen as incapacitating millions of people. The commission, which changed its name to the International Health Board (IHB) in 1916 and later to the International Health Division (IHD) in 1927, would become the most powerful and influential international-health organization during the first half of the twentieth century. By the time it wrapped up business in 1951, it had operated in 80 countries around the globe, fighting hookworm, bilharzia, malaria, and yellow fever. In addition, it had financed the construction of schools of tropical medicine and public health in 21 countries and funded the training of hundreds of health professionals from developing countries. International Health Board program officers, colonial health authorities, and public-health professionals from many countries passed through these schools, becoming part of a community of international-health experts who shared a common set of ideas about how to transform the health of peoples living in colonial and postcolonial settings.

A great deal has been written about the motives of the foundation and its creator, John D. Rockefeller, in creating the International Health Commission. Rockefeller amassed a fortune as founder and chief executive officer of Standard Oil, and some have argued that his philanthropic health work was designed to advance his far-reaching financial interests, focusing narrowly on increasing the efficiency of tropical labor. There is little doubt that Rockefeller was concerned about labor productivity. Yet the record of the IHB's activities hardly supports a narrow, materialist argument. Though many of the IHB's campaigns were targeted at plantation workers, many were not. Rockefeller was a visionary, who saw the promotion of health as part of a larger civilizing mission. His goal for the IHB was to sanitize the world and, in doing so, advance the cause of Western civilization. If this contributed to his fortune, he would not object. Nor would he complain if it countered his image as someone who was solely interested in increasing his wealth. But increasing his personal wealth was not the purpose of his investments in international health.[1]

Hookworm was chosen as the IHB's first vehicle for achieving its goals, because of the foundation's earlier experience in the American South and because the foundation viewed it as a disease of immense economic importance, as well as one for which "relief and control" could be readily achieved through medical intervention. Success in eliminating hookworm could demonstrate the benefits of a biomedical approach to public health and provide a means of advancing broader initiatives in health. The IHB's first director, Wickliffe Rose, described the purpose of its hookworm work in Latin America: "The work for relief and control of [hookworm] . . . is to be regarded merely as an entering wedge toward a larger and more pervasive service in the medical field. It will lead inevitably to the consideration of the whole question of medical education, the organization of systems of public health, and the training of men for public health service."[2] At the same time, neither Rockefeller nor Rose were interested in efforts to address the underlying cause of hookworm or other diseases. Hookworm was a disease of poverty and was strongly associated with particular forms of agricultural and industrial production that brought together large numbers of workers in settings that fostered the transmission of parasites.[3]

From its inception, the hookworm campaign was deeply entangled in the world of colonial medicine. The IHB recruited many of the directors for its hookworm programs from Panama and the Philippines. Tracking the movement of these directors reveals the influence that ideas and practices developed in America's colonial possessions had on the IHB's early hookworm campaigns. The most prominent of the men recruited by the IHB from Panama and the Philippines was Victor Heiser, who was tapped to be the presumptuously titled "director of the East," responsible for the IHB's disease-control programs in 40 countries in East and Southeast Asia. In addition, Louis Schapiro and David Molloy, who worked under Heiser in the Philippines, became directors of the foundation's hookworm programs in Costa Rica and Nicaragua, respectively. Shapiro replaced Henry Carter Jr., who had been hired from the Marine Hospital in Colón, Panama, in 1914.

John D. Snodgrass, after working in Manila and the Culion leper colony, took over the Rockefeller hookworm campaign in Ceylon.

The International Health Board's decision to initially situate its campaign in the British colonial empire further strengthened colonial influences on its hookworm work. Rose traveled to London to obtain the Colonial Office's blessing for the campaign in 1913. He convinced the Colonial Office of the potential value of the IHB's work and won its support for initiating campaigns within British colonies. To facilitate this work, Rose agreed to set up a British Advisory Board, with headquarters in the Colonial Office. The Colonial Office agreed that the IHB's first work would be in the British West Indies. Rose followed his London meetings with a tour of the British West Indies and, subsequently, Britain's colonies in South and East Asia.

The IHB's hookworm work in the British Caribbean followed an organizational model that became known as the American, or intensive, method. First developed in the American South, the method was perfected in British Guiana by the IHB's country director, Howard H. Hector. The method involved mapping large populations, collecting and testing their stools for evidence of hookworm infection, and treating all those found to be infected with a highly toxic worming medicine, administered until the patients were worm free. The treatment was painful and, in a number of cases, fatal.[4] The backing of British colonial authorities allowed the IHB to conduct their campaign without the need to develop strategies for gaining local cooperation. As the campaign expanded to other countries in the region where colonial forms of control did not exist, the campaign's ability to ensure the participation of local populations became more challenging, and the IHB was forced to develop communication strategies to overcome resistance and adapt to local conditions. Thus there was considerable variation in the ways the hookworm campaign was conducted in different locales. Yet the campaign retained the basic elements of the American method that had been established within colonial conditions.[5]

The IHB viewed the local hookworm campaigns as entry points into communities in which it hoped to introduce principles of scientific hygiene. The intent was to sanitize the world. Similar goals had driven the IHB's foray into tuberculosis control in France, where it attempted to apply hygienic principles to eliminate this disease after World War I. To eliminate hookworm, it was necessary to get people to construct and use latrines, wear shoes, and generally improve their sanitation, as the IHB had attempted to do in the American South. These activities were also necessary to prevent reinfection. Yet sanitation activities often took a back seat to treatment in the IHB's international campaigns. In part this was because sanitation

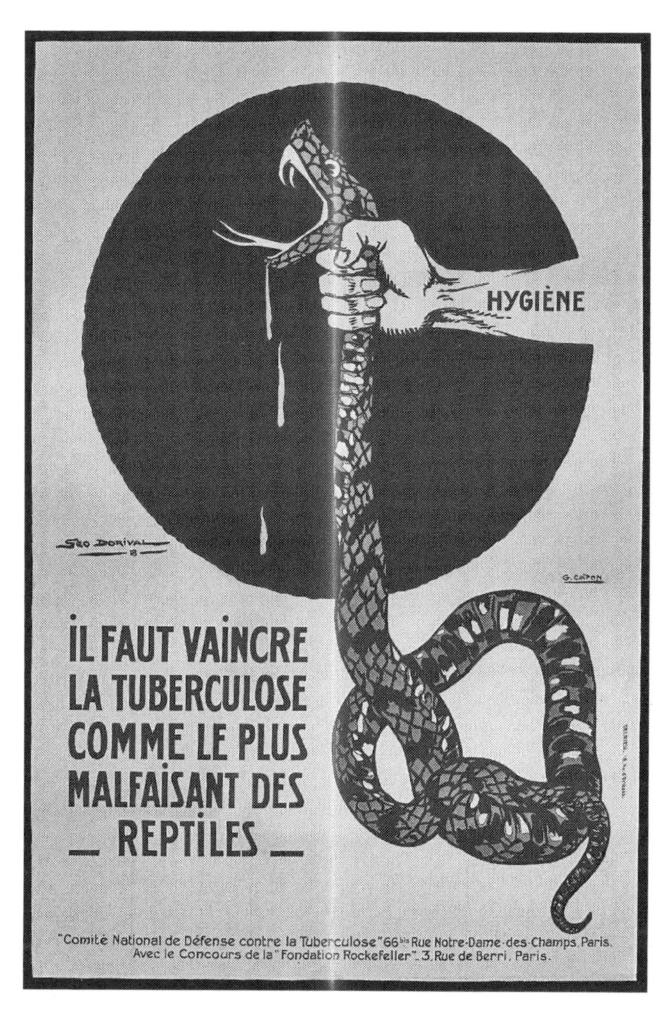

A poster from the International Health Board's anti-tuberculosis campaign in France, ca. 1918

efforts required investments in time and money. But it also was because physicians who believed in the power of drugs to eliminate hookworm ran the campaigns.

In addition, sanitation efforts suffered because program officers failed to appreciate the perceptions of local populations and the conditions under which they lived. In Trinidad and British Guiana, sanitation efforts ran into local resistance to building latrines, both because they were expensive and, in some places, conflicted with local cultural attitudes regarding excreta. In

Costa Rica, local populations complained that the latrines attracted mosquitoes and led to increases in malaria.[6] In Mexico, IHB officers attempted to convince local residents to wear shoes to prevent reinfection. Yet there was a serious disconnect between the local villagers, who had little understanding of what the IHB was trying to do, and the Rockefeller health officers, who had little appreciation for the social and economic conditions of the people they were trying to treat. IHB officials employed educational posters showing enlarged images of hookworms. Villagers had never seen such monstrous creatures and were incredulous that they could enter the bodies of humans through the soles of their feet. Biological notions of infection were also alien to the locals' understanding of the causes of sickness and health. IHB officers additionally failed to understand that most villagers could not afford to purchase shoes. Instead, the health officers insisted that wearing shoes was a behavioral issue, a sign of individual enlightenment. As Andrew Warren, the director of the campaign, observed, "it is the culture and intelligence that causes people to wear shoes."[7]

The Rockefeller Foundation's political interests also undermined the IHB's sanitation efforts in Mexico. In addition to curing and sanitizing local populations, the foundation viewed its hookworm program as a means of building support for the pro-US Mexican government and countering political opposition in the region.[8] It thus focused on treating infected individuals, in order to have an immediate impact, and on rapidly expanding its eradication activities. This approach meant that prevention efforts, which required time and patience, were often not pursued with much enthusiasm. More importantly, the IHB did not conduct followup surveys to verify that their attempts to prevent reinfection had, in fact, worked. Mexican authorities took prevention more seriously than the Americans and, once they took over programs begun by the IHB in the late 1920s, pushed for the construction of latrines. The desire to have a rapid impact, a lack of understanding of local social and economic conditions, and the absence of followup surveys to measure outcomes would become enduring characteristics of international- and global-health campaigns.[9]

The IHB's hookworm-control efforts were was further perfected in East Asia by another Panama alumnus working for the IHB, Samuel Taylor Darling. Darling studied medicine at the College of Physicians and Surgeons in Baltimore, graduating in 1903. In 1905, he accepted a position as an intern at Ancon Hospital in Panama and joined Gorgas's effort to eliminate yellow fever and malaria. Darling was in charge of laboratories and patient services at Ancon Hospital. He also conducted research on malaria and introduced species-specific control measures in Panama. In addition, Darling accompanied Gorgas to South Africa. Following the triumph of sanitation in Panama and the successful completion of the Panama Canal, Darling accepted an offer to join the International Health Board of the Rockefeller Foundation in 1915. He was soon appointed to head a commission to study hookworm disease in the Far East. Marshall A. Barber, who had begun his public-health work in the Philippines before being recruited by Rockefeller, joined Darling. For the next three years, Darling and Barber studied the incidence and consequences of hookworm and malaria in Malaya, Java, and Fiji. All three locations were British colonies, to which planters recruited large numbers of Tamils from southern India to work on sugar and rubber plantations. The laborers were poorly housed and worked on plantations that lacked basic forms of sanitation. They routinely defecated in areas outdoors, near where they were working. Not surprisingly, hookworm was a major source of disability for them.

Barber and Darling's final report revolutionized the diagnosis and treatment of hookworm disease. They introduced the concept of mass treatment, in which a sample of the population was tested to estimate a hookworm index for an entire community. If the index revealed substantial infection, the whole community, infected and non-infected alike, would be treated. The goal was to treat everybody quickly, within a few days. If this was done, very few hookworm eggs would be passed onto the soil, making the disease less infective. The mass-treatment method eliminated the need for testing every individual to determine that person's level of infection. Barber and Darling argued that further sanitation efforts could also be eliminated. In this manner, the cost and effort of controlling hookworm would be reduced "advantageously."[10]

Darling's desire to develop a method that would eliminate the need for sanitation was driven by his negative view of the Tamil laborers among whom he worked. He saw them as incapable of observing sanitary practices and claimed that their superstitious ways and poor sanitary behaviors were a menace to the world. Describing his hookworm work in Malaya in 1920, Darling and his colleagues noted: "Most of the Tamil coolies in the Federated States come from the Madras Presidency in southern India, and are in the main ignorant, superstitious, and servile. Probably not more than 5 percent of them are able to write their names. While it is true that Tamils are hard workers, it is also true that they are almost entirely lacking in ambition. Docile, unstable, and apparently quite incapable of administering their own affairs, with either dispatch or intelligence, they constitute the white man's burden."[11] This view of Tamil workers as ignorant and incapable of caring for their own health echoed those of US colonial officials in the

Philippines. It ignored the fact that plantation owners provided few, if any, facilities that the workers could have used. In addition, the report made no mention of plantation working conditions.

Darling and Barber also conducted clinical trials to test the effectiveness and cost of various purging agents, using subjects found in hospitals and jails. They noted that they had difficulty finding subjects for their trials and that "many patients absconded during the rather prolonged period [when] it was necessary to keep them under observation and treatment." This loss of subjects was not surprising, given the range of aftereffects observed in patients who took the test drugs. These included dizziness, unsteadiness of gait, an inability to rise, a semicomatose state, tingling in the hands and feet, deafness, burning in the stomach, and headache. All of these side effects, except the last two, were more commonly produced by oil of chenopodium than by any other vermicide.[12] Yet Darling and Barber found that oil of chenopodium was the most effective drug for removing worms, costing half as much as thymol, which was widely used by IHB hookworm campaigns at that time. Barber and Darling accordingly recommended the adoption of

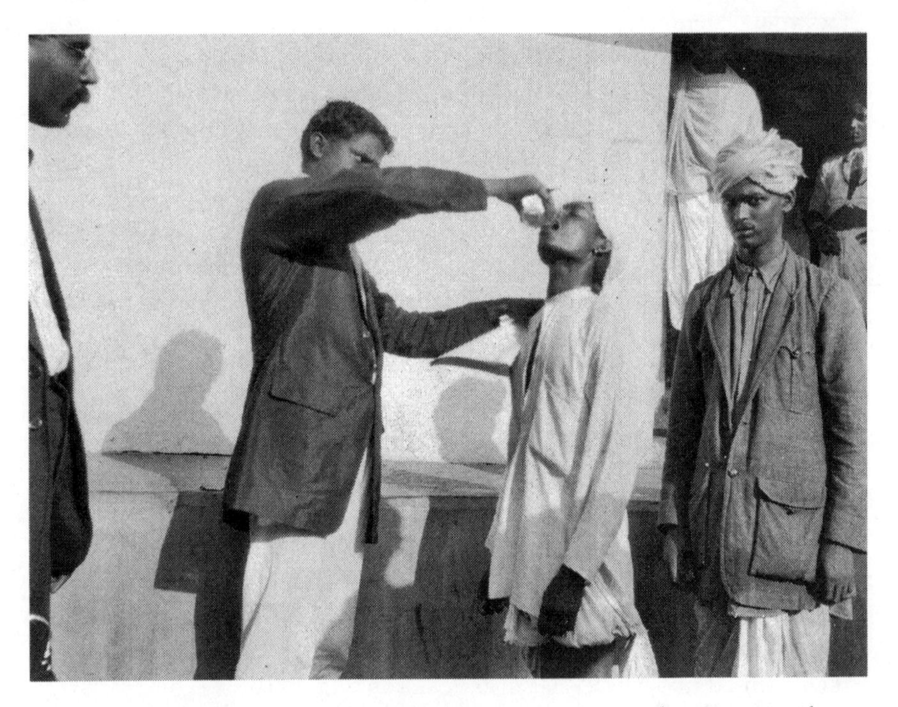

The International Health Board's hookworm campaign in India. Courtesy of Rockefeller Archive Center.

oil of chenopodium, though they emphasized the need for larger trials. The fact that it was the most toxic drug tested seems not to have dampened their enthusiasm.

The results of their hookworm studies were very influential in shaping the IHB's subsequent reliance on the mass-treatment approach to hookworm control around the world.[13] Darling would later become a member of the League of Nations Malaria Commission, extending the network of US colonial health officials who took up positions in international-health organizations. The emphasis on treatment over prevention ultimately undermined the IHB's goal of eradicating hookworm. Even with the use of mass treatment, treating infections without implementing effective sanitary measures did not prevent reinfection. It was recognition of this failing that eventually led to the IHB's disenchantment with hookworm control and their decision to wrap up their existing hookworm programs by the end of the 1920s. The IHB shifted its resources to its campaign against yellow fever, which had begun with the establishment of its Yellow Fever Commission in 1915 and had paralleled the IHB's hookworm campaigns. By the time this occurred, hookworm campaigns had been extended over wide areas of Latin America, China, and South and Southeast Asia. Together, these efforts had established the colonial-disease campaign as a central part of international health.

The Colonial Roots of the Yellow Fever Commission

Like the hookworm campaign, the IHB's yellow fever campaign had strong colonial connections. The Rockefeller Foundation developed an early interest in eradicating yellow fever by eliminating it in the few places where it appeared to exist in the Americas and Africa. They invited William Gorgas to become director of the IHB's Yellow Fever Commission in 1915. Gorgas accepted the position but had to wait until after World War I to take up his new post. Following the war, Gorgas devoted three years to laying the groundwork in Peru, Mexico, and Ecuador for what he expected to be a successful campaign to eradicate yellow fever from the Americas. The campaigns employed methods Gorgas had developed in Panama and Havana and, like the IHB's hookworm campaigns, were directed by colonial old hands, recruited from Panama and the Philippines. In Ecuador, the Guayaquil campaign that Gorgas had recommended in 1913 began in 1918, under the leadership of Michael Connor, a graduate of Dartmouth medical school who had served in the US Army Medical Corps in the Philippines. He had also served as assistant sanitary inspector in the Panama Canal project. Within a year, he had replicated Gorgas's antimosquito methods, visiting

Yellow fever workers in Guayaquil, Ecuador. Courtesy of Rockefeller Archive Center.

372,278 houses; inspecting 1,104,862 water containers; and dispersing 14,300,500 gallons of oil on water surfaces.[14]

In Peru, another Panama veteran, Henry Hanson, ran the campaign.[15] In 1921, the Peruvian government solicited the financial support of the Rockefeller Foundation to eradicate yellow fever, and Hanson was hired to direct the campaign. The campaign was slow to develop, and yellow fever began spreading south. Hanson chose to focus on major towns and cities, rather than on smaller communities where, he claimed, the disease would burn itself out. While he employed a number of methods, including the unsuccessful application of a vaccine that Hideyo Noguchi had developed with Rockefeller Foundation support, Hanson came to depend on the use of small larvae-eating fish, placed in water tanks, to eliminate mosquito breeding. This method was applied with military precision, using 100 sanitation workers. Towns were divided into districts, each in the charge of an inspector and a squad. The workers covered their entire district in seven days, examining every water container to ensure that no mosquitoes were breeding. When mosquitoes were found, the receptacles were emptied, all adherent eggs were removed, and larvivores were placed in the newly resupplied water. On Sundays, Hanson and the inspectors made rapid surveys of the critical towns. Hanson kept a detailed daily record of funds expended, premises inspected, and water containers treated.[16]

Hanson conducted his campaign in a top-down manner, making no effort to educate the public or gain their cooperation through persuasion. His resistance to such undertakings was based on his assumptions about the scientific backwardness of the populations with whom he was working and their inability to understand the need for cooperation. In his personal account of his work in Peru, Hanson wrote, "Superstitious, ignorant, and seemingly satisfied, the rank and file of the natives of Lambayeque, as everywhere in Peru, were resentful of any effort to direct or change their mode of life."[17] Hanson's views were hardly unique. They channeled those of his counterparts in the Rockefeller Foundation's hookworm campaigns in the Caribbean and Latin America and by the US military in the Philippines, as well as by British and French colonial medical officers working in many settings across Africa and Asia. Despite these attitudes, Hanson succeeded in controlling yellow fever in all towns north of the department of Lima by the end of 1922, and no more cases of yellow fever were reported.

Passing the Torch

The disease-control strategies developed by Gorgas and American physicians, who learned their public-health skills in America's colonial possessions in Panama and the Philippines, were passed on to later generations of US public-health workers, as well as to public-health workers from foreign nations, through training sites and schools of public health established by the Rockefeller Foundation. This educational process contributed further to the entanglement of colonial medicine and early international-health organizations. The Rockefeller Foundation invested heavily in the construction of schools of public health, both in the United States and abroad. These schools were intended to be centers for the development and dissemination of the "new public health," based on the application of scientific methods and principles informed by the new sciences of bacteriology and parasitology.

The first of these was the Johns Hopkins School of Hygiene and Public Health, founded in 1916, with William H. Welch, the noted physician and former dean of the Johns Hopkins School of Medicine, as its first director.

The school attracted a wide range of students. While Welch sought to train young medical researchers, the foundation used the school to train its present and future staff members, award advanced degrees to experienced men in the field, and bring in field staff for specialized training courses. The school also served another purpose: it trained international students, who received an education in hygiene and then returned to head up health programs and become faculty at local schools of public health in their home countries.[18] The foundation invested heavily in the training of such students. Between 1917 and 1951, it provided 473 scholarships for Latin Americans working in the medical sciences.[19] Many of these students came to Johns Hopkins.

Just what foreign students believed they got out of these educational experiences is less clear. The curricula of Johns Hopkins' and Harvard's public-health schools were heavy on science and scientific methods for controlling disease, but contained little material on how to work with local populations. C. C. Chen, who graduated in 1929 from the Peking Union Medical College, which was also funded by the Rockefeller Foundation, went to Harvard to study public health in the early 1930s. He later played an important role in the International Health Division's North China Rural Reconstruction Program (chapter 4). Chen claimed that he learned nothing of value at Harvard. Selskar Gunn, who also worked in China for the Rockefeller Foundation, was another critic of what he was taught at Harvard. Both Gunn and Chen built programs that were community based, something that was not part of the educational experience at Johns Hopkins and Harvard at this time.[20] It was not until the 1950s that the social sciences began to take hold in schools of public health in the United States.

Nonetheless, the schools of public health funded by the Rockefeller Foundation became places where models of disease control and public health, forged in colonial settings and replicated in disease-control projects across the globe, intersected with the scientific practices, methods, and knowledge of the new scientific hygiene. This conjunction of colonial field experience with public-health education can be seen in the short course on parasitology that Frederick Soper took at the Johns Hopkins public-health school in 1923, after he joined the IHB. Soper, who would go on to lead hookworm and yellow fever campaigns for the IHB, became a major advocate for disease eradication and eventually was director of the Pan American Health Organization. As part of his parasitology course, he claimed to have taken a class on yellow fever control that described Gorgas's work in Havana. Years later, when Soper was sent to Brazil to combat yellow fever and malaria, he was met by Brazilian counterparts who had also been trained at Johns Hopkins.[21]

As Americans and foreign nationals who were trained in the new schools of public health took up academic and public-health positions around the world, they became part of an international community of public-health workers who shared a common set of ideals and practices. In this way, disease-control strategies forged originally in the crucible of colonial rule during the first decades of the twentieth century were fused with and became part of the science of hygiene. In this form, they were easily transferable to noncolonial settings and became central to the practices of international health. In addition to those at Johns Hopkins and Harvard, the Rockefeller Foundation eventually funded other schools of public health in 21 foreign countries, including the London School of Tropical Medicine and Hygiene.

The International Health Board created other sites where public-health knowledge was passed from one generation to the next. One such location was the IHB's malaria field site in Leesberg, Georgia. Samuel Taylor Darling, who had developed the mass-treatment approach to hookworm control, was appointed director of the Leesberg site in March 1923.[22] Leesberg served as a training location for IHB physicians on their way to take up positions around the globe. It was there that Frederick Soper arrived in summer 1923, for two months of training with Darling before taking up a new post directing a hookworm campaign in Paraguay. Soper went on to apply Darling's methods to hookworm control in several Latin American countries.

Lewis W. Hackett, who would also become a major leader in the war on malaria, working for the International Health Division in Europe, joined Soper in Leesburg. Also at Leesburg that summer were John L. Hydrick, who would develop new methods for rural community-health education in Indonesia; and Alexander F. Mahaffey, who later discovered the virus that causes yellow fever and played a central role in mapping yellow fever in Africa. Mahaffey headed up the Rockefeller Foundation's Yellow Fever Institute in Entebbe, Uganda, and concluded his career in London, as director of colonial medical research at Britain's Colonial Office. He thus represented the movement of IHB personnel back into colonial medicine. Leesberg was one of many nodal points in an expanding network of tropical-disease experts that connected colonial medicine and international health across the first three decades of the twentieth century.[23]

Philippines Redux

Before concluding this chapter on the colonial roots of global health, I want to return briefly to the Philippines, moving ahead a few years to the early 1930s, to examine the early career of Paul F. Russell, who would become an international expert in malaria control. Russell's career in the Philippines further demonstrates the importance of colonial settings for the development of international health.

The IHB had established a hookworm program in the Philippines in the early 1920s. Repeated surveys indicated that the Philippine programs had failed to reduce infection levels over a wide area of the islands. Heiser and his lieutenants routinely ascribed this failing to the indiscriminate defecation habits of the population in the Philippines, combined with the lack of initiative and general unsuitability of Filipino health workers for this kind of work. The hookworm programs in the Philippines were eventually abandoned, as they were elsewhere. In the 1920s, the IHB turned its attention to malaria. Malaria had menaced and sickened American soldiers and administrators from the beginning of the US occupation of the country. Yet efforts to control the disease had been perfunctory. The threats of cholera and plague and the American obsession with Philippine excrement drew attention away from malaria.

In 1914, Philippine health authorities attempted to control malaria in Manila by eliminating breeding sites and fining householders who failed to keep their households free of mosquitoes. Yet they came to view malaria control as futile in the face of a Filipino population whom they continued to view as unable or unwilling to participate in control activities. Lack of confidence in their ability to change the habits of the Filipino population or to get them to participate in control programs, either through the use of netting or screens or by taking quinine tablets on a regular basis, led the IHB to focus its activities on the elimination of *Anopheles* mosquitoes, concentrating particularly on the elimination of breeding sites outside of homes through the use of larvae-eating fish and the chemical larvicide known as Paris green. These methods, it was hoped, would permit malaria to be brought under control without having to obtain the cooperation of the Filipino population.[24] The IHB established control units to apply these methods in various parts of the Philippines during the 1920s. The application of larvicides without the cooperation of local populations, however, failed to control malaria.

Paul F. Russell arrived in Manila in 1929. He has been appointed by the IHB to head laboratory work at the Bureau of Science. The bureau had been set up by American researchers but was largely under the control of Filipino nationals by the 1920s, much to the dismay of Heiser, who viewed these scientists as incompetent. Russell was also directed to take charge of malaria-control activities. Russell, a graduate of the Cornell Medical School, joined the IHB in 1923. Before coming to the Philippines, he had been assigned

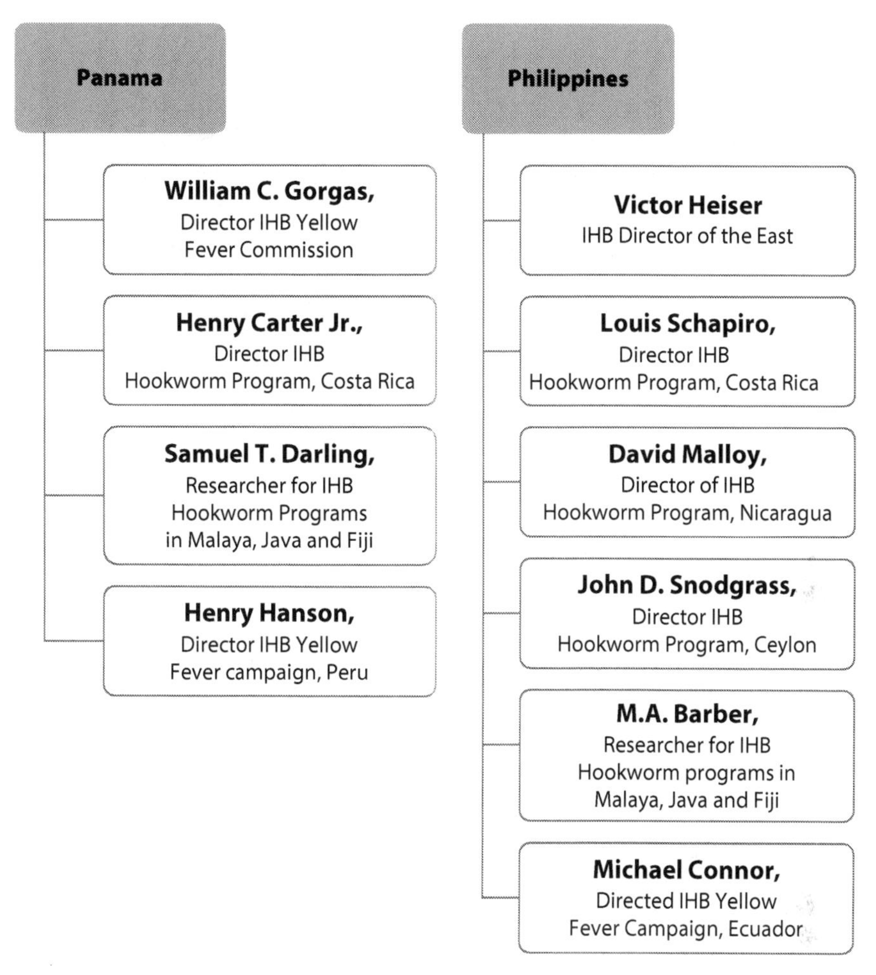

Colonial experience of directors of the International Health Board's hookworm and yellow fever programs

to hookworm work in the Malaysian Straits Settlements, Ceylon, and the American South. He had also received a Master of Public Health degree from the Harvard School of Public Health in 1929.

Russell initially recommended a shift in control strategies away from centralized larval-control programs, which he viewed as a total failure, and toward programs that involved more community participation. He also recommended the use of quinine and netting, as well as educational efforts. In a letter to Heiser, he wrote, "I feel strongly that control work must be locally desired and locally carried out," adding that "the work of a central malaria control division should be advisory, experimental, and instructive."[25] Russell's ideas presaged a growing interest in community participation in international-health programs, an attitude that emerged in the 1930s (chapter 4). Yet his malaria-control work failed to sustain this commitment. Like Heiser and other American health officers working in the Philippines, Russell succumbed to the view that Filipinos were incapable of participating in the protection of their own health. In a 1932 *American Journal of Tropical Medicine* article on "Malaria in the Philippine Islands," Russell wrote that "the difficulties of getting large groups of civilians to take antimalarial drugs systematically are as insurmountable in the Islands as elsewhere." Filipinos could not be trusted with quinine. Larval control was the only way forward. Like Heiser, Russell also came to view his Filipino counterparts as incompetent: "The average [Filipino] physician, whether or not he is called a health officer, dislikes to get his hands, his feet, or his white collar muddy and is rarely qualified for antimosquito control."[26] In the end, Russell's campaign was better organized than previous efforts, but it remained a top-down exercise in applied technology.

Russell would continue his malaria work in India and become a leading figure in the international malaria-control efforts. He was centrally involved in the International Health Division's malaria-eradication efforts in Sardinia after World War II and was one of the prime architects of the World Health Organization's Malaria Eradication Programme in the 1950s and 1960s. A close reading of his later diaries and writings indicates that the attitudes and lessons he learned in the Philippines continued to shape his malaria work.[27]

Russell's career clearly illustrates how colonial contexts shaped the attitudes and practices of a generation of American public-health authorities who would go on to lead future international-health efforts. These authorities would continue the tradition of attacking diseases by applying biotechnologies, while largely ignoring the need to develop basic health services or addressing the underlying causes of ill health.

PART II

Social Medicine, the Depression, and Rural Hygiene

The report of the 1932 Cape Town Conference, which began part I, contained a subcommittee report on rural hygiene. The report laid out five principles for the reorganization of colonial health services. First, it emphasized the need for preventive as well as curative health activities. Second, it called for cooperation between colonial health authorities and other administrative and technical departments, including agriculture, veterinary services, education, and police. Third, it highlighted the importance of improving the economic status of native populations, in order to ensure their health, asserting "that most important in the amelioration of health conditions are measures toward raising the economic status of the population." Fourth, it noted the need to increase education in elementary hygiene, beginning in the primary schools. Finally, it insisted that in both curative and preventive work, native personnel should be utilized to the fullest extent, particularly women.[1]

Three years later, in November 1935, the League of Nations sponsored a followup conference in Johannesburg. Like the 1932 meeting, it was attended by both African colonial officials and representatives of international organizations, including the League's Health Office, the Rockefeller Foundation, and the International Office of Public Health. Once again, the conference stressed the need for a broad-based approach to health, which rested on economic advancement. With regard to the control of malaria, the conference noted that additional research on African malaria was needed, but "it must not be forgotten that, without raising the economic status of the vast bulk of the population of Africa as a whole, there can be no hope of applying successfully on a continental scale the results of research or of markedly improving the position of great populations with regard to malaria as a disease."[2] Similarly, in discussing the problem of typhus in Africa, the conference participants had only one recommendation: "The Conference considers that raising the economic status of the natives is the only practical way of eradicating typhus."[3] Finally, with regard to health services, the conference reaffirmed the need for Africans—and particularly African women—to play a leading role in providing health services, the importance of obtaining the whole-hearted cooperation of all peoples within African countries, and the need for cooperation among departments of government "in the great task of maintaining the health of African populations."[4]

The recommendations of these two conferences reflected a vision of public health that deviated in important ways from that described in part I. In contrast to the military-styled disease campaigns that had dominated colonial medicine and the work of the Rockefeller Foundation's International Health Board during the 1910s and 1920s, the Pan African conference reports championed a broader conceptualization of health and health services, which included a wider range of institutional and local actors and took the social determinants of health seriously.

The two Pan African conferences were not anomalies. They were part of a broad movement of public-health leaders in the nations of Europe and their colonies who believed that improvements in health required both the development of health services and a link between health and patterns of social and economic advancement. This movement began in the late 1920s, with the emergence of social medicine in Europe. Social medicine was a concept that had many meanings, reflecting different national and intellectual traditions during the 1920s, 1930s, and 1940s.[5] For some, it meant eugenics and social hygiene. In the context of the international-health activities in the 1930s, social medicine encompassed a broader understanding of disease etiology, which included social and economic factors. It was also associated with the development of universal health-care systems. This broader vision of health gained ground during the Great Depression in the early 1930s and reached its height with the 1937 Intergovernmental Conference of Far Eastern Countries on Rural Hygiene, held in Bandoeng, Java. It paralleled and, in many ways, shaped the expansion of international-health activities during the interwar period.

Part II explores the development of this movement. It begins in chapter 3, with an examination of the work of the League of Nations Health Organization (LNHO) and how that organization moved from collecting statistics and developing technological solutions to health problems immediately after World War I to pursuing a broader vision of health and health care in the 1930s. The chapter describes the LNHO's efforts to apply this broader vision to the problems of rural hygiene and nutrition in Europe.

Chapter 4 describes how the LNHO and the Rockefeller Foundation's International Health Division implemented this broader vision of health beyond Europe during the 1930s. It also examines the efforts of British colonial researchers to improve nutrition. I argue that the 1930s raised the possibility of significant new approaches to international health, yet these changes remained more in the realm of vision and rhetoric than in actual achievements. In the end, international-health interventions retained a faith in scientific solutions, a limited understanding and valuation of local cultures, and an inability or unwillingness to address the structural conditions that underlay patterns of sickness and ill health around the globe.

In addition, more narrowly focused disease-control programs persisted during this period. For example, French scientists established a series of medical institutes, modeled on the famous Pasteur Institute in Paris, in France's colonies in Africa and Southeast Asia, as well as in several other international cities. Initially focused on extending Pasteur's vaccine research, the overseas Pasteur institutes became involved in local disease-control activities. Yet, following Pasteur's famous dictum, "Whatever the poverty, never will it breed disease," the Pastorians had little interest in extending health services and rural hygiene, or addressing the underlying social and economic sources of ill health. At the same time, throughout Latin America, the Rockefeller Foundation's International Health Division continued its disease-control campaigns to battle hookworm and yellow fever up through the 1930s.

In short, the broader visions of international health that emerged in the 1930s did not replace earlier approaches to public health. Instead, they represented an expansion or diversification of public-health strategies: an opening of new doors, or perhaps old doors that had been partially closed in the wake of the bacteriological revolution. Chapters 3 and 4 explain why this broadening of international-health strategies occurred, and why it achieved less than many had hoped.

CHAPTER THREE

The League of Nations Health Organization

THE LEAGUE OF NATIONS was established after World War I to maintain peace through collective security and disarmament and foster the settling of international disputes through negotiations and arbitration. The League's founders also recognized that maintaining peace required improving the social and economic welfare of the world's populations. Improvements in health were viewed as a critical part of this mission. To this end, the League established a Health Section, directed by a Health Committee, in 1922. The section eventually became known as the League of Nations Health Organization (LNHO).

Early Years

From its creation, the Health Organization was controversial. An International Office of Public Hygiene had been established in Paris in 1907. Its mission was to oversee international rules regarding the quarantining of ships and ports and prevent the spread of plague and cholera. It was also charged with collecting and distributing statistics from health departments around the world. It was an outgrowth of the international sanitary conferences that had been convened at the end of the nineteenth century. The responsibilities of the new LNHO clearly overlapped with those of the International Office of Public Hygiene. This overlap created tensions between the two organizations.

A second set of tensions centered on the Health Committee's definition of health and the extent to which its activities extended to social and economic issues. Would it, for example, make recommendations regarding occupational health, which many felt was the purview of the newly created

International Labour Organization (ILO)? Would it become involved in the health of refugees, or was this the responsibility of the International Red Cross? The charter of the LNHO called for cooperation with these other bodies but did not prevent it from crossing the line between health and social issues.

A third set of differences was over the question of whose interests the LNHO represented. Was its role to serve the administrative needs of member governments, providing statistics and standards with which to plan national health programs, or was it to promote the health of the world's populations, making policy recommendations that would place increased demands on the resources of member states? This tension split representatives from Great Britain and the United States, who believed that the LHNO should serve the needs of member states, from the Health Committee's first director, Ludwik Rajchman, who believed that the organization should work to improve health globally.[1] The United States was not a member of the League of Nations, its participation having been rejected by isolationist politicians in the US Congress. It did, however, accept an invitation to join the League's Health Committee. As a member of the Health Committee, and later the World Health Organization, the United States consistently resisted any proposals that would require it to alter US health policies.

The geographical scope of LNHO activities was another source of tension. The organization's immediate concern was the health and well-being of Europe's war-torn populations. Yet the League viewed itself as an international organization, and Rajchman believed that the LNHO had a responsibility to protect the health of all the world's people. He accordingly attempted to extend its reach to the Americas, Asia, and Africa. These efforts were met with resistance from European colonial authorities, other international organizations, and the United States.

In the Americas, the LNHO found itself partially blocked by the Rockefeller Foundation's International Health Board, which was heavily invested in hookworm- and yellow fever–control activities. In addition, the various countries in the region had their own international-health organization, the Pan American Sanitary Bureau (PASB), which had been formed in 1905 to share information and prevent the regional spread of epidemic diseases. The headquarters of the PASB was in Washington, and the surgeon general of the United States was the permanent president of the organization. Given the Unites States' refusal to join the League of Nations and the sensitivity of many US political leaders to subordinating the country's interests to a supranational organization, the LNHO found it difficult to gain a foothold in the Americas. Hugh S. Cumming was the US surgeon general and director of the PASB from 1927 to 1947. Cumming was also a member of the League's Health Committee. While he supported many of the LNHO programs in Europe, he was very sensitive to the committee's efforts to expand its activities into the Americas. Cumming viewed official League visits to Latin America as "a quite evident effort to lessen the prestige of the Pan American Sanitary Bureau."[2]

Africa and South and Southeast Asia also provided the LNHO with limited opportunities to extend its reach. These regions remained largely divided into colonies controlled by various European powers.[3] Colonial administrations were generally resistant to having the health conditions of their colonial subjects scrutinized by outside authorities. Nonetheless, some progress was made in these areas.

The League established a Far Eastern Bureau in Singapore in 1924 to collect health data from East Asian and Southeast Asian countries. The Rockefeller Foundation initially funded the bureau for five years. In a relatively short time, the bureau formed an epidemiological intelligence network, connecting hundreds of ports from Cape Town to Vladivostok. The Far Eastern Bureau subsequently extended its field activities to collecting and disseminating information about Far Eastern health services, the value of oral immunizations against acute intestinal infections, and certain aspects of plague and cholera. The Far Eastern Bureau also helped organize the Intergovernmental Conference of Far Eastern Countries on Rural Hygiene, held in Bandoeng, Java, in 1937.[4]

In Africa, the LNHO succeeded in organizing two Pan African conferences of health officers to discuss common health concerns. In addition, the LNHO appointed a subcommittee to assist in the study of sleeping sickness and tuberculosis in Central Africa. France, Britain, and Belgium had sent scientific missions to research sleeping sickness, which threatened the economic development of their colonial holdings in the region. The subcommittee was composed of representatives of each of these three countries. Its final report recommended more intercolonial cooperation and presaged the Cape Town meeting in 1932.[5] It was in China, largely free of direct European colonial control or American imperial influence, where the LNHO was able to make its most significant headway in establishing health programs outside of Europe (chapter 4).

Uncertainty about the LNHO mission and the geographical scope of its activities, together with recognition that it needed to establish its credibility among the League's member states, led the Health Committee to initially focus on activities that were widely viewed as having practical value for its members. These included the collection and distribution of health data from around the world. One of the LNHO's most appreciated activities was its Epidemiological Intelligence Service. The committee also worked on establishing standard medical measurements related to vitamins, syphilis diagnostics, and sex hormones, as well as testing medical interventions, such as the recently developed bacille Calmette-Guérin (BCG) vaccine for tuberculosis. It also established standards for public-health education.

In addition, the LNHO served in an advisory capacity to countries that wished to reform their health services. In 1928, the newly elected government of Greece asked for the LNHO's assistance in rebuilding its health system, following a devastating epidemic of dengue fever. The resulting reforms led to the centralization of the country's health services and the creation of a national school of public health. The Health Organization received a similar request from the government of Bolivia. Yet the LNHO's efforts there were impeded by both inter-European politics and the interference of the US surgeon general, who felt that Bolivia should have turned to the Pan American Sanitary Bureau for assistance. The LNHO's Bolivian experience revealed the barriers that stood in the way of the LNHO's involvement in Latin America.[6] At the same time, the LNHO went out of its way to limit discussions to nonsocial aspects of health. In 1922, Rajchman explicitly stated that the LNHO would not be principally concerned with social health but would advise the League's Social Section on issues requiring health expertise.

Finances also drove the LNHO's early conservatism. Chronically underfunded, the Health Committee sought outside financial resources. It turned to the Rockefeller Foundation, which agreed to fund one-third of the LNHO's budget. This financial dependence encouraged the LNHO to operate in ways that were compatible with the Rockefeller Foundation's own narrow view of health as a problem of science and organization and to avoid social and economic issues.[7] The LNHO and the IHB advocated similar approaches to international health up through the 1920s. They also had overlapping personnel. The major distinction between the two organizations was that the LNHO reported to the League's member nations, while the IHB answered only to the Rockefeller Foundation's head office. In addition, the LNHO's approach was more oriented toward data collection and the testing of therapies, while the IHB was concerned with the development of organizational systems for the application of medical science to disease elimination. This alliance would not be the last time that a wealthy private foundation was able to shape global-health agendas.

For Rajchman, the LNHO's conservatism was clearly a question of pragmatism rather than ideology. As the person most centrally responsible for the shift in international-health policies during the 1930s, it is important to understand how he came to take up this role and how his background led him to espouse broader social and economic perspectives on health while, at the same time, advocating technical solutions that reflected his medical background. This tension between a broad vision of health and society and narrow technical solutions would characterize much of international health during the 1930s.

Rajchman was born in Poland in 1881 and received his medical training in Cracow. Active in student politics, he cofounded a resource center on labor conditions, became involved in an association of socialist students, and joined the Socialist Party in 1905. Arrested for his part in a 1906 worker uprising, Rajchman was forced to leave Poland. While Rajchman would retain his socialist sympathies throughout his career, he dedicated the next 15 years of his life to advancing his expertise in scientific approaches to public health.[8] From Cracow, Rajchman went to Paris, where he gained additional scientific training at the Pasteur Institute. The Institute was a major center for the advancement of microbiology, attracting researchers from around the globe.[9]

Rajchman's training and connections at the Pasteur Institute led to his being appointed head of the bacteriology laboratory at the Royal Institute of Health in London, even though he did not speak English at the time of his appointment.[10] While in London, Rajchman reconnected with Polish socialist leaders, who constituted a significant immigrant group in London before the war. He also had contacts with the progressive British Fabian Society. These associations did not distract him from his bacteriological work, but reflected his continued engagement with socialist ideas.

Following World War I, Rajchman returned to Poland, which had regained its independence but had been economically devastated by the war. The threat of epidemic disease was ever present. Rajchman founded the Polish Central Institute of Epidemiology (the future State Institute of Hygiene) on the model of the Pasteur Institute. In his role as the institute's director, between 1918 and 1922 he was involved in efforts to contain a series of epidemics. Spanish influenza was the most well known of these afflictions, but Poland was also hit by outbreaks of dysentery, diphtheria, typhoid, venereal disease, and typhus. The latter disease infected around 4 million Poles.

Rajchman attempted to set up a surveillance system to identify cases throughout the country, but he was hampered by the lack of trained bacteriologists in Poland. To solve this problem, Rajchman approached the Rockefeller Foundation for funds to create a school of public health in Poland. The foundation hesitated, because it viewed the country as politically unstable. But its representative to Eastern Europe, Selskar Gunn, met with Rajchman and was very impressed with the man and what he had achieved. Meanwhile, typhus continued to spread in Europe. In an effort to respond to the crisis, the League of Nations appointed an Epidemic Commission, composed of Rajchman and two British officials, to spearhead control efforts.[11]

Rajchman was subsequently appointed as medical director of the LNHO in 1921. He was chosen in part because he was Polish, and Poland was the epicenter of the battle to contain the typhus epidemic that was threatening Europe. But he also came with strong recommendations from those who had worked with him, including representatives of the Rockefeller Foundation, who praised his efforts to establish a bacteriology laboratory in Warsaw, and the Red Cross, which had worked with him in fighting typhus in Poland. It also helped that by this time, he had become totally fluent in English, French, and German, as well as Polish. The appointment forced Rajchman back to Poland, but it provided him with a much larger stage on which to act, as well as an opportunity to define a new agenda for public-health practices in Europe and beyond, one that combined his knowledge of bacteriology and the principles of hygiene with a commitment to socialist principles and the need to address the underlying social determinants of health.

International Health and the Great Depression

There is general agreement that the Great Depression played a key role in moving the League of Nations Health Organization away from previous models of public health and toward ones that took a broader approach to health. The Depression, which drove millions of people across the globe into poverty, made clear the relationship between economics and health. But there was widespread appreciation of the importance of the social and economic determinants of health and of the need for sanitary reforms prior to the Depression (chapter 1). In many ways, what emerged in the 1930s represented the recasting of a set of ideas about social medicine that had their basis in the writings of Rudolf Virchow and others. Moreover, there is evidence that these ideas had begun to find expression in international-health circles by the late 1920s, stimulated in part by the limited explanatory power of bacteriology. Few medical advances had flowed from the identification of the pathogens that caused the major diseases affecting human populations. In addition, the failure of disease-elimination campaigns to achieve their goals also stimulated health authorities in various parts of the globe to rethink how best to achieve improvements in health.

This rethinking led to a broader inquiry into disease etiology that encompassed the roles of nutrition, housing, working conditions, agricultural production, and the economy. It also encouraged investments in the development of health-care systems. This shift in perspective was manifested in the LNHO's efforts to deal with malaria, rural hygiene, and nutrition in postwar Europe. It subsequently shaped the programs the LNHO developed in other regions of the globe, as well those of the Rockefeller Foundation's programs in China, Mexico, Ceylon, and Java during the 1930s (chapter 4).

MALARIA

Interest in the social basis of health emerged before the Great Depression, in efforts to control malaria in Europe. The LNHO created a Malaria Commission in 1924 to deal with wartime upsurges in malaria in many European countries. The Malaria Commission's members were asked to examine malaria-control measures in various League of Nations countries and make recommendations concerning the most effective strategy for combating the disease. But the underlying purpose of the commission was to resolve a disagreement between supporters of vector control who, like Gorgas, Ronald Ross, and the Rockefeller Foundation's International Health Board, believed that killing mosquitoes was the most effective way to eliminate malaria, and those who preferred to focus on the human host and thus attack malaria parasites through the distribution and use of quinine and by general improvements in health. The human-host approach stemmed from the work of Robert Koch, who had identified the specific causative agents of cholera, tuberculosis, and anthrax. Koch asserted, "Treat the patients, not the mosquito." The human approach had been successfully applied through the mass distribution of quinine in Italy before World War I. The Malaria Commission's first report, issued in 1924, placed great emphasis on the need to protect human populations through a combination of "quininization" and improvements in housing, education, nutrition, and agriculture.[12] It concluded: "The Commission feels bound to reiterate the importance of the general social and hygienic condition of the people in relation to the extent and severity with which malaria shows it[self]. This has been brought to notice in many areas: better housing, an ampler and more varied dietary, and better environmental conditions make for a more intelligent and willing people and for greater individual resistance."[13]

The report's first author was British malariologist Colonel Sydney P. James. James's lack of enthusiasm for vector control was deep seated and stemmed from his experience with a failed early vector-control experiment carried out at a military encampment in Mian Mir, India. Like the American physicians who joined the IHB following their service in US colonial possessions, James came to the LNHO's Malaria Commission from colonial service in India. James's opposition to vector control was also based on his experience in England, where malaria largely disappeared before the initiation of any antimalarial activities. James was strongly committed to the idea that the elimination of malaria followed from broader social and economic development. Visiting Kenya in the late 1920s, he recommended that the best way to eliminate malaria in the so-called native reserves, in which the majority of the colony's African population lived, was "to introduce agricultural, and in some cases industrial welfare schemes aimed at improving the economic and social conditions of the people and their general well-being and standard of living."[14]

The Malaria Commission's appreciation of the role of social and economic advancement was influenced in part by an early Italian success in eliminating malaria through a policy of generalized rural uplift known as *bonifica integrale*, or bonification, begun after World War I. Bonification sought to both eliminate malaria and create new agricultural lands that would support poor farmers, allowing them to improve their economic status along with their overall health and well-being. The League of Nations

The League of Nations Malaria Commission. Wellcome Trust.

Malaria Commission praised this Italian experiment. Bonification achieved its most spectacular success in clearing malaria from the Pontine marshes outside of Rome under the Fascist government of Mussolini, which came to power in 1922. The project drained 200,000 acres of swampland, and the reclaimed land was then cleared of undergrowth. New roads were constructed and new agricultural colonies were founded, each with its own church, hospital, and social club. In addition, the state established rural schools and health centers. By 1939, the population of the bonified areas had grown from 1637 people in 1928 to over 60,000.[15] The Italian success became a model for similar projects in Palestine in the 1920s and, with less success, in Argentina in the 1930s. The late 1920s also saw the creation of malaria societies in Bengal that combined malaria-control activities with improvements in overall sanitation and agriculture.[16] Broader approaches to health and development were thus under way before the 1930s.

Yet it was the Depression that led the LNHO to move more decidedly in this direction. The move was signaled and made possible by the LNHO's partnering with other League organizations, including the International Institute of Agriculture (the predecessor of the post–World War II Food and Agriculture Organization), the International Red Cross, and the International Labour Organization. These partnerships led to the formation of joint, or mixed, committees on rural hygiene, housing, labor conditions, maternal and child welfare, and nutrition during the 1930s. In doing so, they linked health to a broader array of social and economic issues.

Symptomatic of the LNHO's changing vision of health during the Depression was its approach to the problem of tuberculosis (TB). During the 1910s and early 1920s, the LNHO's Tuberculosis Committee focused its activities narrowly, measuring the effectiveness of the BCG vaccine in preventing tuberculosis; in other words, concentrating on finding a technical solution to TB control.[17] By contrast, its 1932 report, while acknowledging the potential contribution of BCG and the role of medical interventions in the fight against TB, concluded: "Society must cleanse itself; it cannot look for success to the practice of traditional medicine, concerned mainly with the treatment of the individual. . . . Nowadays, the general line of advance must be toward improved nutrition and housing, education and social discipline, and greater security in the matter of health and employment."[18]

RURAL HYGIENE

The health of rural populations became a central focus of the LNHO's new approach, first in Europe and then more broadly across the globe. Farming economies were devastated by the Depression. In 1930, 1931, and 1932, world agrarian prices fell 40, 28, and 12 percent, respectively. Despite diminished prices, farmers found themselves unable to sell their produce, due to a lack of demand. Farmers in Eastern Europe were particularly hard hit, and their growing impoverishment raised fears in the West that Bolshevism, the anticapitalist ideology that had arisen with the Russian Revolution of 1917, would spread from Russia to eastern and southern Europe.[19]

Colonial populations were also seriously affected. For the first time, the Depression revealed the extent to which the welfare of colonial populations had become tied to their participation in the colonial economy, which, in turn, was linked to the global economy. Income from the sale of cash crops and animals, or from industrial or agricultural wages, had become essential to their economic well-being. As the Depression eliminated or severely reduced access to these forms of income, colonial populations suffered from increased levels of malnutrition and disease.[20] The results of these changes became especially apparent in the increasingly wage-labor economies of southern Africa, where labor recruiters noted the deteriorating physical condition of African mine recruits during the 1930s. This led to a series of investigations into rural health conditions, which revealed widespread patterns of malnutrition and disease.[21]

In September 1930, the Spanish government requested the LNHO's assistance in organizing a conference on rural hygiene in Europe. The LNHO received a similar proposal from Hungary, which suggested that the conference focus on the role of health centers in promoting rural health. After World War I, health centers had emerged in a number of European countries as multipurpose institutions encouraging a broad-based approach to health. The Health Committee agreed to organize the conference. After several preparatory meetings and discussions, the conference was held in July 1931 in Geneva. It brought together representatives from 24 countries, the International Institute of Agriculture, the International Labour Organization, the International Red Cross, and the Health Committee of the LNHO. It included public-health administrators, agricultural experts, engineers, hygienists, and insurance experts and emphasized that improvements in rural hygiene required cooperation among these various groups.[22]

The conference report represented a careful balance between two approaches to public health. On the one hand, it defined rural hygiene in the broadest terms, linking it clearly to problems in agriculture, inequalities of land ownership, and rural poverty. This approach harkened back to nineteenth-century models of public health and sanitation, only it focused on rural rather than urban health. It also echoed the conclusions of the League's Malaria Commission in linking health to overall rural development.

On the other hand, in making its recommendations, the report emphasized technical problems and solutions and trod lightly on the politically contentious issues of land ownership and poverty. This is not surprising. The term *hygiene*, which we tend to associate with cleanliness, had a more specific technical meaning for public-health workers in the 1920s and 1930s. The 1934 edition of *Webster's International Dictionary* defined hygiene as "the science of the preservation of health; sanitary science; a system of principles or rules designed for the promotion of health." *Webster's* secondarily defined hygiene as "the branch of medical science that deals with the preservation of health." The emphasis here was on *science*. Hygiene was the application of the new bacteriological knowledge that had emerged at the turn of the twentieth century. Thus, when public-health officials spoke or wrote about "rural hygiene," they focused on the technical aspects of health related to the prevention of disease, be it clean water supplies, the avoidance of overcrowding, or nutrition.

The balance between a broad vision of health and a focus on hygiene can be seen in the conference report's discussion of housing. That section began with the statement: "Good housing is a fundamental requirement for rural hygiene. It is influenced by social and economic conditions and in its turn exerts a strong influence on these conditions, resulting in better health and a general elevation of the standard of life."[23] But the report quickly shifted gears and identified a series of technical problems: too small houses, too few bedrooms, insufficient cubic space per occupant, inadequate toilet and sanitary facilities, lack of ventilation, insufficient exposure to sunlight, and minimal protection from mosquitoes, flies, and dust. In order to ameliorate these conditions, the report recommended the creation of standards and regulations for ensuring the proper construction and maintenance of houses. It also called for educational activities aimed at teaching farmers the basics of "good housing." It did not call for higher wages and agricultural prices, or land reform that could produce larger rural incomes.

The conference report also made a series of recommendations concerning the provision of both curative and preventive rural health services. It concluded that health authorities "should ensure that the entire population benefits from an effective medical assistance" and that there should be "one duly qualified medical practitioner" for every 2000 persons, with the goal of reducing this to one practitioner for every 1000 persons.[24] By balancing broader approaches to health with technical recommendations, the conference reflected the realities of the postwar political environment, in which there was considerable disagreement about the role of international organizations, the limits of rural reform, and the very definition of health.

NUTRITION

Nutrition was another area in which the LNHO attempted to promote a broader vision of health while retaining a focus on technical aspects of the problem. Since the discovery of the role of vitamins and minerals in human health in the 1910s and 1920s, there had been growing interest in the role that nutrition played in determining the health of individuals and populations. Part of this interest stemmed from recognition that the bacteriological revolution had yet to provide much in the way of medical weapons for attacking the causes of diseases it had revealed. Improving nutrition provided a way of fighting disease by strengthening the health of individuals. Yet unlike bacteriology, nutritional science had clear connections with broader social and economic processes related to the production and consumption of food. Focusing on nutrition therefore had the potential to cross disciplines and encourage collaboration among a wide range of professionals, including economists, botanists, agriculturalists, and engineers. Whether one decided to make these connections or to focus narrowly on dietetics always prompted tension within nutritional investigations during the 1930s, including those undertaken by the LNHO.

During the 1920s, the LNHO chose to concentrate on the development of nutritional standards, as part of its larger mission to establish health standards for procedures (such as the Wassermann tests for syphilis), medicines (such as insulin), and vaccines. Later in the 1920s, the standards program was extended into epidemiology, with the standardization of disease categories and statistical procedures.[25] In the wake of the Depression, however, the LHNO broadened its approach to nutrition. In 1931, the LNHO appointed British scientist Wallace R. Aykroyd as its first nutritionist. Aykroyd had received his medical training in Dublin and taken up his first appointment in Newfoundland, where he investigated the effects of vitamin A deficiency and, oddly enough, beri-beri, a disease more often associated with poor Asian laborers living on processed white rice, but one that also affected fishing communities in this region of the world. In Geneva, Aykroyd joined a team of researchers who were primarily concerned with the effect of mass unemployment on public health and spearheaded efforts to broaden discussions of the sources of malnutrition to include social and economic forces.[26]

In 1932, the LNHO and the ILO issued a joint memorandum titled "Economic Depression and Public Health." The memorandum reviewed statistics for overall mortality and for TB mortality and morbidity in a number of European cities. It found that there was little evidence that the Depression had increased any of these health indicators. The memorandum also reviewed German data on nutrition among unemployed workers and their families, however, and concluded that there were serious qualitative deficiencies in their diets; these deficiencies were accentuated by inadequate unemployment allowances.[27] The memorandum, like the reports of the Malaria Commission and the European Conference on Rural Hygiene, shed light on the broader economic causes of ill health.

Over the next three years, the LNHO put forth a series of recommendations regarding nutritional standards. These recommendations met with resistance from those who opposed the League's role in establishing health requirements for member countries. Predictably, the British representative, George Buchanan, wrote in 1933 that it was not, in his opinion, "the business of an international conference of experts, who did not represent their governments, to say how health administrations in different countries should be reformed, reorganized, or rationalized. The work of the Health Organization on this type of question should be limited to supplying carefully considered and authoritative intelligence reports."[28]

Despite this resistance, the LNHO continued to push for improvements in nutrition. In 1935, Aykroyd and Etienne Burnet published a 152-page report titled "Nutrition and Public Health." Much of the report summarized existing knowledge about normal nutritional requirements and dietary standards. In addition, it reviewed the public-health consequences of various dietary deficiencies. Yet it also drew attention to the larger issues related to food supply. At the beginning of the report, the authors stated that "nutrition is an economic, agricultural, industrial, and commercial problem, as well as a problem of physiology."[29] Under a chapter on "Food Supply," they included a section on "Points of Contact between Nutrition and Economics" and another section on "Nutrition and Poverty." The latter section referred to studies from a number of countries that correlated dietary patterns with socioeconomic status. The report also contained a section on "Colonial Dietary Problems." It concluded, "It is now clear that 'colonial' populations in general are under nourished, and governments have begun to recognize this fact and taken steps to deal with the situation."[30]

That same year, concerns over nutritional problems led to the creation of a Mixed Committee of the League of Nations on the Relation of Nutrition to Health, Agriculture, and Economic Policy. The committee included representatives of the LNHO, the ILO, and the International Institute of Agriculture and brought together specialists in health, labor, agriculture, finance, social welfare, cooperation, and administration. The Mixed Committee's report, published in 1937, focused on "Europe and countries with a Western civilization," excluding Asia, Africa, and tropical counties in general, due primarily to what it termed "methodological problems." Yet it acknowledged that malnutrition was a major problem in many of these areas and chided colonial governments for their lack of attention to them, noting, "It is notorious that a large proportion of them live in a stage of continual malnutrition; but up to the present very little systematic information on this point has been collected."[31] This criticism stirred Great Britain to initiate a survey of nutritional conditions in its empire.

The report attempted to strike a balance between a technical scientific approach to nutritional studies and a desire to highlight the broader societal forces shaping access to adequate diets. It recommended that governments work to provide scientifically based nutritional information to all segments of society. But it also encouraged governments to increase the availability of food by making credit available to food producers and regulating food prices, stating, "Governments should also ensure that those without adequate incomes have access to proper foods." The report also observed:

> Food consumption levels depend primarily on two factors—the amount of real income at the disposal of the family or individual, and the intelligent use of that income to meet food requirements. . . . Where income is adequate, the problem is one of education; but where resources fall below the limits of reasonable living standards, adequate nutrition cannot be secured even by the most scientific expenditure of those resources.[32]

The Mixed Committee trod gently here, stopping short of proposing that national governments employ any specific strategy for increasing the ability of the poor to acquire adequate nutrition. Instead, it listed a series of measures that had been tried by various countries, including family allowances, systems of social insurance, and the establishment of minimum-wage levels. It came down strongly, however, on the need to focus particular attention on the nutritional status of children.

In making its recommendations, the report stressed that nutrition was a responsibility of state governments. But it emphasized that one size did not fit all and that it was up to each state to decide what policies to follow in line with its own specific conditions, a clear acknowledgment of the resistance of some countries, particularly the United States and Britain, to having international bodies dictate which public-health policies they should follow. It recommended that each nation should form a nutrition committee—

staffed by experts not only in physiology and nutrition, but also including agriculturalists, economists, consumer representatives, and teachers—so the problem of nutrition could be seen from different angles, including health, labor, agriculture, economics and finance, social welfare, cooperation, and administration.[33] Despite its recognition of the need to address broader social and economic forces, the Mixed Committee's final recommendations nonetheless gave more weight to technical questions related to the production and dissemination of nutritional knowledge and standards. Only three of its fifteen recommendations addressed broader economic concerns.

Beginning in the late 1920s, the League of Nations Health Organization shifted directions, developing a more broad-based social approach to improving health and nutrition in Europe that was formed in partnership with other international organizations, including the ILO. This approach marked a break with the past. Yet it was tempered by a political environment that limited efforts to address the fundamental structural causes of disease and malnutrition. For Western governments, broad-based social reform raised the specter of Bolshevism. It was also influenced by the continued dominance of a scientific vision of hygiene that focused attention on the proximate causes of ill health and on technical solutions to health problems. As we will see in the next chapter, this tension would also shape the efforts of international-health organizations to improve the health of populations outside of Europe, particularly in China during the 1930s.

CHAPTER FOUR

Internationalizing Rural Hygiene and Nutrition

THE LEAGUE OF Nations Health Organization's commitment to improving rural hygiene and nutrition and to addressing the social determinants of health in Europe carried over to its work in other parts of the world, particularly China, during the 1930s. Similarly, the Rockefeller Foundation's International Health Board, now renamed the International Health Division (IHD), undertook new initiatives in China, Mexico, Ceylon, and Java that shared a commitment to improving rural health services and sanitation and, in some cases, a broader vision of health and development. Also during this period, British colonial researchers, responding to increased evidence of malnutrition among that nation's colonial subjects, initiated a series of inquiries that broadened the scope of nutritional research in ways that were similar to the LNHO's nutritional inquiries in Europe.

In many ways, these various interwar interventions represented a break from the previous, narrowly focused disease campaigns (chapter 1). Yet in some cases, they retained a set of attitudes and practices that reflected earlier conceptualizations of the health of colonized populations. In addition, like the LNHO's activities in Europe (chapter 3), they reflected an ongoing tension between the application of technical solutions and advocacy of broad-based social and economic reforms.

The LNHO in China

Ludwik Rajchman, the director of the LNHO Health Committee, traveled to China in 1924 as part of a Far East tour. He was struck by China's lack of sanitation and poor medical services and saw innumerable possibilities for improving health conditions in the country. In an unofficial report, he recommended that the League undertake a multifaceted program in China, developing assistance projects and sending technical experts on public health, economics, state finances, and public transport.[1] The recommendation was not immediately acted on, but in 1929, Chiang Kai-shek assumed leadership of the Chinese government and established a new Ministry of Health. He requested assistance from the Rockefeller Foundation's International Health Board and invited several eminent health experts, including Rajchman, to come to China. Rajchman spent two months in China, traveling, observing, and meeting with officials. He came away even more convinced of the opportunities and needs for public-health improvements, but also with the recognition that such improvements required economic progress that would take time to achieve. In addition, Rajchman believed that health could be the leading edge for a wide range of technical collaborations with China. He and the country's minister of health developed a plan for antiepidemic measures and maritime quarantines.

Rajchman's broader vision of rural reconstruction in China had to wait until 1933, when the League's Council appointed him to serve as the technical agent responsible for coordinating cooperative relations between China's National Economic Council and the League of Nations. The National Economic Council had been created in 1931 to oversee the country's economic reconstruction in the wake of a series of disasters, including civil war with Mao Zedong's Communists, devastating floods, Japan's invasion of Manchuria and Shanghai in 1929, and the effects of the worldwide economic depression. China was a member of the League of Nations, and the new National Economic Council turned to the League for technical advice, establishing an agreement for a program of technical cooperation. This led to a series of missions by experts in health, agriculture, water conservancy, road construction, and education, who traveled to China for varying periods of time. Importantly, the League's health interventions were part of an integrated approach to health and development.

Not all League members were supportive of the League's involvement in China. The French and British representatives to the League viewed Rajchman's China forays with suspicion. These two European powers had substantial investments in Asia and were concerned about the League's interference in what they saw as their spheres of colonial influence. Moreover, Rajchman's view of the relationship between China and the League was one of equals. He had a high regard for the Chinese officials who staffed the country's various ministries, many of whom had been trained in the West. He also believed that the League should respect Chinese Nationalist aspirations and not try to dictate policies to them.[2] He employed the term "technical collaboration," rather than "technical assistance," in his dealings with the Chinese. This was a viewpoint that often clashed with the more paternalistic views held by the British and French, as well as by officers of the Rockefeller Foundation's International Health Division prior to 1930.[3] The Japanese also opposed the League's involvement in China and Rajchman's role as a technical agent. The Japanese viewed Rajchman as an unquestioned supporter of Chinese interests and viewed his proposals for Chinese development, as well as his efforts to raise European capital to support them, as impediments to Japan's expansionist goals in China. An economically strong China would be difficult to manipulate.[4]

Despite this opposition, Rajchman was able to move forward with his proposals for rural reconstruction. The various programs initiated through Rajchman's negotiations, however, did not always reflect his progressive vision of technical cooperation. For example, William Campbell was tapped to head an agricultural project as part of the League's technical cooperation agreement. Campbell had worked for the Indian Civil Service in Ceylon for more than two decades. He was one of a number of British colonial officers who found their way to the League of Nations in the 1920s and 1930s, paralleling the earlier migration of US colonial physicians to the Rockefeller Foundation's IHD. Like a growing number of technical advisors in China, Campbell had no prior experience working in the country. He did not view this as a handicap, however, asserting that he was an expert in cooperatives in "backward countries." Campbell was thus part a growing body of international-health advisors who defined themselves in terms of their technical expertise rather than their local knowledge.[5]

Like the public-health officers in the Philippines and Panama, Campbell viewed local populations as unable to manage their own affairs, a view that Rajchman explicitly rejected. Campbell noted in his book on practical cooperation in Africa and Asia that "undeveloped, ill-educated, and poverty-stricken masses in less fortunate countries" were "resigned and apathetic," strikingly different from the "inhabitants of highly developed, well-educated, and sophisticated countries."[6] While Campbell believed in the power of agricultural cooperatives, he did not see them as necessarily evolving from below. He asserted that "in undeveloped countries the need for it [cooperation] is so great and the prospect of its spontaneous birth so small that Government is justified in taking a hand in fostering it." At the same time, he was critical of Chinese efforts to create their own cooperatives. He complained that the Nationalist government was moving too fast to create cooperatives. In this he failed to appreciate the political motivations that shaped the government's cooperative policies—an example of his general ignorance of Chinese society. For the Chinese government, cooperatives were a means of extending its authority over rural areas of the country. The failure to appreciate local motivations for development programs characterized the work of a generation of international-health experts over the next half century. Campbell's cooperative movement also failed to appreciate the nature of rural village economies. He viewed cooperatives as purely agricultural in nature, whereas local village economies often involved a range of economic activities, including small-scale manufacturing.[7]

Not all of the League's advisors shared Campbell's views of Chinese development. Some, like Carlos Dragoni, an internationally prominent agricultural expert from Italy, appreciated the need for China to carve its own path to the future and not to be dictated to by outside advisors. He also believed in land redistribution and reclamation as means to increase the access of many Chinese to land.

Of all the experts who went to China as part of the League's program of technical cooperation, Andrija Štampar most closely shared Rajchman's broad vision of public health. Štampar was a Croatian physician who graduated from medical school in Vienna in 1911. He worked for a while as a hospital physician and then as a district health officer. At the end of World War I, he moved to Zagreb to take up an appointment as health adviser to the Croatian Commission for Social Welfare. He became intensely involved with health policy, publishing a book and a series of outspoken articles in the journal *Jugoslavenska njiva*. Štampar's articles and book expressed his commitment to the ideals of social medicine. In 1919, at the age of 31, he was appointed head of the Department of Public Health in Belgrade, in the newly constituted country of Yugoslavia.[8]

The country had been devastated by World War I and was in the midst of a period of reconstruction. Štampar began his work with few resources, but he was able to attract a group of young doctors committed to transforming health conditions in the country. Henry Sigerist, director of the Johns Hopkins Institute of the History of Medicine, visited Yugoslavia for an international history of medicine meeting in 1939. He noted that Štampar approached his task from the broadest possible perspective, recognizing that it made no sense to distribute drugs if people needed food. Štampar focused on improving not only peoples' health, but also their living and working conditions. He oversaw the creation of an extensive health system, with over 300 new health units across the country. In 1930, as the political atmosphere in Yugoslavia shifted toward Fascism, Štampar's progressive approach to public health fell into disfavor, and he was forced out of his position. He shifted his energies to international health, taking up a full-time position with the LNHO.

Štampar was sent to China as a health advisor in 1933. Coming from a country that suffered from the effects of war and was in the midst of its own program of economic reconstruction, Štampar recognized the need for linking health with broad-scale social and economic reform. His first assignment was to serve on a commission sent to survey health conditions in Kiangsi Province, a region that had been the site of a protracted civil war between the Communists and governmental forces. The commission's report pointed to the burden imposed on rural farmers by the system of land tenancy and taxes that ensnared them in poverty. It called for land reform that would prevent the concentration of land in the hands of a few wealthy landlords. The commission also proposed that a central welfare center be created, joining social and economic reforms with health services. While the government had no interest in engaging in land reform, it did create a welfare center in Nanchang, which linked agriculture, the cooperative movement, and health.[9] In his report, titled "Health and Social Conditions in China," published after his return from that country in 1936, Štampar stated: "Public health policy must be intimately connected with a programme for general social improvement. Education is important, since, unless the farmer can read pamphlets, and is given a rudimentary scientific attitude, it is very difficult to reach him by propaganda. Of perhaps even greater importance is the removal of social grievances, such as the sense of exploitation by the landlord."[10]

Any objective evaluation of the League's activities in China would have to conclude that it achieved less in terms of serious social and economic reform than Rajchman or Štampar viewed as being essential to improving health. Some of their recommendations were followed, but the Chinese government was slow to move on others, and in the end the Japanese invasion in 1937 stymied some reforms that might have occurred, given more time. Nonetheless, the League's activities in China need to be viewed in relationship to the earlier activities of the LNHO. Rajchman and Štampar clearly laid out a blueprint for health reform that was bold in both its scope and its recognition of the need for international organizations to work cooperatively with local governments and populations in developing regions of the globe. It marked a clear departure from colonial health interventions, as well as those implemented by the IHD, even if certain remnants of earlier attitudes regarding the backwardness of local populations could be seen in the statements of some League consultants. Importantly, the LNHO's China program was part of a larger shift in international-health aid during the 1930s and was mirrored by the Rockefeller Foundation's own China program and its work in Mexico, Ceylon, and Java. The shift was also evident in the nutritional work of British colonial researchers.

The Rockefeller Foundation in China, Mexico, Ceylon, and Java

CHINA

The Rockefeller Foundation had been involved in China since before the war and had funded Peking Union Medical College (PUMC), which opened in 1921. The PUMC focused on training physicians in Western medicine and was narrowly biomedical in its scope, though one of its directors, John B. Grant, was actively involved in developing primary health-care centers.

The foundation's China program in the 1930s was based on a much broader vision of health. The China program was the brainchild of Selskar M. Gunn, a foundation vice president and director of the IHD's European programs. Gunn was not a physician and was not wedded to a narrow, biomedical view of health. Moreover, he had spent time with Štampar in Yugoslavia and had come to appreciate the need for health efforts to be integrated with social and economic reform. Gunn's China program was intended to bring together health services, agricultural development, economics, mass education, preventive medicine, and industrial reform in an integrated program of rural reconstruction.

The program represented more than an extension of the foundation's earlier commitment to advancing health in China. It also reflected its growing belief that the fledgling social sciences and social research could play an important role in shaping societies as a whole. The trustees of the foundation adopted a resolution in 1929, stating that henceforth, "the possibilities of social experimentation were to be constantly in mind." By 1930, the foundation was displaying interest, through the IHD, in experimenting with holistic programs of community development in southeastern Europe. This holistic view, in which improvements in health were inseparable from broader social engineering designed to ensure the uniform and systematic advance of all essential phases of life, was given its most extensive trial in China.

The China program was sold to the foundation's trustees as an opportunity to provide the social sciences with something that had hitherto been lacking, namely, a "laboratory" where experiments could be carried out under controlled conditions. John B. Grant, who was the foundation's representative at the PUMC, went further and suggested, somewhat naively, that a "demonstration of principle can take place in China long before it will occur in this country due to the absence of vested traditional interests in the former." The Committee on Appraisal and Planning, reporting to the Rockefeller Foundation's trustees in December 1934, agreed that "China might become a vast laboratory in the social sciences, with implications that would be international in scope."[11] Gunn argued that China was an ideal laboratory for conducting an experiment in rural reconstruction, in that it existed outside the realm of formal European rule and was sufficiently "backward" to avoid struggles with vested economic and political interests.

The depiction of China as a laboratory for the social sciences harkened back to earlier colonial references to Africa as a giant laboratory for the natural and medical sciences.[12] At the same time, it emphasized how much the China program was viewed as a grand experiment, a kind of testing ground, which had implications for rural reconstruction in Europe. Above all, the Rockefeller Foundation's view of China as a vast social laboratory was commensurate with the foundation's view of the world and its ability to shape it. China was a place where social variables could be controlled, or so they thought. Events, including the Japanese invasion and the Communist rise to power, together with the political influence of landed interests, would soon conspire to reveal just how removed from reality the foundation's view of the world—and its ability to manipulate it—really were.

While the foundation saw the China program as its social experiment, the program itself built on already-existing programs.[13] Gunn's social-reform vision was stimulated in part by his encounter with Y. C. Yen, a Yale-educated Chinese promoter of education. Yen had founded the Mass Education Movement, a literacy campaign that was based on a simplified lexicon of Chinese characters that could be imparted in only 96 classroom hours. By mid-1930, Yen claimed to have provided basic reading skills to 40 million Chinese. He then expanded his educational activities to include scientific agriculture, cooperative marketing, public health, and governmental integrity. As Yen put it, "We are trying to develop not just rural education alone, or rural health, or economic improvement, or citizenship, each individually and by itself, but to coordinate and correlate all these various aspects of community improvement and social upbuilding for the remaking of the Chinese *hsien* [county] and the Chinese citizen."[14] The Chinese government gave Yen a county with a population of 400,000 in which to run his experimental program in rural reconstruction.

The public-health activities promoted by Yen were directed by Harvard-trained physician C. C. Chen. Chen developed a program that was based on the utilization of local resources and the concurrent improvement of social and economic conditions. At the village level, health stations were staffed with village health workers—local people who had received basic training in preventive treatment and simple medical procedures. They were responsible for first aid, vaccinations, and reports of births and deaths. They also oversaw improvements in sanitary conditions, especially the construction of drinking wells and latrines.[15]

While Gunn and Grant based the Rockefeller Foundation's China program on Yen's rural-reconstruction model, they emphasized the scientific-research aspects of the program and expanded it to include a wider range of disciplines and university programs. These included the Institute of Rural Administration at Yenching University, the Mass Education Program in the Institute of Economics at Nankai University, the College of Agriculture at Nanking University, the Department of Preventive Medicine at Peking Union Medical College, and the North China Industrial Institute. The result was the North China Council for Rural Reconstruction, which was established with an initial grant of US$300,000 in 1936.[16]

What actually transpired on the ground has yet to be fully studied. What is clear is that fundamental reforms were difficult in the face of vested class interests, the same "political interests" that Gunn and Grant had claimed did not exist in China. China's landed classes were a formidable barrier to the agricultural reforms envisioned by Gunn. Chiang Kai-shek's government recognized that existing land-tenure arrangements were oppressive, and that they provided an opening for his Communist opponents. Yet any effort by his government to reallocate land would threaten the interests of the landed classes, on whom he depended for much of his political support. In the absence of serious land reform, the overall impact of the North China Rural Reconstruction Program was bound to be limited. More devastating to the program was the outbreak of the Sino-Japanese War in July 1937. While this war did not lead to an immediate cessation of program activities, it seriously undermined its efforts, as the foundation's main experimental areas were located in war zones, so all fieldwork had to be relocated and begun again. Also, the universities that had provided the institutional core of the program were either under Japanese control or forced to relocate. By 1939, despite Gunn's continued optimism about the rural reconstruction program's achievements, the foundation wound down its commitments to the program. In the end it had invested US$1 million on the experiment, or approximately US$15 million in today's dollars.[17]

MEXICO'S RURAL HYGIENE SERVICE

In the late 1920s, as the IHD's hookworm program in Mexico was winding down, the IHD became involved in developing multipurpose cooperative-health units, initially on its own, and then, by the early 1930s, in cooperation with the Mexican Department of Public Health, or Departamento do Salud Pública (DSP). The units were intended to address the most-pressing health problems of the regions in which they were located. Henry Carr, who headed the initiative for the IHD, claimed that the health units were a more effective structure for delivering health care than the itinerant campaigns that the IHD had funded up until that point.[18] The IHD's initial unit was established in the state of Veracruz. It focused on diagnosing and treating hookworm, testing antihelminthic drugs, constructing latrines, and giving lectures on hookworm prevention. Dental services, prenatal care, and routine vaccinations were added later.

The DSP established its own health unit in the city of Veracruz, under the leadership of Miguel Bustamente, who had trained at the Johns Hopkins School of Hygiene and Public Health. The DSP unit was more comprehensive than the IHD unit. It provided medical and veterinary services, as well as prenatal care, and it addressed a wider range of diseases, including typhoid, malaria, yellow fever, tuberculosis, and venereal disease. It also focused on plague prevention, provided rabies vaccinations, and conducted restaurant and housing inspections.

Integrating the two units into a "rural hygiene service" (which, despite its name, included urban units) was not easy. Bustamante's vision of the health units was more comprehensive and community based than Carr's. Bustamante saw the health units as a source of community empowerment, something that medical professional Carr saw as peripheral to the role of these units. Bustamante also supported the use of untrained, nonmedical personnel to carry out public-health activities. Again, this ran counter to Carr's and the IHD's methods of operation. Yet the IHD recognized that it needed to find an alternative platform for advancing its sanitary activities in Mexico. For its part, the DSP faced financial problems and viewed a partnership with the IHD as a means of funding its health units. It was, as Anne-Emanuelle Birn has noted, a marriage of convenience. Over time, a number of other health units were built in different regions of Mexico. They tended to reflect Bustamante's vision more than Carr's, drawing the IHD into a more comprehensive, community-based system of health care.

Of particular importance to the success of the units was the incorporation of women into the provision of health-care and public-health services.

Women were directly targeted as healers, health workers, patients, and family-health intermediaries. Women served as a "gendered bridge" between doctors and the sanitary units on the one side, and peasants and town folk on the other. Female nurses performed house visits and provided expectant mothers with personal hygienic knowledge. The Rural Sanitary Service also trained a large number of traditional midwives, teaching them scientific birthing techniques.[19]

The IHD acknowledged that poverty influenced the development of public-health problems in Mexico. But it insisted that this was not the primary factor. As an annual report of one of the health units noted, "The lack of knowledge of hygiene determines health, and it is precisely the poor man who needs this knowledge so that he may live healthily in spite of his economic situation."[20] Unlike the IHD's China experiment, the Rural Hygiene Service did not engage in efforts to integrate health with broader social- and economic-development efforts. It adhered more closely to the Rockefeller Foundation's early commitment to advancing hygienic science. Yet it moved away from the disease-campaign model and toward building up basic health services.

HEALTH UNITS IN CEYLON

Another example of the IHD's support of rural health units occurred in Ceylon. The program began in the late 1920s but took off during the 1930s. As in Mexico, the health units in Ceylon grew out of the IHD's work with hookworm. The International Health Board had instituted a hookworm-control program in Ceylon in 1915, using the curative model (chapter 2). The campaign initially focused on plantations and successfully treated large numbers of workers. As occurred elsewhere, however, many of the workers became reinfected in the absence of sanitary reforms on the part of plantation owners, undermining the campaign's early successes. The plantation campaign was deemed a failure and terminated in 1920. The IHD also attempted to control hookworm outside the plantations but met with similar problems. More importantly, Sinhalese villagers questioned the usefulness of a campaign that focused only on hookworm, while ignoring the presence of other, more pressing health problems, including malaria, typhoid, smallpox, and maternal and child health.

Johns Hopkins–trained IHD officer William Jacocks, who led the hookworm campaign, became convinced that the IHD could not achieve any results unless the campaign addressed the basic relation between sanitary conditions and health and focused on the wide range of health problems experienced by villagers. He was able to convince his superiors that the IHD had to develop a more broadly based approach to health, which centered on the creation of a system of health units dedicated to improving the general health of the population. According to historian Soma Hewa, the country was divided into health units, which served approximately 100,000 people each through a network of hospitals, dispensaries, and clinics.[21] The health units represented a shift in policy instigated by local pressure from below, a rare development in the field of international health both at the time and during much of its history.

Each health unit was responsible for carrying out a range of functions. These included the collection and study of vital statistics, health education, control of acute communicable diseases, hookworm treatments, immunizations against smallpox and typhoid, antimalarial work, maternity and child welfare, school hygiene, household sanitation, water-supply improvement, the disposal of night soil and other wastes, and control of food-handling establishments. While the health units were modeled after programs run by the Rockefeller Foundation in the American South, they were to be directed by the Ceylon Department of Medical and Sanitary Services.

The first unit was established in the Kalutara District in 1926. Public-health and sanitary personnel, rather than physicians, staffed it. The health units were prevented from treating patients, except in the case of hookworm and malaria. Prevention rather than cure was the central mission of the health units. Another distinguishing feature of the health-unit model was the role played by native Ceylonese in running the health facilities and providing health education to the local population. The attention that health workers gave to building community support was also important. Local villagers were encouraged to support the health unit by contributing to the building and maintenance of health facilities. In many ways, the health units presaged the later development of primary health-care systems in Ceylon (now Sri Lanka) and elsewhere. By the end of World War II, there were 602 health centers and 20,488 health clinics divided among the country's 63 health units.

RURAL HYGIENE IN JAVA

The Rockefeller Foundation's rural hygiene program in Java during the 1930s was yet another example of the growing interest among early international-health organizations in making general improvements in rural hygiene instead of focusing narrowly on the elimination of specific diseases. The foundation's emissary to Java was John Hydrick. Hydrick was from

South Carolina and had graduated from Wofford College in 1908. He received a Rhodes Scholarship to study at Oxford and then pursued a medical degree at Jefferson Medical College in Philadelphia. He joined the IHB in 1913 and was posted to Trinidad to learn hookworm control before being stationed in Brazil in 1920. In 1923, the IHB sent him to Baltimore to take a special course in hygiene at Johns Hopkins. That summer he also attended the malaria training camp in Leesburg, Georgia, directed by Samuel Taylor Darling and attended by Lewis Hackett and Frederick Soper (chapter 2). Hydrick thus followed a career path that was common to many IHD officers.

Hydrick was seconded to Java in 1924 to launch a hookworm campaign. The Dutch initially assigned Hydrick to Serang, an insurrectionary district in West Java. Perhaps they hoped that a successful campaign would reduce local political resistance, a strategy that had also been used by the Mexican government in the 1920s. The hookworm campaign did not go well. Rumors abounded among colonized populations over the nature of the campaign, with some claiming that Hydrick and his staff intended to steal the souls of Muslims and convert them to Christianity. As happened elsewhere, deaths from overdoses of oil of chenopodium, used to purge worms from the body, further escalated distrust. These tensions, coupled with strict financial limitations imposed by the Dutch government and Depression conditions, led Hydrick to adopt a particularly frugal model of health prevention that sought to ease tensions with rural populations and achieve compliance with Western standards of hygiene through the cheapest possible means.[22]

Hydrick's strategy was worked out in an experimental health center he established in Poerwokerto. It was described in his book, *Intensive Rural Hygiene Work and Public Health Education of the Public Health Service in Netherlands India*.[23] Hydrick was convinced that villagers had to understand proper sanitary behaviors and willingly adopt them for any change to occur in sanitary conditions. Sanitary principles had to be internalized. He noted that with a stronger colonial state, sanitary behavior could be achieved through coercion, but in the absence of state control, other methods had to be employed. The strategy he developed made use of readily available local materials and locally recruited health educators to promote the adoption of hygienic behavior by rural Javanese villagers. The strategy also encouraged villagers to participate in their own health care, building latrines and protecting their water supplies, rather than have these things done for them.

To disseminate the health information, Hydrick trained a cadre of educators recruited from the villages in which they worked. These hygienic technicians (*mantri*) were chosen as much for their personalities as for their knowledge of sanitation. Above all, they needed to be polite, modest, and sensitive to the needs of the people among whom they worked and able to clearly and patiently communicate the principles of hygiene to them. The *mantri* were trained to teach simple, inexpensive hygienic techniques. They taught school children how to make toothbrushes and fingernail brushes from the fibers of coconut husks; to drink from individual, homemade bamboo cups; and to build bamboo piping to funnel water for hand washing and bathing.

The IHD and the Dutch colonial government deemed the experimental program to have been successful in transforming rural health behaviors in Poerwokerto. Anthropologist Eric Stein's recent research, however, which included interviews with program participants,[24] suggests that the program retained elements of a top-down, development-assistance approach that had characterized the foundation's earlier international-health work, and that it was less successful in transforming rural health conditions than either Rockefeller propaganda or recent historical accounts might suggest.[25] Stein notes that the local villagers whom Hydrick tapped to be *mantri* were not chosen from all classes of villagers. Instead, many represented village elites, who had been used before by the Dutch as smallpox vaccinators, tax collectors, labor contractors, and farm administrators. They were, in effect, elements of Dutch colonial rule. They may have been locals, but they were not necessarily peers.

Dutch officials had introduced land reforms in the early years of the twentieth century. These were designed to benefit an elite class of village landowners (*haji*) and avoid what the officials saw as the pitfalls of communal ownership: the creation of smaller and smaller plots of land, and a decline in agricultural production. The land reforms resulted in the accumulation of land among a small percentage of local farmers and the creation of a larger class of landless peasants who were expected to earn a living working on the lands of these elites. In reality, however, those deprived of land were only able to survive by finding extra work as casual wage laborers outside the agrarian economy, in the areas of petty trade, transport, public works, or handicrafts. Opportunities in these nonagrarian sectors were limited. By the 1930s, only two-fifths of the working populations in some locations were cultivators. While colonial officials maintained that the landed and the landless existed together in rural villages in a state of communal harmony, the landless villagers lived in a state of increasing desperation and were dependent, in many ways, on the good will of the village elites for survival.[26]

It was from these same elite farmers that Hydrick drew his health educators. Their ability to motivate villagers was, in many cases, less a product of their persuasive powers than their inherent local authority. Village elites occupied a role as patrons who could provide work and other benefits to their subordinate clients. But they were also feared for the power they wielded over the social order.[27] A former hygienic *mantri* told Stein that people would sometimes flee from their houses or refuse to answer the door when he arrived for educational home visits. Stein also points out that despite Hydrick's stress on locally available, affordable solutions to sanitation problems, many villagers were unable to adopt some of the measures, due to their extreme poverty. Banyumas villagers explained that it might take a full week to dig a latrine pit and complete the latrine cover. The villagers noted that "the value of such intensive labor investments—which might have been used instead to secure family income through other productive activities—was offset by the ephemerality of the makeshift latrines, which, if they survived heavy monsoon rains, filled to capacity every several years and had to be replaced."[28]

Thus Hydrick's experiment, like the rural hygiene programs in China, marked a departure from earlier IHB interventions in that it focused on general improvements in health, relied heavily on local villagers to carry out its program, and emphasized low-tech solutions to health problems. Yet, also like the China program, it was limited by its inability to address the structural causes of ill health. In the Java case, it does not appear that Hydrick had any intention of addressing the social determinants of health or that he understood the conditions that contributed to the poverty of the people among whom he worked.

Finally, Hydrick's approach—for all of its rhetoric about local participation and efforts to train villagers to protect their own health—was based on principles of scientific hygiene developed in the West. He made no effort to actually solicit the views of local villagers or to incorporate their cultural understandings of health into his educational program. His program was designed to be a more effective, low-cost instrument for inculcating the mass of unknowing natives with Western hygienic knowledge.

The Java experiment in rural hygiene is often viewed as a precursor to later village health-worker programs associated with the primary health-care movement in the late 1970s. It certainly contained elements that were part of that later movement. It also shared a utopian vision of bottom-up transformation that was often out of touch with local politics and local social and economic inequalities. Also, like many later efforts, the Java experiment failed to address the underlying social determinants of ill health.

Colonial Malnutrition

The broadening vision of health that marked the investments of the LNHO and the IHD in building local health-care systems can also be seen in British efforts to illuminate and address a rising tide of colonial malnutrition during the 1930s. Like the earlier nutritional work of the LNHO and ILO in Europe, British colonial researchers began to acknowledge the broader social and economic forces shaping nutrition but, in the end, continued to focus on technical solutions to these problems.

British interest in colonial nutrition began in the 1920s, with the comparative work of Boyd Orr and John Gilks on the nutritional status of the Kikuyu and Masai in Kenya. Their study highlighted the impact of different dietary regimes on the overall health and development of the two groups. This work also had wider importance for nutritional studies, because it assumed the universality of nutritional processes. Unlike much of colonial medicine at the time, it did not begin with an assumption of racial differences. This meant that Orr and Gilks's findings could be readily taken up by nutritional researchers in Britain, and their work contributed to nutritional science more generally. Further research into the nutrition of the Masai and the Kikuyu was carried out between 1927 and 1931, and the Medical Research Council published a final report in 1931. While British researchers largely took up the report as a contribution to dietetics, it made larger claims regarding the prevalence of malnutrition among so-called native tribes. Moreover, it asserted that the problem was not confined to Kenya, but existed in many colonial dependencies. Finally, it argued that colonial malnutrition had serious economic implications.[29]

Meanwhile, beginning in the mid-1920s, British and Indian researchers working at the Pasteur Institute in Coonoor, India, conducted studies on the problem of beri-beri and its relationship to the consumption of machine-milled rice. It is useful to trace the history of this research, for it reveals the widening of colonial nutritional research during the 1930s, going from the study of diets to a broader effort to understand the social determinants of malnutrition. The work at Coonoor was directed initially by Colonel Robert McCarrison, who began his nutritional research before World War I, working on the problem of goiters in India. In 1918, McCarrison founded the Beri-Beri Enquiry Unit in a single-room laboratory in Coonoor. In 1924, he published an article on "Rice in Relation to Beri-Beri in India," which focused attention on the effects of different forms of processing on the nutritional value of rice and, in turn, on the epidemiology of beri-beri.[30] This research led him to conclude that beri-beri was not simply the product of vitamin B deficiency in processed rice, which was consumed as a main dietary staple. He claimed that rice that was excessively processed and devoid of vitamin B did not produce beri-beri. Therefore some degree of vitamin B must be present. McCarrison thus played down the connection between beri-beri and the increasing spread of rural rice mills, which produced highly milled and polished rice. Instead, he argued that there were other, as-yet-unknown factors that explained the occurrence of beri-beri in different populations. This early work, therefore, did not address the wider social and economic forces that might shape the epidemiology of beri-beri.

In 1925, the Beri-Beri Enquiry Unit was converted into the Disease Deficiency Inquiry and, three years later, became the Nutrition Research Laboratories, with McCarrison as its director. McCarrison continued to conduct detailed laboratory studies on the effects of different diets on the health of rats and pigeons. Yet he began to articulate a wider vision of the relationship of nutrition to agricultural production and called for the joining of nutritional and agricultural research. In testimony before the Royal Commission of Agriculture in India in 1928, he stated, "The solution to the problem of malnutrition is thus, to a great extent, one of improvement in agriculture."[31]

McCarrison retired as director of the Nutrition Research Laboratories in 1935. He was replaced by Wallace R. Aykroyd, who had resigned from his position as health secretary for the LNHO—yet another example of the movement of health experts back and forth between international-health organizations and colonial medical services. In Coonoor, Aykroyd was charged with evaluating Indian nutritional problems and developing recommendations for improving them. To this end, he expanded the laboratory, and his team began a systematic analysis of Indian foodstuffs, conducting a series of nutritional surveys. His surveys revealed widespread problems of malnutrition among different populations. This led Aykroyd and his Indian colleagues to return to the question of the nutritional value of different forms of rice that were consumed widely in south India and, particularly, the relationship between diets based on highly milled white rice and the occurrence of beri-beri.[32]

Aykroyd and his Coonoor colleagues, B. G. Krishnan, R. Passmore, and A. R. Sundararajan, published their research findings in 1940 in a report titled *The Rice Problem in India*. Much of the report was technical in nature and focused on measuring the vitamin B content of various varieties of processed rice in India. It showed, as earlier investigations had, that the vitamin B content of highly milled rice was considerably less than that of home-pounded rice. Yet it also contained a chapter on the "Economic and

Sociological Aspects of the Milling Problem in South India." This chapter reported that an estimated 70 percent of the population of Madras consumed milled rice rather than home-pounded rice. The report concluded that there were multiple factors involved in the widespread adoption of a less nutritious form of rice and focused attention on the problems of poverty, landlessness, the heavy debt burdens of small-holder farmers, and changes in the processing and marketing of rice in south India:

> The rice mill has spread everywhere and the people have come to regard it as a labor saving convenience; small-holders are often forced by poverty and debt to dispose of their paddy to merchants and do not retain sufficient for their own use; a large portion of the poorer sections of the community lacks accommodation for the storage and pounding of paddy; paddy cannot be purchased in small quantities for pounding; when home pounded rice is put on the market as an article of commerce it is dearer than machine-milled rice because of the labor cost involved.[33]

These conclusions clearly moved the study of malnutrition in India away from the laboratory and the study of dietetics, aspects with which McCarrison had begun his work at Coonoor, to the study of the political economy of rice production and consumption. Yet one needs to be careful not to read too much into the conclusions reached by Aykroyd and his Indian colleagues.

Historian Sunil Amrith has suggested that Aykroyd's report was implicitly critical of British economic policies in India: "[The report] focuses on the negative impact of the regional colonial economy, involving the import of rice by the densely settled parts of eastern India . . . in exchange for the export of labour and skills to the frontier lands of Burma, Malaya, and Ceylon."[34] While it is likely that this regional colonial economy did play an important role in shaping rice-consumption patterns in Madras, it is not at all clear that Aykroyd was aware of these broader economic patterns or that he viewed his report as being in any way critical of British economic policies in India. Aykroyd and his colleagues acknowledged that changes introduced by the British were contributing to a greater consumption of milled rice. But the changes they cited were advances in transport technologies:

> The development of transport and the improvement of roads enables the paddy grower to bring his grain to the mill. The ubiquitous motor-bus has loosened the bonds that attach the villager to his own plot of ground and traditional manner of life. Cheap electric power is obtainable over wide areas and other sources of power—the steam and combustion engines—are familiar and obtainable. . . . All of these factors have played a part in the abandonment of the ancient practice of hand pounding.[35]

These changes were benefits of colonial rule, and Aykroyd and this coauthors did not ascribe the poverty and indebtedness of Indian villagers to British colonial policies. Like many colonial authorities, anthropologists, and international-health professionals of his day, including John Hydrick, Aykroyd viewed the economic conditions he was seeing as characteristic of an Asian economic system that had operated largely unchanged since the distant past. The prevailing attitude was that colonial rule had introduced modern technologies, but not the oppressive economic forces that had encouraged changes in food-consumption patterns. Viewed in this way, the more destructive impacts of colonial policies were often unseen or unappreciated.

In making recommendations for reducing the consumption of highly milled rice, Aykroyd at no point suggested changes in British colonial policies related to land use, food production, labor patterns, or the commoditization of rice. He proposed instead that agricultural research should focus on the production of high-yielding rice rather than strains that are more nutritive, since changes in milling and preparation could improve the nutritive value of locally grown rice. Second, he suggested that educational programs should be developed to discourage consumers from abandoning the process of parboiling rice, since this step helped preserve nutrients, and to limit the amount of water used in washing and cooking rice, since excessive water reduced the vitamin content of the rice. Third, he did not recommend policies for discouraging the continued spread of rice mills. Instead, he called for legislation that would limit the milling of rice past a certain point: "A minimum content of 1.5 micrograms/gramme of vitamin B1 is suggested on the basis of a standard milling." Finally, he wrote that efforts should be made to encourage consumers to partially substitute millet for rice in their diets and to add additional vitamin- and mineral-rich foods to what they ate.

In short, the report's recommendations were technical in nature and did not address any of the broader economic forces described in its chapter on the economic and sociological aspects of the milling problem in south India, let alone the widespread political economy of labor and food in South and Southeast Asia. Like many of the activities that characterized international health in the 1930s described in this chapter, the report was important for the larger perspectives it brought to the table, not for the solutions it proposed. It represented the ongoing tension between an increased recognition of the broader structural causes of health problems and a need to provide technical, achievable solutions for them.

Aykroyd and his colleagues wrote their report the same year that the British Economic Advisory Council's Committee on Nutrition in the Colonial

Empire published its findings, based on responses to a circular that surveyed nutritional conditions in the British colonies. The committee had begun its work in 1936, meeting four times between 1936 and 1939. It issued a draft report in 1938 that, like Aykroyd and his colleagues' *Rice Problem in India*, balanced the need for education to encourage colonial populations to consume balanced diets with the need to address the underlying problem of undernutrition caused by the Depression, pricing and marketing policies that discriminated against colonial producers, and the commercialization of food production. All of these processes were undermining the ability of local populations to maintain adequate nutrition levels. As Michael Worboys has shown, however, the report underwent significant revisions before its publication in 1939.[36] The roles of the economy and poverty remained in the final version but were greatly toned down, overshadowed by new technical sections on dietetics, food processing, and storage. The report focused heavily on the role of native ignorance as a cause of both poverty and malnutrition. Nutrition had now been transformed from being conceptualized as a major structural problem to being viewed as a technical problem.[37]

The Bandoeng Conference

Interest in rural reconstruction and hygiene outside of Europe culminated in August 1937 with the Intergovernmental Conference of Far Eastern Countries on Rural Hygiene, held in Bandoeng, Java. Indian and Chinese delegates to the League of Nations Assembly proposed the conference in 1932, following up on the successful European Rural Hygiene Conference in Geneva in 1931. The Health Committee and the League's Far Eastern Bureau in Singapore organized the Bandoeng conference. Nearly all of the countries invited to the conference were colonies of Britain, France, or the Netherlands. While colonial subjects represented some of these colonies at the conference, the majority of representatives were European colonial officials. The only independent countries among the participants were Japan, China, and Siam. Representatives of the League of Nations Health Organization, the Rockefeller Foundation, the Far Eastern Association of Tropical Medicine, the International Institute of Agriculture, and the League of Red Cross Societies also attended the Bandoeng conference. The makeup of the conference, and the dominant role of colonial officials in it, reflected the continued entanglement of international-health organizations and colonial medicine.[38]

The conference was organized around five topics: (1) health and medical services, (2) nutrition, (3) sanitation and sanitary engineering, (4) rural reconstruction and collaboration by the population, and (5) measures for combating certain diseases in tropical areas. Each topic was assigned to a committee, which reported back to the full conference. Discussions on each of these topics highlighted underlying social and economic problems. Yet at almost every turn, the conference participants backed away from recommendations that would have radically changed the status quo in order to improve rural hygiene in their countries.

In discussing nutrition, for example, the conference report began by acknowledging that the diets of the mass of the population in the East were "thoroughly deficient" in terms of the standards set by the Technical Commission on Nutrition of the LNHO. It also acknowledged that this problem was "largely bound up with economic conditions." In addition, it argued for cooperation between nutritionists and agriculturalists. Yet in making recommendations, the report called for further study by each country of its own nutritional conditions, including dietary surveys; the creation of national nutrition committees; the establishment of local dietary standards; and the sharing of nutritional information among countries. It also recommended that countries study the relationship between income and the cost of "minimum adequate diets, well balanced diets, etc. as drawn up by nutrition workers." There was no critique of economic conditions and their causes and, not surprisingly, no mention of the impact of colonial economic policies on nutrition. The closest the report came in this regard was to rehearse concerns about the spread of rice milling and its negative impact on the nutritional status of rice eaters in the region. But here again, it did not go so far as to suggest that the practice should be stopped or limited, but only that governments should make a thorough investigation of the nutritional, commercial, economic, and psychological aspects of the problem. Most of the technical reports prepared for the conference and attached to the conference report focused on specific dietary recommendations.[39]

Discussions of rural reconstruction focused largely on the problem of obtaining the collaboration of local populations. Here again, the participants put forth bold proposals for change, which were tempered in the conference's actual recommendations. As defined in the report, rural reconstruction meant the total transformation of peasant lifeways, requiring the cooperation of "representatives of public assistance, health insurance, agricultural associations, the medical profession, and scientific agriculturists, architects, hygienists, and engineers."[40] Yet reform was not possible without the cooperation of rural populations: "Whether in the matter of housing or diets, sanitary arrangements or treatment of diseases, it is necessary to work through the people, and avoid, as far as possible, imposing systems from above."[41]

To ensure local cooperation, the conference recommended the creation of village committees. These committees would oversee water supplies, the cleansing of public roads, tanks, and wells; the construction and repair of minor roads, drains, and bridges; and the maintenance of proper village sanitation. The report also stressed the need for women to be involved in all aspects of rural reorganization. By emphasizing the need for local participation and not imposing systems from above, the conference amplified the lessons learned from the LNHO and the Rockefeller Foundation's China rural development programs, as well as those of other programs described above. It marked an evident departure from pre-Depression forms of international health.

Yet if one looks more closely, it is clear that there were limits to how far broad-based rural reconstruction or popular participation could go. At several points, the report stressed the need for rural development to be overseen by experts: "There should be an organization of control—namely a committee of management consisting of Government experts, representatives from villages, and non-officials—to advise village committees as to the programme to be undertaken and to see that the programme is carried out." Moreover, the village committees themselves should be composed of representatives "nominated by the Government," with a chairman elected by themselves or nominated by the government. This vision of rural government, overseen by outside experts and selected by governmental officials, was some distance away from true popular participation. These limitations were hardly surprising, given the strong representation of colonial authorities at the meeting.

Second, the report contains very little in the way of direction as to just how village committees and management committees would go about making improvements in nutrition, sanitation, health services, or disease control. The sections of the report that focused on these issues were largely silent about implementation, other than making general statements about the need for education and the setting of standards. Nor were any models cited. Hydrick's work on sanitation was not mentioned in the sanitation section of the report, though conference attendees were given the opportunity to tour his work site at Poerwokerto. This lack of specificity reflected an unwillingness on the part of conference organizers to be proscriptive in advising countries on how to reform their rural health conditions. Sweeping recommendations for change would have been particularly troubling to the colonial authorities who dominated the meeting. A lack of specificity and the absence of models would also characterize the recommendations of the 1978 Alma-Ata conference on primary health care (part VI), which some historians have compared with the Bandoeng conference.[42]

The interwar period opened up new possibilities for understanding and responding to international-health problems. While the early postwar initiatives of the LNHO and the Rockefeller Foundation's IHD focused on technical solutions to disease problems and the measurement and setting of nutritional, sanitary, and housing standards, both organizations, as well as many of the colonial authorities with whom they worked, gradually shifted their gaze to the underlying social and economic determinants of ill health. They proposed integrative models of rural development, which linked health initiatives with broader investments in agricultural development, improvements in housing, the development of local health services, and land reform. Efforts to apply these models in various international settings met with some success, though they were hard to measure, as the onset of World War II short-circuited many of these efforts.

These rural hygiene and reconstruction initiatives were also different from earlier health initiatives in the degree to which they acknowledged the role of local populations as partners in efforts to improve their own health. In China, Ceylon, Mexico, and India, the increasing role of local health authorities in research and in the designing and implementation of health programs contributed to this view of technical cooperation, as opposed to the top-down health interventions that characterized the early work of the IHD, the LNHO, and colonial medical authorities.

At the same time, it is important to recognize the limits of these progressive approaches. Continued belief in the power of medical science to solve health problems, combined with the inability of rural reconstruction or nutritional programs to bring about significant changes in land ownership or patterns of regional economic development, or to empower rural villagers, meant that the progressive visions of people like Rajchman, Gunn, and Štampar failed to fully materialize. In addition, there were limits to how far Western technical advisors were willing to go in devolving decision-making authority to local populations who, for many, were still viewed as needing outside assistance and unable to improve their own health. Finally, the continued dominance of colonial authorities in many parts of the globe contributed to this lack of progress.

That said, the 1930s must be seen as a time of great promise for an alternative vision of global health—a vision that addressed the underlying determinants of health; was committed to building broad-based health services, not just disease-control programs; focused on preventative as well as curative services; and sought to involve local populations in the planning and implementation of health services.

The coming of World War II disrupted efforts to implement this alternative vision and created an economic and humanitarian crisis. In Europe, China, and Southeast Asia, the war displaced millions of people, who faced disease and starvation. The economies of many countries were shattered. The war also transformed the landscape of international organizations. The League of Nations and its Health Organization were disbanded in 1946. As World War II came to an end, a new set of international organizations were formed to respond to the crisis conditions created by the war and to plan for a new world order. Among these were the United Nations, the Food and Agriculture Organization, the United Nations Relief and Rehabilitation Administration, the World Bank, and a new World Health Organization. The question for many was, would the broader vision of health and development that had begun to emerge in the 1930s survive within this changed environment?

PART III

Changing Postwar Visions of Health and Development

In July 1946, representatives from 50 member states of the newly formed United Nations met in New York City for an international-health conference. The purpose of the conference was to finalize the constitution for a new international-health organization that would replace the now-disbanded League of Nations Health Organization. A draft of the constitution had been produced by the Technical Preparatory Committee (TPC), which had met during March and April of that year. The final text of the constitution approved in New York City began with a preamble that included the following assertions:

- Health is a state of complete physical, mental, and social well-being and not merely the absence of disease or infirmity.
- The enjoyment of the highest attainable standard of health is one of the fundamental rights of every human being without distinction of race, religion, political belief, economic or social condition.
- Governments have a responsibility for the health of their peoples which can be fulfilled only by the provision of adequate health and social measures.[1]

The goals expressed in the preamble to the World Health Organization's constitution echoed a set of principles that had shaped international health and development efforts during the 1930s. These ideals has survived World War II and influenced early postwar efforts to improve the health and well-being of the world's war-torn populations, defining the goals of a number of postwar organizations, including the United Nations Relief and Rehabilitation Administration (UNRRA), the Food and Agriculture Organization (FAO), and the World Health Organization (WHO). The founders of each of these organizations were committed to a vision of postwar development that linked improvements in health and nutrition to broader societal changes. But this commitment did not last long.

By the early 1950s, WHO was moving rapidly toward a set of strategies and interventions that focused narrowly on the application of newly developed biomedical technologies to eliminate diseases one at a time. WHO, in effect, returned to an era of international-health interventions associated with the disease-eradication campaigns of the Rockefeller Foundation's International Health Board (chapter 1). While the scale of WHO operations were much greater and the weapons available to combat diseases more powerful, the strategies employed were very similar. As before, they were designed by experts who met in cities in Europe and the United States to discuss and make plans for improving the health of peoples living in Africa, Asia, and Latin America. They also shared a confidence in the superiority of Western knowledge and a disregard both for the ideas and practices of those for whom the campaigns were created and for the ability of targeted populations to meet their own health needs. They also paid little attention to the underlying social and economic determinants of health or the need to develop basic health services. Chapter 5 describes how a broad vision of health and development permeated early international-health planning in the late 1940s, while chapter 6 examines why this vision dissipated so quickly in the 1950s.

CHAPTER FIVE

Planning for a Postwar World

The Legacy of Social Medicine

THE WORLD HEALTH ORGANIZATION'S early commitment to promoting health in its broadest sense was part of a wider effort among world leaders to improve the economic and social welfare of the world's populations in a postwar world. As World War II was coming to an end, leaders across the globe expressed the need to create a new world order in which international cooperation would lead to a world free from want and secure from the violent upheavals of war. This desire was driven by both hopes and fears. European leaders looked to the creation of new international organizations as a way of avoiding the chaos that had followed the end of World War I, as well as a means of garnering financial resources to fund Europe's reconstruction. In the United States, the fear that the US economy would drop back into recession with the decline of wartime industrial production was a major incentive for participating in the reconstruction of the postwar world, blunting, if not completely overcoming, isolationist sympathies and fears that US interests would be subordinated to those of international organizations.

Leaders in both Europe and the United States looked at countries in Africa, Asia, and Latin America, many of which were emerging from colonial rule, as sources for the production of raw materials and potential new consumer markets for products manufactured in Europe and the United States. Efforts to advance the social and economic development of these so-called developing, or Third World, countries, including improvements in health, were closely linked to these economic interests.[1] Development became a justification for intervening in the economic and social lives of millions of peoples across the globe.[2] Development also became a major industry, employing large numbers of experts from the global north who proposed policies for transforming or modernizing newly emerging nations. These proposals supported the political and economic interests of donor nations, as much as those of the countries receiving foreign assistance.[3]

Nationalist leaders in colonial Africa, Asia, and the Caribbean viewed the new internationalism as an opportunity to end colonial rule and advance their own agendas for social and economic advancement. Many of these leaders, educated in European and American universities and groomed by colonial authorities, shared the postwar vision of global economic development expressed by European and US leaders and looked forward to joining the international community as equal partners. Their hopes had been raised by the Atlantic Charter, signed by President Franklin D. Roosevelt and Britain's Prime Minister Winston Churchill in 1941. The charter laid out a vision for a postwar world, which included commitments to "bringing about the fullest collaboration between all nations in the economic field with the object of securing, *for all*, improved labor standards, economic advancement, and social security" [emphasis added]. More importantly, it declared that all countries had the right to self-determination, a point that was not missed by nationalist leaders in Africa, Asia, and the Caribbean, even though Churchill would later disavow the idea that this statement applied to Britain's imperial holdings, at least in the short term.[4]

Overshadowing all of these concerns was the atomic bomb and the potential horrors of nuclear war. The international leaders who met after World War II to plan for a new world order did so amid an emerging nuclear arms race between the United States and the Soviet Union. The threat of nuclear war raised the stakes for building global cooperation toward a peaceful, prosperous world. To achieve a new world economic order, leaders from many countries agreed that it was necessary to create international organizations that would ensure the free flow of trade, the financial stability of member nations, the reconstruction of Europe, and the economic growth of the developing world. It was with these goals in mind that representatives from 44 nations met in July 1944 at Bretton Woods, New Hampshire, to establish the International Bank for Reconstruction and Development, or World Bank, and the International Monetary Fund (IMF). The World Bank would advance loans designed to stimulate economic growth, while the IMF would foster global monetary cooperation, secure financial stability, and facilitate international trade. The delegates to the conference also signed the General Agreement on Tariffs and Trade (GATT), designed to lower tariffs and other trade barriers in order to encourage free trade. These new organizations, while initially focused on problems of economic growth and financial stability, would, by the 1980s, become major actors in the field of international health.[5]

Improving the health of the world's populations was widely viewed as an essential part of creating a new, prosperous world order. The heavy burden of disease in the developing world was a barrier to progress, while improvements in health were increasingly presented as an essential prerequisite for economic development in the postwar world. In the 1940s and early 1950s, many global leaders, like their counterparts in the 1930s, viewed health as an integral part of a broader approach to social and economic development.

In Britain, the importance of improvements in health as part of an integrated approach was captured in the 1942 *Report of the Inter-Departmental Committee on Social Insurance and Allied Services*, commonly known as the Beveridge Report. The report laid out a plan for the provision of universal social insurance to provide income security as part of a "war on want." The report became the basis for the creation of a National Health Service in Britain, which assured medical coverage for every person, regardless of their means, through a system of compulsory contributions collected by the state. The National Health Service would become a model for universal health coverage in other countries. Yet the Beveridge Report was not primarily about health. Rather, it treated health care as one element among the "allied services" necessary for a comprehensive scheme aimed at maintaining employment and income. As economist Philip Musgrove has noted, health care was viewed as a means of protecting or restoring people's capacity to work; hence the report's emphasis on postmedical and rehabilitative care.[6]

The link between health and social and economic development was also captured in wartime commitments by Great Britain and France to improving the health and well-being of their colonial subjects. Up until World War II, the colonial possessions of both countries had been run on a self-supporting model, in which charges to metropolitan treasuries were to be kept to an absolute minimum. Now there was realization that a different financial model was necessary. The British government passed two Colonial Development and Welfare Acts (in 1940 and 1945), driven by the rising discontent of workers—expressed in a series of labor strikes across the empire in the 1930s and early 1940s—and the growing belief that further exploitation of colonial resources required an investment in the social welfare of colonial populations. The 1940 act established a Social Services Department within the Colonial Office to oversee efforts at social advancement in the colonies. Funds from this act contributed to a range of economic and social-welfare activities. About 40 percent of the funds from the act went toward education, health services, housing, and water supplies.[7] Ghana, for example, received £878,858 between 1940 and 1948 for agricultural research, the construction of educational institutions, urban water supplies, and electrification.[8] French colonial authorities followed suit with the creation of the Fonds d'Investissement pour le Développement Économique et Social, or FIDES, in 1946. Both the British and French initiatives linked advances in economic production in the colonies with an extension of health and education services. While they fell far short of meeting the development needs of the populations they were intended to serve, they embodied the principle that improvements in health were part of a plan for broad-based social and economic reform.

The Bhore Committee's proposals for the postindependence reorganization of India's health system also echoed a prewar commitment to integrating health and development. The committee, created in 1943, was composed of a group of conservative Indian physicians and civil-service appointees, headed by Sir Joseph Bhore. It was advised, however, by an external group of physicians who were strongly supportive of social medicine. Among these was John Ryle, the first professor of social medicine at Oxford, and Henry Sigerist, a Zurich-trained physician who was the director of the Institute of the History of Medicine at the Johns Hopkins School of Medicine. Sigerist was a strong advocate of universal health insurance and an admirer of the Soviet Union's system of socialized medicine. Another non-Indian advisor, who also served on the committee, was John Grant. Grant had been centrally involved in the Rockefeller Foundation's North China Rural Reconstruction Program and was a proponent of primary health care and the need for integrated approaches to rural health and reconstruction (chapter 4). These advisors played a major role in shaping the recommendations of the Bhore Committee.[9] The committee was also influenced by earlier proposals for a national health service developed by the Indian Congress Party's National Planning Committee, as well as by the Beveridge Report and the national health systems of Canada, New Zealand, and the Soviet Union.[10]

Not surprisingly, the Bhore Report, published in 1946, put forth a broadly constructed vision of health that highlighted the roles of social, economic, and environmental factors in shaping patterns of sickness and health. The committee asserted: "At the outset, we must ensure the conditions essential for healthful living in town and countryside. . . . The elimination of unemployment, the provision of a living wage, improvement in agricultural and industrial production, the development of village roads and rural communications, as distinct from the great national highways now projected, are all so many facets of a single problem calling urgently for attention."[11] In both its diagnosis and proposed prescription, the report echoed the language of the League of Nations' reports on rural hygiene (chapter 3). The Bhore Report also insisted that public health was a fundamental responsibility of the state, presaging the language of the preamble to WHO's constitution. Finally, the report called for the creation of a national health service, involving a countrywide network of district health centers connected to specialized urban health centers.[12]

In the United States, investing in health had been crucial to efforts to promote economic growth in the 1930s and was reflected in the New Deal, a series of domestic social and economic programs enacted between 1933 and 1938 under the Roosevelt administration, in response to the Great Depression. The US Farm Security Administration (FSA), which was created to help address the economic plight of rural farmers during the Depression, established a network of 3000 rural physicians and 1200 cooperative health plans in 41 states.[13] In addition, the government established the Tennessee Valley Authority, a rural-development program designed to promote improvements in agriculture, forestry, and health, as well as to provide electricity to 3 million people. Both program were examples of this earlier, integrated approach to health and development.

After the war, President Harry Truman proposed a system of national health insurance as an essential part of US economic growth. Truman also believed that advancing the health of populations living in the developing world was essential to their economic development and, thus, to the global economic growth envisioned by participants at the Bretton Woods Conference. Investments in health would not only create healthier workers and consumers, but could also open up new lands for agricultural production and contribute to the economic advancement of developing regions of the world. As George Marshall, Truman's secretary of state, noted in 1948, "Little imagination is required to visualize the great increase in the production of food and raw materials, the stimulus to world trade, and above all the improvement in living conditions, with consequent social and cultural advances, that would result from the conquest of tropical diseases."[14]

While recognizing the importance of health to development, a number of prominent economists and public-health authorities in the United States maintained that health interventions needed to be part of a coordinated program of social and economic development, which included improvements in agriculture, education, industry, and social welfare, as well as health. Otherwise, improvements in health alone could cause rapid population growth that could undermine development efforts. This concern led some to argue for population-control programs (part V).

Charles-Edward A. Winslow, a Yale professor of public health and one of the most prominent postwar leaders in international health, addressed this issue in his book, *The Cost of Sickness and the Price of Health*:

> The interrelationships involved make it abundantly clear that the public-health programme cannot be planned in a vacuum, but only as a vital part of a broad programme of social improvement. . . . It is not enough then, for the health administrator to develop the soundest possible programme for his own field of social endeavor. . . . He must also sit down with experts on agriculture, on industry, on economics, and on education and integrate his specific health programmes as part of a larger programme on social development.[15]

Paul Russell articulated a similar position in his book, *Man's Mastery of Malaria*. Russell had spent years combating malaria in India, the Philippines, and Latin America and was a member of the WHO Expert Committee on Malaria during the early 1950s. In his concluding chapter, "Malaria Prophylaxis and Population Pressure," he argued that having "physicians, malariologists, and sanitarians integrate their activities with those of agriculturalists, demographers, social scientists, economists, educators, and political and religious leaders is of the utmost importance. For only thus can there be joint planning of social reorientation that will result not in bigger populations but in healthier communities."[16]

Early Visions of United Nations Organizations

An integrated vision of health and development shaped the early goals and activities of the United Nations organizations that were created after World War II to address the immediate- and long-term health and nutrition needs of the world's populations. This vision can be seen in the early histories of three of these organizations: the United Nations Relief and Rehabilitation Administration (UNRRA), the Food and Agricultural Organization (FAO), and the World Health Organization (WHO).[17]

UNRRA AND THE WELFARE OF EUROPE'S CHILDREN

The United Nations Relief and Rehabilitation Administration was created to address the immediate plight of women and children in the liberated areas of Europe. It grew out of the Office of Foreign Relief and Rehabilitation Operations (OFRRO), created by President Roosevelt in 1942 and under the leadership of Herbert Lehman, who, at that time, was governor of New

York. OFRRO was to do more than ship supplies of food and medicine to aid the children of Europe. It was intended to help rebuild the shattered economies of liberated territories as part of an integrated approach to improving the health and welfare of Europe's children. OFRRO's health section was dominated by officers loaned from the Rockefeller Foundation's International Health Division and the US Public Health Service. Its first secretary was Selskar Gunn, who had been the director of the International Health Division's European programs and designed the IHD's North China Rural Reconstruction Program in the 1930s (chapter 4). The choice of Gunn to run OFRRO reflected the strong legacy that the 1930s had on postwar planning in the 1940s.

Gunn asked Martha May Eliot, a Harvard-trained physician who had headed the Division of Child and Maternal Health in the Children's Bureau, to produce a plan for addressing the plight of Europe's children. The Children's Bureau was a Progressive Era–institution that had been created prior to World War II as part of the US Department of Labor. It was concerned with all aspects of child welfare, including health, housing, education, and nutrition. The Children's Bureau had already developed international programs in Latin America, and Eliot made a major survey of British wartime programs for children in designing OFFRO's program. Postwar programs, she concluded, must be driven by coordinated and comprehensive planning for children, bringing together health and welfare workers. She accordingly developed a strategy for helping Europe's children that included a broad infrastructure for the protection of maternal and child health. The proposed program would recognize the special needs of a mother and child to appropriate medical care, nutrition, and clothing, and the right of every child to the security of family life.[18]

In 1943, OFRRO was folded into a new, larger, multilateral organization, the United Nations Relief and Rehabilitation Administration. Lehman was appointed its first director. UNRRA continued OFRRO's commitment to a broad vision of reconstruction and rehabilitation in Europe and China. While supplying relief aid was its immediate objective, it devoted millions of dollars to resurrecting health services and stimulated agricultural and industrial development.[19] Over time, however, these larger goals proved difficult to achieve (chapter 6).

THE FAO'S ECONOMIC APPROACH TO THE WORLD FOOD CRISIS

The Food and Agricultural Organization, which was created to oversee the postwar development of agriculture and food production, also adopted a comprehensive approach to improving the world's food supply and advancing worldwide nutrition. Recognition of growing food shortages during World War II and a desire to begin work on postwar reconstruction led the Roosevelt administration to call for a conference to discuss possible "international agreements, arrangements, and institutions designed to promote efficient production and to ensure for the world adequate supplies of food and other essential agricultural products."[20]

The conference was convened in Hot Springs, Virginia, from May 18 to June 3, 1943. Experts in agriculture and nutrition from 44 nations attended the conference. The FAO's founders adopted a broad-based economic framework for understanding and responding to the problem of food shortages. This framework explicitly linked adequate diets to wider political and economic transformations. The final declaration of the conference stated:

> The first cause of malnutrition and hunger is poverty. It is useless to produce more food unless men and nations provide the markets to absorb it. There must be an expansion of the whole world economy to provide the purchasing power sufficient to maintain an adequate diet for all. With full employment in all countries, enlarged industrial production, the absence of exploitation, an increasing flow of trade within and between countries, an orderly management of domestic and international investment and currencies, and sustained internal and international equilibrium, the food which is produced can be made available to all people.[21]

Yet there were disagreements about how the proposed organization should meet its responsibilities. Latin Americans and those from other non-European countries pleaded for a powerful organization able to realize a far-reaching regulation of global food markets. They had a substantial interest in restructuring international trade, in order to regain their former role as food exporters for the industrialized world. Supporters of a strong organization argued that, in the long run, simply distributing food surpluses from the industrialized world more efficiently could not, by itself, prevent regional famines. Representatives from Britain and the US State Department expressed opposition to this model, however, and proposed that the organization have a more limited role, similar to the prewar Institute for International Agriculture, which had focused its attention on collecting statistics.

When the FAO held its first meeting in Québec in 1944, it was unclear which direction the organization would take. The delegates chose to elect Boyd Orr, the famous British nutritionist and LNHO advisor, as the FAO's first director-general. In 1936, Orr had published a groundbreaking study that examined the relationship among food, health, and income and established the foundation for a social approach to nutrition. He believed that the new FAO should take an activist position in promoting global nutrition, a stance that put him at odds with the British government, which had purposely not chosen him as an official British delegate.

Orr was convinced that simply increasing production would not solve the world's food shortages. It also required a profound transformation in established patterns of distribution, without which a growth in food supplies would lower food prices and eventually undermine production. He therefore proposed the creation of a World Food Board (WFB) at the FAO's second conference in Copenhagen in 1946.[22] The WFB would ensure stable prices and food supplies by setting up and operating buffer stocks and regulating trade. In other words, it would stabilize prices for commodities, just as the newly formed IMF stabilized currencies. The WFB would also create a credit facility to ensure that poor farmers had access to agricultural inputs. Ensuring a more consistent demand for food, Orr argued, also provided a stimulus to world economic growth. These proposals replayed arguments that had been made by the League of Nations Mixed Committee on Nutrition and clearly linked improvements in nutrition and food consumption to global economic processes. The conference approved Orr's proposals in principle and set up a 16-member Preparatory Commission to work out the details.[23] But the commission walked away from Orr's proposals and set the FAO on a more narrow approach to food and nutrition (chapter 6).

WHO AND THE LEGACY OF SOCIAL MEDICINE

An examination of the planning and early activities of the World Health Organization also reveals an early commitment to an integrated view of health and development. The United Nations Charter committed its member nations to promoting "social progress and better standards of life in larger freedom" and to employing international machinery to further "the economic and social advancement of all peoples."[24] Accordingly, an Economic and Social Council was created as part of the United Nations to oversee social aspects of its mission.

In January 1946, this council met and proposed that a conference be held to discuss the creation of a new international-health organization to replace the LNHO. While the United States and Britain initially objected to the creation of such an organization (chapter 6), the conference was held in New York City in June 1946 and was attended by all UN member states, nonmember states, and a series of international organizations, including representatives from UNRRA, the Pan American Sanitary Bureau, the FAO, the International Labor Organization, the Red Cross, and the Rockefeller Foundation. The conference established the Technical Preparatory Committee (TPC) to create a constitution for the proposed health organization.[25]

The TPC included a number of long-time supporters of social medicine, such as Andrija Štampar, René Sand, and Karl Evang. Štampar, the Croatian physician who had headed up the League of Nations' health programs in China in the 1930s, chaired the TPC and would be elected president of the first World Health Assembly two years later. The broad definition of health that appears in the preamble of WHO's constitution was part of Štampar's proposal for the new constitution in 1946.[26] René Sand was an activist Belgian physician and major figure in the world of social work. He founded the Association Belge de Médicine Sociale in 1913, became the first academic chair for the history of medicine and social medicine at the Free University of Brussels in 1945, and would publish *The Advance of Social Medicine* in 1952. Norwegian physician Karl Evang was a socialist and believer in the importance of social reform for the advancement of health, the need to develop primary health care, the idea that health was a basic right, and the obligation of states to provide for the health of their populations.

Two other TPC members, while not necessarily strong advocates of social medicine, were nonetheless sympathetic to some of its ideas. US Congressman Thomas Parran had been involved in New Deal politics before World War II. He had served on the House of Representatives' Committee on Economic Security, which drafted the Social Security Act, and had subsequently been appointed to the post of surgeon general by President Franklin Roosevelt when Hugh Cumming stepped down. Parran was an early and strong supporter of national health insurance, a position that earned him the animus of the American Medical Association and ultimately led to his removal as head of the US Public Health Service. Parran chaired the international-health conference in New York City in 1946. Brock Chisholm was a Canadian psychiatrist who had led the Canadian military medical services during World War II and risen in the ranks of the Canadian government to become deputy minister of health in 1944. Chisholm chaired the subcommittee that prepared the draft constitution for WHO and was responsible for the wording of much of the preamble. He was subsequently appointed chair of the Interim Commission (IC), which was created at the New York conference to lay the groundwork for the new World Health Organization, and led the IC for two years. In 1948, he was chosen to be the first director-general of the World Health Organization at the initial meeting of the World Health Assembly. As would be clear from his subsequent publications, as well as the appointments he made to expert committees and other administrative posts within WHO, Chisholm shared with Štampar, Evang, and Sand a strong belief in the importance of the social and economic determinants of health and the development of health services.[27]

At the first World Health Assembly in 1948, the delegates voted to prioritize several problems. Malaria, maternal and child health, tuberculosis, venereal disease, nutrition, and environmental sanitation were the first top-priority "clusters." Public-health administration and parasitic diseases came next.[28] The high priority given to eliminating specific diseases seemed to belie WHO's definition of health as more than the absence of disease and signaled a more general movement back toward the kind of disease-elimination campaigns that had characterized the early work of international-health organizations up through the 1920s. Yet the broader vision of health held by WHO's founding architects was also represented in the early activities of this organization.

In 1950, Chisholm chose Milton Roemer to head WHO's new Social and Occupational Health Section. Roemer had received his MD in 1940 from New York University. He had also earned a masters degree in sociology from Cornell University, having written his thesis on "Sociological Factors Influencing American Medical Practice." Roemer's thesis and subsequent work in public health were influenced by Henry Sigerist. Roemer served as a medical officer in the Farm Security Administration under Frederick Mott after leaving medical school. The FSA was a New Deal agency founded in 1937 and dedicated to assisting rural families who had been hit hard by the Great Depression. Roemer joined Mott in writing *Rural Health and Medical Care*, which was the first systematic analysis of rural health-care needs and services in the United States.[29] The book, in addition to detailing the dearth of medical services available to rural families, made a strong case for the creation of a system of compulsory national health insurance, along the lines proposed in 1945 by President Truman.[30]

At WHO, Roemer quickly set about applying the experience and perspectives he had gained from organizing health units in rural America to creating health demonstration areas in El Salvador and Ceylon. As historian John Farley points out, the goal of the demonstration areas was not to attack individual diseases, but to show "what may be accomplished in general social and economic improvement from a comprehensive health service organized along modern lines."[31] The demonstration areas were similar to the rural reconstruction programs run by the League of Nations and the Rockefeller Foundation in China in the 1930s. They linked health-service improvement with advances in nearly every aspect of village life, including agriculture, industries, education, social welfare, and housing. Medical services were both curative and preventative, with emphases on improved nutrition, health education, sanitation, and latrine construction. The demonstration area in Ceylon was established in a rice- and coconut-growing area with a population of 66,000 people, north of Colombo. The inhabitants of the area suffered heavily from infections diseases, and over half of the deaths there occurred among children under age six. Roemer's program relied heavily on local health workers and was intended to help local people help themselves.

Another indication of the continued influence of social medicine on the early work of WHO was the central role played by advocates of this approach in a number of expert committees that met during the early 1950s. The Expert Committee on Professional and Technical Education of Medical and Auxiliary Personnel, which first convened in 1950, was chaired by René Sand and included Andrija Štampar and John Grant. Among the committee's conclusions was the recommendation that medical doctors should be trained from both a medical and a sociological point of view. It noted that "it had become evident to all observers that the factors necessary for building up a healthy society include economic and social security, nutrition, housing, and general public-health education" and "that particular emphasis should be laid on the social aspects of public health training and the study of economic, social, and working conditions as they affected health."[32] The second meeting of this committee, in 1952, included Štampar and Sigerist, with René Sand present as a consultant to the committee. Its report asserted, as a "fundamental principle," that "the character of medicine at any time and in any place is mainly determined by two factors—the stage of development of scientific and technical knowledge, and the social, economic, and cultural structure of a given civilization." It further stated that "the practice of medicine today is a co-operative effort of many groups of health workers, including physicians, dentists, nurses, social workers, technicians, public-health engineers, chemists, and many others."[33]

Similarly, the first session of the Expert Committee on Public-Health Administration, in 1951, was chaired by Evang and included Štampar. The committee's report opened by citing Article 25 of the United Nations' Universal Declaration of Human Rights: "Everyone has the right to a standard of living adequate for the health and well-being of himself and of his family, including food, clothing, housing, and medical care and necessary social services, and the right to security in the event of unemployment, sickness, disability, widowhood, old age, or other lack of livelihood in circumstances beyond his control." The report went on to note that "the potentialities of world health are great if WHO can become the spearhead of a movement for the social and economic betterment of the underdeveloped countries while also serving as a channel for the exchange of ideas on health administration among all countries."[34]

In defining public health, the committee drew on Charles-Edward A. Winslow's definition in his *Evolution and Significance of the Modern Public Health Campaign*.[35] This definition included, among the various duties of a public-health program, "the development of social machinery to ensure to every individual a standard of living adequate to the maintenance of health." The committee also asserted that "health is a part of individual and social welfare in general; health administrations should, therefore, recognize the interdependence of all agencies concerned with community welfare." Among the list of "other services that contribute to health," the Expert Committee's report included

1. social welfare,
2. social security measures,
3. education,
4. food production and distribution,
5. reservation of land,
6. veterinary services,
7. labor standards,
8. recreation,
9. transport and communications,
10. youth movements,
11. irrigation,
12. environmental sanitation and personal cleansing services,
13. conservation of natural resources, and
14. population and family planning.[36]

World War II is often viewed as a watershed in the history of international relations and global development. And in many ways it was. It gave rise to commitments on the part of global leaders to build new forms of international political and economic cooperation. It also spurred efforts to generate economic development, not only to war-torn countries in Europe and Asia, but also, more broadly, to emerging nations throughout Africa, Asia, and Latin America. Yet the ideas that shaped the early vision and work of postwar international-health and development organizations were strongly influenced by ideas and approaches that had been developed in the 1930s. Many of the leaders who had been involved in the League of Nations Health Organization and the Rockefeller Foundation's International Health Division in the 1930s took up positions in the newly formed United Nations organizations that emerged after the war.

The influence of these leaders and ideas was short lived, however. In the early 1950s, postwar development efforts took a distinct turn away from these early ideas and became focused again on more-narrow technical approaches to health and development: the world of technical assistance. The world of international health returned to the earlier disease-campaign approach that had dominated health activities in the 1910s and 1920s. Why this happened is the subject of chapter 6.

CHAPTER SIX

A Narrowing Vision

International Health, Technology, and Cold War Politics

THE WORLD LEADERS who met to plan out postwar development hoped to advance a broad-based agenda of social and economic change. Yet they were confronted with immediate crises caused by epidemic diseases and acute food shortages among war-torn populations in Europe and Asia. These crises called for rapid responses that could not wait for the creation of integrated programs. At the same time, they had at their disposal new technologies, which had emerged during the war, that promised to bring about dramatic improvements in health and social welfare without widespread social and economic changes. These new technologies inspired optimism about the ability of planners to rapidly transform societies and led to a new "can do" culture of growth and technical assistance. Finally, world leaders were faced with a changing political environment in which broad-based approaches to health and development were viewed with suspicion, particularly by right-wing politicians in the United States. With the midterm elections in November 1946, a group of right-wing Republicans, including Senator Joseph McCarthy of Wisconsin, took control of both the US House and the Senate. They were isolationists, opposed to foreign aid and any increase in federal expenditures. They were also suspicious of the United Nations, as well as strongly anti-Communist. They vociferously attacked President Truman's proposals for national health insurance and the agencies associated with it, including the US Public Health Service and its head, Thomas Parran, who had been the US delegate to the TPC. All of these factors worked against the fulfillment of the expansive vision of health and development that had survived after World War II.

Epidemics of plague, typhus, syphilis, tuberculosis, and malaria threatened the postwar populations of Europe and Asia. The memory of massive postwar typhus and influenza epidemics that had ravaged Eastern and Central Europe after World War I underlined the need to respond quickly and stimulated disease-control campaigns. Even before the first World Health Assembly was held in 1948, the Interim Commission (IC) was faced with a cholera crisis in Egypt. The IC acted as an international clearinghouse for the procurement and distribution of vaccine supplies from around the world and was credited with having played a critical role in ending the epidemic.[1]

Food shortages were also a pressing problem. The war had disrupted agricultural production, and food supplies appeared to be inadequate to meet the needs of many of the world's populations. A League of Nations report on *Food, Famine and Relief, 1940–1946* concluded:

> Approximately 100 million people in Europe will receive less than 1500 calories a day, and of these many, particularly in Germany, Austria, and Hungary, are already receiving 1000 or less. In India the cereal crop is short by about 8 million tons, and a large part of the city population is existing on rations of 1000 calories a day or less. . . . In China acute local famines are reported, but transport obstacles make adequate relief extremely difficult. In Japan as well famine conditions will develop unless large imports materialize.[2]

The Bengal Famine in 1943, which killed an estimated 3 million people, raised the specter of a worldwide food catastrophe. These populations needed immediate assistance in the form of food and medical supplies. Women and children were seen as being particularly vulnerable.

Epidemics and famines caused many world leaders to view the immediate postwar years as a time of crisis. This perception worked against efforts to build comprehensive, integrated programs that combined advances in health with broader social and economic transformations. Instead, it favored short-term, immediate solutions. The perception of crisis would shape global-health planning over the next half century.

New Technologies

Medical advances during World War II, including the discovery of new medical and public-health technologies, facilitated rapid responses to postwar disease threats and made broad-based, integrated approaches to health and development seem unnecessary. These new technologies promised to greatly reduce the global burden of disease, with or without social and economic reforms. They also led to the emergence of a culture of development and technical assistance that encouraged the application of quick-fix solutions to complex problems, while, at the same time, reinforcing the faith of development planners in Europe and the United States in their ability to improve the lives of people living in Africa, Asia, and Latin America.

One of the most important medical advances during the war was the discovery and application of the pesticide DDT. American representatives of the Swiss firm J. R. Geigy AG brought an insecticide that bore the trade name Gerasol to the attention of the US Department of Agriculture in 1942. Tests over the next year showed that the compound, known as dichlorodiphenyltricloroethane, or DDT, had amazing insect-killing powers. More importantly, once it was sprayed on the walls of human habitations, it retained its toxicity for mosquitoes and other household pests for months. This residual effect meant that spraying only needed to be done every six months or, in some cases, just once a year. Long-acting pesticides significantly reduced the cost of vector control, making it possible to extend disease-control programs to cover entire countries or regions. DDT offered the likelihood of controlling and, perhaps, even eradicating some of the most debilitating and life-threatening diseases that threatened human populations, including malaria, yellow fever, plague, and typhus.

A series of successful antimalaria campaigns using DDT convinced many public-health authorities that the new pesticide could eliminate the need for other approaches to malaria, including efforts to improve the general social and economic well-being of at-risk populations, a strategy that had been supported by the Malaria Commission of the League of Nations in the 1920s and 1930s. In 1944 and 1945, the Malaria Control Demonstration Unit of the Public Health Subcommission of the Allied Commission tested the use of DDT in house spraying to control malaria in the Pontine marshes outside of Rome.[3] The success of these trials led the Board of Scientific Directors of the International Health Division of the Rockefeller Foundation to conclude that DDT had made earlier methods of malaria control superfluous.[4]

This conclusion ignored other factors that contributed to the success of this campaign. As historian Frank Snowden has shown, DDT was not the only weapon used in this battle. Quinine was widely distributed, and Paris green was employed as a larvicide. Equally important, the inhabitants of the region had been exposed to decades of malaria-control efforts. They were both informed about the disease and responsive to control efforts. Finally, the campaign itself generated widespread employment, which may have contributed to an overall improvement in health. But the Rockefeller Foundation chose to focus instead on the contributions of DDT, which was promoted as a new magic bullet in the war on malaria.[5]

Experimental programs in *Anopheles* eradication using DDT were subsequently carried out in Sardinia and Cyprus in 1945–1946. The Rockefeller Foundation, the United Nations Relief and Rehabilitation Administration, and the US Economic Cooperation Administration funded the Sardinia program, while the British Colonial Office funded the Cyprus experiment. Though unsuccessful in eradicating the malaria vectors, both programs succeeded in eradicating malaria by reducing the numbers of *Anopheles* mosquitoes below the threshold at which transmission could occur and by treating all cases, so as to eliminate the human reservoir of infection. Large-scale control programs relying exclusively on house spraying with DDT were also begun in Ceylon, India, and Venezuela in 1946.[6] DDT was also employed after World War II to control plague outbreaks in a number of countries, including Italy, India, Peru, and Senegal.

DDT was but one of a number of wartime medical discoveries that raised new hope that science, medicine, and technology could overcome diseases that had plagued human populations for centuries. New antimalarial drugs were developed after the Japanese cut off access to quinine supplies from Java. The US antimalaria research program, much of it situated at Johns Hopkins, tested thousands of synthetic compounds and ratified the use of chloroquine by the end of World War II.[7] Penicillin, which had been discovered in the 1920s, was adapted for mass production and use during the war. Moreover, the penicillin in use by 1945 was four times the strength of the original drug. Similarly, prewar sulfa drugs were brought into large-scale production and distribution during the war. Streptomycin, which was effective in curing TB, was discovered in 1943. New vaccines for influenza, pneumococcal pneumonia, and plague were also developed during the war. Finally, portable x-ray machines greatly facilitated postwar efforts to combat tuberculosis.

In short, World War II provided a new set of tools, making it possible—for the first time—to greatly reduce the global burden of disease through the application of biomedical technologies. The war had also brought advances in many other areas of technology, all of which contributed to a new faith in the ability of technology and science to transform the world. Whether it was nuclear weapons, atomic energy, drought-resistant plants, or penicillin and DDT, the postwar world was one that technology promised (or threatened, depending on one's perspective) to transform.

The Culture of Development

Faith in the power of these new technologies fed the emergence of a culture of development and technical assistance, which further discouraged the pursuit of broad-based approaches to health. This attitude was especially strong among US public-health officials in the 1950s. The attitude of "know-how and show how" was reflected in President Truman's 1949 inaugural address, in which he laid out four points that would guide US foreign policy. Point four was, "We must embark on a bold new program for making the benefits of our scientific advances and industrial progress available for the improvement and growth of underdeveloped areas." Truman stated: "More than half the people of the world are living in conditions approaching misery. Their food is inadequate, they are victims of disease. Their economic life is primitive and stagnant. . . . For the first time in history humanity possesses the knowledge and the skill to relieve the suffering of these people." He went on to observe: "The United Stares is preeminent among nations in the development of industrial and scientific techniques. The material resources we can afford to use in the assistance of other people are limited. But our imponderable resources in technical knowledge are constantly growing and are inexhaustible."

Confidence in the United States' ability to use its technological know-how to transform the developing world was expressed that same year in a US Department of State report on malaria-control programs: "The most dramatic results from the employment of a very small number of skilled men and very small quantities of scientifically designed materials have been achieved in the field of medicine. In many areas of the world one trained public-health doctor or a group of two or three working with local people able to follow their guidance have been able to rout one of man's oldest and deadliest enemies."[8]

Faith in technology and the optimism it inspired about the ability of the United States to transform the world carried with it certain assumptions about the peoples and societies that were to receive this technology and be transformed. Thus Truman's inaugural address described the underdeveloped world as "primitive and stagnant," and the above passage from the report on malaria-control programs privileged the skills and knowledge of the outside expert while placing local populations in a position of dependence, in need of guidance and assistance.

These attitudes echoed those of an earlier generation of international and colonial health authorities in the Philippines and elsewhere. Framing resource-poor countries as underdeveloped and in need of assistance justified outside intervention and empowered those who were intervening to define the goals of development. These goals, not surprisingly, reflected a Western industrial vision of modernity. Underdeveloped countries would become developed by achieving "high levels of industrialization and urbanization, technicalization of agriculture, rapid growth of material production and living standards, and the widespread adoption of modern education and cultural values."[9] They would also become developed by changing their reproductive behavior, creating smaller nuclear families that resembled families in the United States and Western Europe (part V). That the people receiving assistance might have their own visions of modernity and traditions of development was simply ignored.

It would be a mistake, however, to assume that the categories of developed and underdeveloped and the concept of development were primarily the constructions of postwar planners in Europe and the United States. Many of the leaders of resource-poor countries also adopted them. They touted development as a desirable process, justified political and economic policies in terms of the needs of development, and constructed 5- and 10-year plans to show how development would be achieved. The goal of development was also invoked by local populations and opposition leaders, who criticized governmental authorities for not delivering on their economic promises. Governmental leaders were blamed for failing to achieve development or, in some cases, such as in the Congo in the 1960s, for having "stolen" development.[10]

The Rise of the Social Sciences

The expansion of the social sciences—particularly political science, sociology, demography, anthropology, and economics—was critically important to the growth of this confidence in technical assistance and its ability to transform the developing world. The 1950s saw a burgeoning faith among social scientists in their ability to predict outcomes and to control for a whole range of social, economic, and cultural variables. Keynesian economists took the lead during and immediately after World War II in employing demand-side economic theories to predict policy outcomes. Development economics emerged after the war as a subdiscipline of economics, focused on formulating policies that would contribute to the economic advancement of developing countries.[11] The immediate postwar period also saw the rise of demography as a predictive science (chapter 9).[12] In the United States, motivational research firms employed behavioral scientists to predict consumer decision-making patterns and discover how these could be manipulated to increase the consumption of particular products. The universalizing claims of formalist disciplines made it easy to make the cultural leap from selling laundry detergent to housewives in Toledo, Ohio, to convincing African farmers to accept new fertilizers, take antimalaria tablets, or use contraceptives to limit their fertility. Focus groups designed originally to test consumer preferences for various products and marketing techniques were adopted by European and American development planners to promote social and economic changes in Africa, Asia, and Latin America. The social sciences provided the theoretical underpinnings to support and rationalize the efforts of countries in Europe and North America to transform the developing world through technical assistance.

Within this context, anthropologists, in particular, succeeded in inserting themselves into the practices of international technical assistance. In 1951, the director of the Smithsonian Institute for Social Anthropology (ISA), George Foster, met with the acting head of the Health Division of the Institute of Inter-American Affairs (IIAA). The IIAA had been created to oversee US development assistance to Latin American countries during the war.[13] Foster, an anthropologist trained at the University of California–Berkeley, proposed to utilize the ISA's anthropologists working in Latin America to assist IIAA health programs in meeting their goals. These anthropologists, Foster asserted, could collect data on the cultural ideas and practices of local populations. This data could then facilitate local cooperation and preempt possible resistance to innovations being introduced by the IIAA's technical-assistance programs. The IIAA's Health Division accepted the proposal and the ISA's anthropologists began conducting research linked to technical-assistance programs in eight project sites.

The results of these studies were written up in a report titled *A Cross-Cultural Anthropological Analysis of a Technical Aid Program*, which caught the attention of Henry van Zile Hyde, the permanent head of the IIAA's Health Division, who agreed to pay the salaries of Foster's team for a year as part of a broader project to evaluate the IIAA's health programs.[14] In this report, Foster laid out the problems facing technical-assistance programs and the role that anthropologists could play. His introduction to the report reflected the growing faith that social scientists had in their ability to identify cultural patterns and overcome obstacles to changes in health practices around the globe: "This paper illustrates the manner in which the social sciences, on the operational level, may contribute to the success of specific technical aid programs by analyzing and explaining the behavior of the peoples involved, and by pointing out to administrators the means by which customary action patterns may most easily be modified, and the points at which particularly strong resistances will be encountered."[15]

Foster would go on to serve as a consultant to WHO and an advocate for applied anthropology. It is worth noting that for Foster, the cultural problems facing technical-assistance programs resided in the culture of aid bureaucrats and fieldworkers, as much as in the culture of the people whose lives they were trying to change. As he noted later in life, the value of studying the culture of development was less appreciated by the technical-assistance community than the insights anthropologists provided into the lives and ways of local populations.[16] That said, Foster played a major role in integrating the social sciences, and particularly anthropology, into the practices of international health during the 1950s, 1960s, and 1970s.

Cold War Politics

The postwar political environment in which international-health and development interventions were forged also encouraged the shift toward narrow, technical approaches to health in the 1950s. Despite the existence of widespread agreement about the need to create a new world order, based on international cooperation in a number of fields, there were disagreements over what that new order should look like. These disagreements often pitted leaders in continental Europe, and particularly the Soviet Union, against Britain and the United States. They also divided political leaders within the United States.

Right-wing politicians, who gained ascendancy in the United States during the late 1940s and 1950s, objected to the creation of any organization, or the implementation of any international policy or program, that could potentially threaten US political and economic interests at home or abroad. They reacted strongly to plans that appeared to advance a "socialist" agenda and pushed hard for programs that promised to further the development of free-market economies, Western-style democracy, and capitalist forms of production. Within the United States, fears of Communist expansion in Europe and China, where the Communists finally succeeded in overthrowing the Nationalist government of Chiang Kai-shek in 1949, had a chilling influence on efforts to advance broad-based health and development agendas, let alone a system of national health insurance. Right-wing politicians viewed large-scale, government-funded projects that integrated health, education, and rural development, such as the Tennessee Valley Authority and the Farm Security Administration, as examples of a Soviet-style, command-economy approach to development. Those who supported such programs faced criticism and censure. When President Truman nominated the Tennessee Valley Authority's director, David Lilienthal, to head up the new Atomic Energy Commission in 1946, Lilienthal's appointment was held up for weeks while senators examined his political record. Senator Robert Taft, a Republican, claimed that Lilienthal was too soft on issues related to Communism and the Soviet Union.[17] These attitudes also limited the ability of postwar international organizations to carry out their early commitments to broad-based, integrated approaches to health and development and instead encouraged reliance on narrowly focused technical programs, made possible by advances in technology and science during World War II.

US fears about the spread of Communism were matched by Soviet concerns over what they saw as an aggressive American economic imperialism, which threatened Soviet economic and political interests. Growing tensions between the United States and Soviet Union gave rise to the Cold War. The United States and its Western allies wrestled with the Soviet Union, Eastern Europe, and China for global dominance and influence over emerging nations in Africa, Asia, and Latin America, the so-called Third World. For the US government and its allies, technical-assistance programs quickly became weapons in their battle against Communist expansion. These programs were ways of relieving the poverty and suffering that were viewed as the breeding grounds for Communism. Of particular value were "impact programs." These were programs designated by the US Special Technical and Economic Mission as having a rapid, positive effect on local populations, as well as building support for local governments and their US allies. Malaria-control programs using DDT were viewed as one of the most effective impact programs. The apparent speed with which malaria could be brought under control with DDT, together with its short-term effects on other household pests, made malaria-control particularly attractive for those who saw tropical disease–control as an instrument for winning the support of Third World populations; in other words, for "winning hearts and minds," a phrase associated with US efforts to draw Vietnamese peasants away from the Communists in the 1950s, 1960s, and 1970s.[18]

The political benefits that had been achieved by malaria control were laid out in the 1956 International Development Administration Board's report to President Dwight Eisenhower on malaria eradication:

> As a humanitarian endeavor, easily understood, malaria control cuts across the narrower appeals of political partisanship. In Indochina, areas rendered inaccessible at night by Viet Mihn activity, during the day welcomed

> DDT-residual spray teams combating malaria. In Java political tensions intensified by overcrowding of large masses of population are being eased partly by the control of malaria in virgin areas of Sumatra and other islands, permitting these areas to be opened up for settlement that relieves intense population pressures. In the Philippines, similar programs make possible colonization of many previously uninhabited areas, and contribute greatly to the conversion of Huk terrorists to peaceful landowners.[19]

The report claimed that the governments of developing countries also perceived eradication as a means to gain popular political support: "The present governments of India, Thailand, the Philippines, and Indonesia among others, have undertaken malaria programs as a major element of their efforts to generate a sense of social progress, and build their political strength."[20] As Indian malariologist Dharmavadani K. Viswanathan pointed out, "No service establishes contact with every individual home at least twice a year as the DDT service does unless it be the collection of taxes."[21] Within the cold war environment, programs that promised to make an immediate impact were prioritized over programs that would make longer-term, fundamental change in peoples lives.

The Soviets and their allies also engaged in technical-assistance programs during the 1950s, focusing particularly on the Middle East, Africa, and South and Southeast Asia. Like Western aid, Soviet-bloc assistance targeted specific areas and did not encourage integrated development programs. In contrast to Western technical assistance, however, the Soviets sought to support longer-term development, directing their aid toward basic industries, such as steel and textiles, and overhead investments, like irrigation works, power, roads, and communications.[22] They also supported scientific and educational facilities. Much of the Soviets' technical assistance came in the form of military aid. In the area of health, while the Soviets and East Germans had developed primary health-care services in their own countries, they built hospitals in Africa, serving as symbols of socialist scientific achievement. Thus neither side in the Cold War supported a return to the broad-based integrated approaches to health and development that had emerged in the 1930s.

The Changing Strategies of Postwar Health and Development Organizations

The histories of UNRRA, the FAO, and WHO reveal how the crisis environment of the immediate postwar period, the availability of new medical technologies, the culture of development, and the emergence of cold war tensions combined in different ways to reshape the policies of international-health and development organizations in the late 1940s and 1950s.

UNRRA AND THE POLITICS OF RELIEF AID

The need to respond to crisis conditions, as well as political pressures from the Right and Left, forced UNRRA to narrow its goals and ultimately led to its dissolution.[23] Responding to the acute food shortages faced by populations in Europe provided little opportunity for UNRRA to build new infrastructures to support the health and nutrition of women and children, as originally envisioned by Martha Eliot. Nor was there time or resources to rebuild health systems, resurrect agricultural production, and revitalize industrial production. UNRRA struggled to provide needed relief supplies across Europe, particularly to the worst-affected areas of Eastern Europe. UNRRA also found itself faced with the problem of dealing with millions of displaced persons, many of whom had found shelter in camps scattered across Europe. The camp populations needed to be protected from disease and hunger and ultimately repatriated to their countries of origin, some of which no longer existed. There was little time to design, let alone implement, enduring programs, such as welfare and training programs.[24]

UNRRA officials also found it politically difficult to engage in broader planning efforts. Wealthier European countries were suspicious of UNRRA, seeing it as an effort on the part of the United States to make economic inroads in Europe. The Soviet Union was particularly critical of UNRRA for not doing more to aid the populations in Eastern Europe, which it increasingly viewed as falling within its sphere of influence. The political Right in the United States ironically attacked UNRRA for devoting too much of its work to supplying aid to Soviet-dominated Eastern Europe. Republicans in the US Congress also opposed any efforts at institution building on the part of UNRRA. They viewed UNRRA primarily as a solution to the problem of large agricultural surpluses in the United States. Institution building did nothing to advance food exports.[25]

UNRRA officials also faced local conditions that made building health and welfare programs difficult. In Greece, UNRRA personnel arrived immediately after the withdrawal of Axis powers in 1945. Their operations were delayed by a civil war that broke out between Communist-led partisans and the British and American–supported government in Athens. By 1946, however, the UNRRA mission was employing thousands of Greek workers and professional staff, along with some 700 foreign staff, primarily

British and American; 191 of these were health professionals. James Miller Vine, UNRRA's director in Greece, proposed to build a comprehensive "first best" health system. Yet the challenges of working with state and local administrations, which UNRRA authorities viewed as disorganized, corrupt, and incompetent, soon led UNRRA officials to dial back their enthusiasm for broad-based social engineering. It should be noted that these characterizations of Greek officials ignored the fact that these officials had alternative ideas about the methods and goals of health reform in their own country.

In the end, UNRRA focused on the most serious disease threats, devoting the bulk of their energies to controlling malaria with DDT, even though an early US Army report warned that "personnel must not consider an effective insecticide like DDT as a substitute for sanitation and preventive medicine." The supposition that Greek officials opposed serious health reform led UNRRA officials to quickly adopt this technological fix, employing airplanes to spray wide areas of infected territory. The success of the aerial malaria campaign convinced many in UNRRA that malaria could be controlled in the absence of social and political reform. This lesson would be widely accepted in the years to come. Greece also provided another lesson. Shortages in DDT supplies led the control program to curtail spraying. UNRRA officials feared that malaria would make a quick comeback, but it did not. A few cases occurred, but these were identified and treated, and Greek health authorities were able to maintain control without reverting to new rounds of aerial spraying. This experience would inform the design of WHO's malaria-eradication program in the 1950s and 1960s (chapters 7 and 8).[26]

Much of UNRRA's work in southern Europe ended up focusing on disease-control programs. Efforts to reconstruct postwar Italy also fell afoul of local administrative resistance, and the agency reverted to meeting the immediate food needs of mothers and children and combating diseases (particularly malaria), working with personnel from the IHD and Alberto Missoroli of the Institute of Hygiene in Rome. These efforts kept malaria in check in Allied-controlled areas. In addition, they succeeded in eliminating malaria from Sardinia using DDT. As in Greece, these results were achieved without the development of an effective sanitary organization. Sardinia, like Greece, became an example of how vertical, military-style disease campaigns, using effective technologies, could defeat major disease threats without social reconstruction.[27]

Finally, UNRRA's workers were often overwhelmed by the sheer number of people who needed their help and the depth of their destitution, particularly in the hundreds of refugee camps they operated. While many relief workers developed a kind of cultural relativism, attempting to understand the needs and actions of local populations on their own terms,[28] they sometimes lost sight of the refugees' humanity. They came to see them as logistical problems to be solved, rather than as individuals and families with complex needs. In such circumstances, the burden of just getting through each day delayed efforts to build for the future. The memory of one UNRRA worker captured the dehumanizing experience of working in the camps. This particular camp worker was faced with the problem of distributing food vouchers to long lines of refugees filing into the camps, their arms filled with their few remaining possessions. He found a simple solution: he stuck the vouchers into their mouths.[29]

Continued concerns about UNRRA's work in Eastern Europe, and the growing belief by some US congressmen that it was not the most effective instrument for dealing with the distribution of US food surpluses, led the United States to withdraw its funding from the organization in 1947, forcing UNRRA to close its doors. US officials increasingly came to see international organizations as ineffective and instead opted to channel funds to Europe through bilateral assistance. In the same year that United States withdrew its support from UNRRA, it launched the Marshall Plan, which would funnel US$13 billion in direct economic aid to Europe.

UNRRA's leaders had viewed its role as providing both immediate relief and support for rebuilding war-torn economies. But they found it politically and operationally difficult to do the latter and ended up focusing most of UNRRA's energies on providing emergency relief.[30] When UNRRA was disbanded, many of its aid workers were absorbed by other organizations, including the United Nations International Children's Emergency Fund (UNICEF), which took over many of UNRRA's responsibilities, as well as much of its funding.

THE FAO: FROM THE WORLD FOOD BOARD TO TECHNICAL ASSISTANCE

The Food and Agriculture Organization had also adopted a broad-based stance on the problem of food shortages and elected John Boyd Orr, a strong advocate for social approaches to nutrition, as the organization's first director-general. In 1946, the delegates to the FAO's second conference in Copenhagen reaffirmed their support for a broad-based approach to solving the world's food problems by approving Boyd Orr's proposal for the creation of a World Food Board (WFB), which would regulate global food prices and supplies. But, like UNRRA, the FAO found it difficult to carry out this vision.

When the FAO's Preparatory Commission met to work out the details of the proposed WFB later that year, it became clear that not everyone on the commission was happy with Boyd Orr's proposal. While it received strong support from France, Austria, Poland, Greece, and other food-producing countries, many nations were nervous about giving an external agency control over their commodity policies. Britain, as a major food importer, was particularly concerned that the WFB would be unable to adequately regulate prices and feared that this would lead to rising food costs.

There was also concern that growing cold war tensions would prevent the Soviet Union, one of the world's largest grain producers, from participating in the WFB. The Soviet Union had refused to take part in the FAO meetings, and how could food prices be regulated without Soviet cooperation? That said, it was US opposition that prevented the creation of the WFB. The United States had little to gain from Boyd Orr's proposals, but it would have had to contribute the bulk of the funding to support the new institution. The United States also opposed any organization that would have the power to exercise control over its foreign or domestic policies. That same year, the US Congress declined to ratify the charter for the International Trade Organization.[31] The US State Department also objected to Boyd Orr's schemes, because they conflicted with its existing national food-aid programs, which were used to deal with domestic surpluses and, increasingly, as an instrument to support US foreign-policy objectives.[32] The United States preferred to use bilateral intergovernmental arrangements to regulate food supplies.

An American, Norris E. Dodd, who was the US undersecretary of agriculture at the time, succeeded Orr as director-general of the FAO in 1948. Dodd, an Oregon farmer, was sensitive to the problem of food prices. Yet he was also aware that there was limited support for the global regulation of prices, as proposed by his predecessor. In addition, the world's food situation had changed. European reconstruction and the Marshall Plan had greatly reduced Europe's food crisis. Dodd recognized that the focus of the world's food crisis had shifted to developing countries, and that the solution lay less in regulation than in the provision of technical assistance. Dodd also was aware that the FAO had a very small budget, US$5 million, which limited what it could do. There would be no Marshall Plan for developing countries. Finally, Dodd realized that political opposition in the United States to any efforts to control global food markets would prevent their creation.[33]

Dodd therefore turned the FAO's attention away from global regulation strategies, which linked food and nutrition to the global economy, and concentrated on providing technical assistance that was aimed at increasing agricultural production in developing countries. The availability of new fertilizers and pesticides developed during and immediately after World War II and, later, hybrid seeds, provided additional incentives for the organization to focus on providing food-strapped countries with technical assistance. In effect, the FAO, like UNRRA, narrowed the focus of its activities.[34] The United States continued to use its influence to keep the FAO focused on the narrow task of providing technical assistance through the 1950s. When increased funding, made available through the United Nations' Expanded Programme for Technical Assistance (EPTA) in 1950, provided Dodd with an opportunity to enlarge the FAO's mission, the United States was able to use its voting power within EPTA to block this plan.[35]

Meanwhile, the US government developed a bilateral strategy for providing food aid to the world in a way that benefited US economic and political interests. In 1954, the US Congress passed the Agricultural Trade Development and Assistance Act, better known as Public Law 480. This act provided a way of dealing with the massive food surpluses that the government purchased from US farmers in order to stabilize food prices. These surpluses cost the government millions of dollars per day for storage. The solution to the problem was the export of food surpluses, through concessionary sales, to countries experiencing food shortages. To overcome dollar shortages in nations receiving food surpluses, the act allowed those countries to purchase the food with local currencies. By targeting food sales to pro-American governments in the developing world, the United States converted its surplus food problem into a foreign-policy instrument.[36] In the ensuing years, US administrations increasingly used food-aid programs, now termed "Food for Peace," as a political tool. This was done most blatantly under President Lyndon B. Johnson, who, for example, insisted in keeping India, which maintained relations with the Soviet Union, on a short tether by linking food assistance to India's political behavior. Under both Johnson and President Richard Nixon, local currency generated by the sale of American concessionary food aid to South Vietnam was used to support counterinsurgency activities.[37]

WHO RATIFICATION AND US POLITICS

The United States played an even greater role in redefining the mission of the World Health Organization. From the outset, the United States lobbied against the creation of an international-health organization. This opposition reflected the country's longstanding resistance to participating in institutions that would be able to exercise influence over America's health-care system. The United States had been a founding member of the Pan American Sanitary Bureau (PASB). But it had successfully balanced its participation in the organization with its desire to restrict interference with domestic health policies by virtually controlling the PASB's agenda. The bureau's constitution made the US surgeon general its permanent president. Moreover, it permitted the president of the PASB to establish the "official" agenda of the bureau's annual conferences. The conference actually had two agendas—one set by all the member states, and one set by the president—but only the agenda formulated by the president could generate official policies or initiatives. The agenda set by the whole membership provided opportunities for wide-ranging discussions of numerous topics, but none of these discussions could result in policies that were binding to PASB member states. The United States could thus participate in the PASB without fear of this organization influencing US health policy.[38] By contrast, the proposed WHO constitution allowed each country to have one vote, regardless of its financial contributions to the organization. Moreover, any WHO member could introduce a proposal to the World Health Assembly for discussion. The United States could not control the agenda directly.

US opponents of the new organization had little success in shaping WHO's constitution.[39] Advocates for a broad-based approach to health, and the idea that health was a right, dominated the drafting committee. Even the US delegate, Thomas Parran, was a supporter of universal health insurance. Yet the constitution had to be ratified by a majority of member nations before it could be enacted. Many realized that without US ratification, the proposed new organization would lack the financial resources it needed to perform its duties, regardless of how many other countries ratified the constitution. The new health organization would, in effect, be stillborn. In the end, the US Congress ratified the constitution, but in a manner that allowed it to shape WHO's future agenda.

US ratification of WHO's constitution was not at all certain. Many US congressmen on the political Right were critical of the proposed constitution, with its insistence that member countries were responsible for the health of their citizens, and that health was a basic human right. They were also sensitive to the constitution's broad definition of health. Early discussions leading to the creation of WHO had included references to social insurance; as a result, the Republicans' response to the new organization became entangled in their attacks on national health insurance, "socialized medicine," and the war on Communism. Hugh Cumming, director of the PASB and a former US surgeon general, actively campaigned against ratification. He referred to WHO's constitution as reflecting "the dominance of stargazers and political and social uplifters" and described the organization as "an entering wedge for Communism into Pan America."[40] Cumming had been a member of the LNHO Health Committee, but he was deeply opposed to what he saw as Rajchman's "socialist" policies and had opposed efforts to hold a rural hygiene conference, along the lines of the Bandoeng Conference, in the Americas. He urged Latin American governments to also reject WHO.

Supporters of the new organization, including President Truman, insisted on the importance of US participation in it. The US State Department issued a statement making the argument that "a broad gauged international organization in the field of health is absolutely essential," and that, through WHO, "the United States, which is one of the countries far advanced in medical science and public health, can play an important role in improving the health conditions of more backward states." Also advocating ratification were a number of powerful private institutions, including the Rockefeller Foundation's Institute of Medical Research, the Milbank Memorial Fund, the Nursing Council, and the Johns Hopkins School of Hygiene and Public Health. The American Medical Association was in favor of the creation of WHO but predictably objected to anyone coming to the United States to tell Americans how to practice medicine. Thomas Parran made a case for ratification, pointing to the trade benefits that would result from the creation of WHO: better pharmaceutical standards and improved health services in developing countries would increase US exports of drugs and health supplies. The United States would invest millions of dollars in building tertiary health systems in developing countries that would consume large quantities of US-manufactured medical equipment and pharmaceuticals. The control of diseases would also increase world trade, and US scientists would enjoy greater opportunities for research. Similar arguments would be presented in support of future efforts to convince the US Congress to support technical-assistance and aid programs across the globe.[41]

The debate was heated, but the US Congress finally ratified WHO's constitution, with only three days left before the first World Health Assembly was to meet in June 1948. Congress, however, attached two critically important qualifications to US membership. First, the United States reserved the right to withdraw from the organization at any time, despite the fact that the right of withdrawal was not granted in WHO's constitution. Second, Congress placed dollar limits on its obligations to WHO. The initial limit for 1949 was US$1,920,000. All member states had to make compulsory contributions, based on their ability to pay. The United States, being the richest country, was given the highest contribution assessment. At first it was responsible for 39 percent of WHO's total budget, but this was later dropped to 35 percent. The immediate consequences of Congress setting the United States' contribution limit at US$1.9 million was that WHO's total budget for 1949 had to be reduced from US$6.5 to US$5 million, making it difficult for the new organization to carry out its plans. In subsequent years, Congress would increase this limit, but it continued to keep a lid on WHO's regular budget. The longer-term consequence of this limiting process and the insistence on America's right to walk away from the organization was an implicit reminder that the United States controlled WHO's purse strings, and that if the United States was unhappy with the direction taken by WHO, it could take its contributions and leave.

Finally, Congress required that US delegates to WHO had to be medical-school graduates with at least three years of active practice, a bone thrown to the American Medical Association, which had objected to the absence of practicing physicians among the US delegation to the TPC. Congress also stipulated that a delegate could not serve "until such person had been investigated as to loyalty and security by the Federal Bureau of Investigation."[42] Restrictions on who could represent the United States at WHO, combined with US control of the organization's purse strings, made it highly likely that WHO would adopt policies that reflected a narrow, biomedical approach to health and avoid large-scale, integrated development programs. The Republican-dominated US Congress would later extend its "loyalty" condition to American staff members working for WHO. In 1953, Milton Roemer, who, in 1948, had faced charges of consorting with Communists by the Board of Inquiry on Employee Loyalty of the Federal Security Agency, was forced to resign from WHO after the US State Department revoked his passport when he refused to sign a loyalty oath, which the United States required of all Americans working for the United Nations. The removal of Brock Chisholm's lieutenant, who was the architect of WHO's early experiments in rural health and reconstruction, had a chilling effect on efforts to move the organization along the pathway laid out in the preamble to its constitution.[43]

Cold war politics would continue to play an important role in forcing WHO away from its original vision. To begin with, they led to the early withdrawal of the Soviet Union and its East European allies from WHO in February 1949. The Soviets claimed that the organization served to advance US interests, an assertion that was not totally mistaken, given that the United

States exercised a great deal of control over WHO's agenda. The Soviets also expressed dissatisfaction that WHO was not doing more to help Eastern European countries that had suffered greatly during World War II, a policy the Soviets also attributed to US influence. More broadly, the Soviets felt that WHO should be doing more to advance health-care services along Soviet lines and to address the underlying social determinants of health.

The withdrawal of the Soviet block, along with Nationalist China, which had left WHO for undisclosed reasons, led to a US$1 million shortfall in the organization's budget. It also meant that the voices of Eastern European countries, which had been central to the advancement of social medicine in the 1930s, were no longer heard in the World Health Assembly and thus were unable to counter the narrower medical approaches to health advocated by US delegates. While Nordic countries continued to lobby for more-progressive approaches to health, they had little success in shaping WHO's agenda.

Early on, WHO found itself caught between the restraints of a tight budget and the need to show that it could make a difference in improving the health of the world's populations. The first few meetings of the World Health Assembly laid out an ambitious agenda and established expert committees to discuss and make recommendations on medical education, public-health administration, environmental sanitation, the provision of health services, and maternal and child health. WHO lacked the financial resources to move forward on all fronts, however. Choices had to be made.

Under Chisholm's leadership, much of WHO's early work was focused on setting up its administration and working out the organization's relationships with regional health organizations, the most problematic of which was its relationship to the PASB. The US government was opposed to the subordination of the PASB to WHO and insisted that the PASB remain largely autonomous, representing the interests of countries in the region (read the United States). The two organizations eventually reached an uneasy state of détente, which ensured cooperation at a distance.[44] When it came to making choices regarding the prioritization of problems and approaches to health, Chisholm was committed to preserving the principles laid down in WHO's constitution, which he had played an important role in drafting. While recognizing the power of new medical technologies to reduce the world's burden of disease, he insisted that improvements in health needed to be based on efforts to also address the social and economic forces that shaped health outcomes.

Yet such a broad-based vision was difficult to implement. WHO's expertise lay in a more narrow range of health activities. It lacked both the expertise and the resources to engage in efforts to improve the underlying social and economic conditions of health. To do this, WHO would need the cooperation of other organizations, including the ILO, the FAO, the UN Development Programme (UNDP), UNICEF, and the United Nations Educational, Scientific and Cultural Organization (UNESCO), all of which had interests in health but had their own mandates to fulfill. The ILO was primarily concerned with labor and social security issues, the FAO with agricultural production and consumption, the UNDP with economic development, UNICEF with the problems of women and children, and UNESCO with education. This postwar division of labor among UN agencies made the formation of broad-based approaches to health and development difficult to achieve.[45]

The challenges facing cooperative efforts were evident in early interactions between WHO and the FAO around questions of food supply and nutrition. Both organizations were concerned with improving nutritional standards across the globe. The FAO was charged with developing broader strategies for increasing agricultural production and ensuring overall food security. WHO, on the other hand, was concerned with combating deficiency diseases, such as pellagra and kwashiorkor. In other words, WHO's mandate was the medical consequences of malnutrition. In theory, the two organizations had much to gain from working together to form a comprehensive approach to food security and nutrition. To this end, they formed a Joint WHO/FAO Expert Committee on Nutrition, which met regularly during the 1950s. The early sessions of the joint committee acknowledged the need to view the problem of nutrition broadly, involving experts from multiple fields. These statements echoed those of the LNHO and ILO's Mixed Committee in the 1930s (chapter 3).

Yet the activities of the WHO/FAO joint committee were much more limited than such statements suggested. This was largely because the joint committee was composed of experts in nutrition from *both sides*. There were no representatives who specialized in issues related to agricultural production, marketing, and the distribution of food supplies. As a consequence, most of the committee's recommendations related to the development of demonstration projects aimed at improving nutritional practices and reducing nutritional-deficiency diseases, such as kwashiorkor, by improving knowledge about healthy foods and addressing local communities' farming practices. These efforts largely ignored the structural causes of malnutrition.[46] Instead, severe malnutrition was presented as a cultural/behavioral problem for which technical solutions, such as the provision of dried skim milk, were the answer.[47]

Another area in which there was early cooperation between the FAO and WHO was in malaria control. The FAO viewed malaria control as important for the expansion of food production. By eliminating malaria, large tracts of land could be opened up for farming, and farmers could be more productive. The two organizations worked to identify malaria-control demonstration areas in locations that were deemed agriculturally important. Over time, however, this effort at interorganizational cooperation broke down over policy issues related to the use of pesticides. The FAO wanted to use pesticides to protect plants as well as people. WHO saw this dual usage as a recipe for producing pesticide resistance in insects posing public-health problems, undermining WHO's efforts to protect human health. While Chisholm might favor integrated approaches to health and development, WHO was not positioned financially, politically, or in terms of its expertise to do more than articulate the importance of this approach.[48]

Meanwhile, the world faced a growing burden of disease for which biomedical responses were now available. Tuberculosis, yaws, malaria, syphilis, smallpox, and typhus were widely viewed as critical global-health problems. WHO officials estimated that there were 250 million cases annually of malaria worldwide in the early 1950s, causing 2.5 million deaths.[49] Tuberculosis ravaged postwar populations in Europe, with an estimated prevalence in Warsaw of 500 deaths per 100,000 individuals. The Bhore Committee stated that there were 2.5 million active cases of TB in India, leading to a half-million deaths per year.[50] The Indian Council of Medical Research found that in the late 1950s, the average prevalence of sputum-positive pulmonary tuberculosis was about 400 per 100,000 people, amounting to an absolute number of 1.5 million infectious cases, with similar prevalence rates in rural and urban areas.[51] During the same period, there were an estimated 50 million cases of smallpox worldwide. Many in WHO argued that reducing the global burden of these major diseases was both possible, given the new tools at their disposal, and morally imperative. The world could not wait for advances in social and economic development before beginning to reduce the global burden of disease.

Finally, WHO needed to establish its leadership in international health, where it increasingly faced competition from other organizations, the most important of which was the newly created United Nations International Children's Emergency Fund. Reducing or eliminating the global burden of disease was viewed as a way of asserting this leadership role. A brief examination of WHO's relationship to UNICEF reveals a great deal about the way interorganizational politics shaped international health in the 1950s.

UNICEF AND WHO: AN UNEASY PARTNERSHIP

UNICEF emerged from the ashes of UNRRA in 1946. When the United States announced that it was withdrawing its financial support from UNRRA at the organization's fifth council meeting in Geneva,[52] Ludwik Rajchman, the former director of the LNHO and the Polish delegate to UNRRA, proposed that the remaining funds from UNRRA be used to help children, who were the primary victims of World War II. The proposal was supported by the British and won approval from other UNRRA members. Most of the participants at the meeting expected that the new organization would be a temporary measure. But Rajchman had other plans. He viewed UNICEF as a means for creating international cooperation for a better world. Supporting children was a cause that everyone could agree on. While UNICEF was initially established to assist children in Europe and China, Rajchman hoped to use it to meet the needs of children around the world. Rajchman also viewed UNICEF's mission as more than providing for the immediate needs of children. Its function was to strengthen permanent child-health and -welfare programs in the nations receiving assistance from UNICEF, so they could improve the situation of their countries' children on a long-term basis. He wanted the organization to support self-help and capacity building. In this sense, UNICEF built on the early vision of UNRRA.[53]

The new organization was officially established as a part of the United Nations in December 1946. Its relationship to the parent organization, however, was different from that of other UN agencies in that its funding was dependent on voluntary contributions from member nations, rather than on fixed contributions. This made its funding base less stable, but also potentially much greater than that of other organizations. The arrangement quickly worked to UNICEF's advantage, as the United States pledged US$40 million to get the organization up and running in 1947.[54] The initial WHO budget was only US$5 million. America's contribution reflected the power that children had for attracting financial support from the US Congress. UNICEF also benefited from the creation of a UN Appeal for Children, aimed at raising voluntary contributions from private citizens around the world. The initial appeal raised less than was hoped, especially in the United States, where the appeal was seen as threatening the ability of other private humanitarian groups to solicit voluntary contributions.[55] But over time, UNICEF—using celebrities like Danny Kaye and Peter Ustinov to represent its cause, and tactics such as having children collect pennies for UNICEF as part of their Halloween trick-or-treating—would thrive, maximizing its ability to raise funds both from governments and private donations.

Rajchman chose Maurice Pate as UNICEF'S first executive director. Pate was a US businessman whom Rajchman knew from his work with the American Red Cross, providing postwar relief for Poland. Pate had also served with Herbert Hoover, whom President Truman had chosen to assess the social and economic needs of Europe immediately after World War II. Rajchman chose Pate not only because of his organizational skills and commitment to helping children, but also because he was a Republican, and Rajchman viewed this as politically important within the changing environment of US politics. Rajchman also tapped Martha Eliot, who had headed the US Children's Bureau before the war and served under Herbert Lehman in UNRRA, to develop plans for meeting the relief needs of European countries.

It did not take long for Rajchman, the physician, to focus UNICEF on issues related to improving the *health* of children. He appointed a medical advisory board, made up of physicians, to help direct the organization's health interventions. Eliot, meanwhile, drew up plans for addressing the epidemic of rickets that affected an estimated one-third of Europe's children. Rickets was a debilitating nutritional-deficiency disease, caused by a lack of vitamin D and calcium. To relieve the problem, UNICEF mounted a campaign to provide powdered milk to at-risk children as an emergency measure. In the longer term, the campaign also supported the local production of dairy products in affected countries, an example of Rajchman's desire to use UNICEF to promote longer-term solutions to children's needs.[56]

Rajchman moved UNICEF further into the health field in 1948, when he responded favorably to a request from the Danish Red Cross to join it in a campaign to combat tuberculosis in Europe using BCG vaccinations. The Red Cross had begun vaccination programs in Denmark and neighboring regions of Germany in 1947. It quickly received requests from Poland and Yugoslavia—and later Germany, Czechoslovakia, and Hungary—for assistance in fighting TB. In March 1948, UNICEF joined with the Danish Red Cross and other Scandinavian relief organizations in creating the International Tuberculosis Campaign, or "Joint Enterprise." In doing so, Rajchman argued that the campaign should not be limited to Europe, but instead should be worldwide. While uncomfortable with expanding so far abroad, the Scandinavian agencies agreed to the proposal in order to get UNICEF on board. By 1951, the campaign had established vaccination programs in Europe, North Africa, the Middle East, South Asia, and Latin America. It had tested 30 million people and vaccinated 14 million of them.[57] The

International Tuberculosis Campaign (ITC) was the first coordinated global effort to control a disease that affected large portions of humanity. While the campaigns against hookworm and yellow fever in the 1920s covered wide areas of the globe, they were not part of a single, coordinated campaign. In 1948, the *New York Times* called the ITC "the greatest medical crusade in history."[58] It was not insignificant that UNICEF, not the World Health Organization, launched this first campaign.

UNICEF's incursion into disease control did not go unnoticed by WHO. WHO's Interim Commission openly opposed UNICEF's infringement on WHO's territory, but the first World Health Assembly, in 1948, took a somewhat more muted position, recommending that a Joint WHO/UNICEF Committee be formed to coordinate the health activities of the two organizations. Chisholm and others recognized that TB was a public-health emergency. They could not very easily object to the "Joint Enterprise." Nor was WHO in a position to replace UNICEF in running the campaign. WHO did not have the funds to support such a large endeavor. UNICEF did.

Still, Chisholm was hesitant to become involved in the TB campaign. Members of WHO's Expert Committee on Tuberculosis had expressed doubts about the efficacy of BCG. The vaccine had been deployed across the globe, but there was little solid research indicating that it was effective in preventing infection. Establishing case-control studies in the middle of an epidemic did not appear to be ethically acceptable. So WHO signed on to help run a campaign while having only limited knowledge of whether the intervention being employed would work. This was not the last time, however, that WHO would venture into a global disease campaign with doubts about the efficacy of the proposed intervention (chapter 7). That WHO ignored these reservations in order to become engaged in the TB campaign speaks to both the pressure it was under to reaffirm its leadership role in world health in the face of UNICEF's challenge, and to the extent to which the postwar culture of crisis drove health and development organizations to adopt quick-fix solutions to international-health problems.[59]

In the end, WHO was forced to accept UNICEF as a partner in health matters, agreeing to provide technical advice and leadership in the development of programs, while UNICEF provided material support. Accordingly, the Joint WHO/UNICEF Committee took over control of the ITC in 1951. Rajchman was not totally happy relinquishing leadership in the campaign to WHO, but he recognized the need to reach an accommodation with the new health organization, which he supported. The terms of the partnership would be repeated in future campaigns and would serve both organizations well. Joint campaigns would be organized against yaws, leprosy, malaria, and trachoma during the early 1950s. But the partnership would continue to be fraught.[60]

TB CONTROL IN SOUTH INDIA: A MISSED LESSON

Before concluding this chapter, I want to look closer at the history of the joint WHO/UNICEF BCG campaign in south India and the lessons it provided (or, perhaps more accurately, should have provided) for future disease campaigns. India was the first developing nation to agree to be included in the campaign. The country was viewed by the ITC's leadership as a critical test case for determining whether BCG could be an effective instrument for preventing TB around the world.

The campaign did not get off to a good start. Indian leaders on the ground in Madras, where the ITC began its campaign in early 1949, were quick to disparage its use of BCG. A. V. Raman, a former sanitary engineer, published a series of editorials criticizing the use of BCG in India in his monthly publication, *Peoples' Health*. Raman had earlier taken the Indian government to task for favoring technical solutions over sanitary reforms, clean drinking water, proper housing, and sewage. In doing so, he echoed the conclusions of the Bhore Committee's report (chapter 5). For Raman, BCG was an example of a quick-fix technical approach. This argument would become a recurrent element in Indian critiques of disease-eradication campaigns, as well as other biotechnical solutions to complex health problems, including more-recent attempts to distribute Vitamin A capsules to prevent child mortality (chapter 16).

Raman, however, was equally concerned about the uncertainty surrounding the efficacy and safety of BCG, and he accused the campaign and the Indian government of using Indian children as guinea pigs in the ITC's grand experiment. On February 4, 1949, Raman addressed a scientific society in Madras, where he restated his opposition to the use of an unproven vaccine, declaring, "I strongly protest in the name of India . . . against our boys and girls being made a sort of cannon fodder and treated like guinea pigs for the sake of experimentation."[61] Raman was not opposed to Western medicine. He was in favor of the use of antibiotics to treat TB. He also drew on evidence published in Western scientific journals to support his argument that there was uncertainty about the effectiveness and safety of BCG. From the beginning, these questionable aspects regarding the use of BCG had been a central weakness of the campaign, and Raman exploited them to the fullest.

Raman's attack gained momentum, and local health authorities and politicians began withdrawing their support for the project. In June 1949, the provincial government in Madras informed the Union Government that, considering the significance of the opposition to the BCG campaign, it preferred to train only one team, instead of the three that were originally planned. When the negative arguments died out, more teams could be trained. In autumn 1949, the Madras provincial government opted out of the Union Government's plans to intensify BCG vaccinations.

Opposition in Madras ebbed in early 1951, in part because Raman had to close down his journal for financial reasons. In addition, the ITC had produced its own media attack, supporting the campaign, and intensified its educational activities, aimed at winning local approval for the use of BCG. Until then, the campaign had paid little attention to the need to build popular support. It had assumed, as earlier campaigns had, that the value of the intervention would be self-evident. As opposition waned, there was growing optimism among the ITC leadership, including Halfdan Mahler, WHO's senior representative in India and a future director-general of WHO, about the campaign's ultimate success.

Yet resistance reemerged four years later. In August 1955, the *Indian Express* reported that, according to the Madras minister of health, the BCG teams were creating "mass hysteria" wherever they went. Opposition radiated out from Madras to other provinces. In many parts of the country, tuberculin testing plummeted from highs of above 90 percent to, in some cases, zero. Opposition was reported in Uttar Pradesh; Andhra Pradesh; across the Bay of Bengal in Rangoon, Burma; and to the north in Punjab. At the center of the second round of protests was Chakravarti Rajagopalachari, the first Indian governor-general, former chief minister of the Madras government, and a man whom Mahatma Gandhi called the "keeper of my conscience." His protest was part of his wider critique of the modernist pretensions of India's government under Prime Minister Nehru. Rajaji, as he was known, was a friend of Raman and reiterated many of Raman's arguments about the dangers and inappropriateness of BCG. As historian Christian McMillan noted, "Rajaji and his supporters opposed what they perceived to be a scheme hatched by foreign experts and an increasingly powerful strong central government bent on imposing its will on Indian bodies."[62]

Rajaji's opposition movement, like that of Raman, eventually receded, and by October 1955, UNICEF reported that active opposition was fast dying down. In the end, resistance to BCG in India was largely a Madras story, despite Rajaji's wider influence. Nonetheless, it was important. It provided lessons about the risks involved in launching disease-elimination campaigns without first perfecting the technology needed to eliminate the targeted disease and without giving adequate attention to communicating and building cooperation with local populations. It also demonstrated the potential for people on the ground to mobilize resistance, bringing campaigns to a halt, if only temporarily, when these conditions were not met.

By the mid-1950s, all of the international organizations involved in one way or another in efforts to improve the health and well-being of the world's populations had turned away from integrated approaches to health and development, concentrating instead on more-focused strategies as part of a growing world of technical assistance. For WHO and UNICEF, the next 25 years would be dedicated to eradicating malaria and smallpox. But, as we will see in the next two chapters, it is clear that the lessons of the Madras BCG campaign were only partially learned by those who mounted efforts to eradicate these two diseases.

PART IV

The Era of Eradication

In spring 1955, the eighth World Health Assembly convened at the Palais des Arts in Mexico City. The assembly ratified several proposals prepared by WHO's Executive Board. It acknowledged the importance of health issues related to the uses of atomic energy and created an Expert Committee on Atomic Energy in Relation to Public Health. It also made a special appropriation for expanded programs in polio control and tightened up international sanitary regulations related to quarantine procedures. Yet the centerpiece of the assembly's proceedings was the decision to increase funding for malaria-control efforts, with the goal of eradicating the disease from the face of the earth. This decision led to the launching of the Malaria Eradication Programme. Four years later, in 1959, the World Health Assembly passed a similar proposal to eradicate a second disease, smallpox, though that campaign would not get off the ground until 1966, when the assembly agreed to intensify the smallpox-eradication efforts.

WHO engaged in a range of health activities during the 1950s and 1960s, including programs related to environmental health, the training of health professionals, the collection and publication of health statistics, maternal and child health, and health-services development.[1] WHO also launched campaigns against yaws, venereal disease, leprosy, trachoma, and tuberculosis. Yet the campaigns against malaria and smallpox dominated international-health activities, drawing substantial support from a range of international and bilateral aid organizations for nearly a quarter of a century and eclipsing all other health programs in terms of commitments of funding and personnel.

In the end, the two campaigns had very different outcomes. Malaria eradication failed to achieve its stated goals, though it was successful in eliminating malaria from 26 countries. In 1969, WHO acknowledged this failure—which dealt a major blow to WHO's reputation and authority—and terminated the campaign. The fallout haunted WHO for years afterward, throwing into question the value of vertically organized disease campaigns and seriously undermining all efforts to control malaria.

Smallpox eradication, on the other hand, had a very different conclusion and legacy. Intensified vaccination programs quickly reduced the number of endemic countries, allowing program resources to be concentrated. The last case of smallpox was identified in 1977; three years later, WHO declared that the disease had been eradicated. The smallpox campaign was not only a dramatic success, but arguably the high point of WHO's efforts to improve the health of the world's populations. The aura of this success played a central role in resuscitating disease-elimination programs and in shaping international-health strategies for the next three decades, as a cadre of smallpox-eradication veterans took up leadership roles in existing and emerging international-health organizations.

Despite these different outcomes—and the common view that malaria was the "bad" campaign, while smallpox was the "good" campaign—the two eradication efforts had a great deal in common in terms of how they came about, the ways in which they were organized, and how they were conducted. They were both products of the culture of development and technical assistance that dominated international-health efforts in the 1950s, 1960s, and 1970s. Most importantly, in terms of the longer history of global health, they represented a return to the strategy of disease elimination, based on large-scale campaigns that had dominated international health at the beginning of the twentieth century. Like those earlier initiatives, the attacks against malaria and smallpox were based on the application of public-health technologies through campaigns that were initially conceived and planned outside the countries most affected by the two diseases. While both campaigns were forced to adapt to local circumstances in varying degrees and depended on the cooperation of local health authorities and populations, they strove to establish uniform protocols that could be applied universally. Finally, both campaigns were based on the view that well-administered control programs, using powerful medical technologies, could eliminate diseases without transforming the social and economic conditions in which these diseases occurred.

It is important to look carefully at the histories of these two eradication campaigns, not only because they dominated the international-health landscape for such an extended period of time, but also because, like the joint WHO/UNICEF disease-control programs that preceded them, they involved a partnership of countries from across the globe, unified in an effort to advance world health, only on a much larger scale. Tens of thousands of health workers from multiple nations had a role in these global efforts. Their participation expanded the construction of networks of cooperation that further instantiated the concept of global health.

Building these networks did not come easily. Disagreements between WHO and national governments, as well as between WHO headquarters and the various regional organizations, were frequent, and every move seemed to have to be carefully negotiated. Nonetheless, the two campaigns served to substantially advance the concept of world, or global, health. Finally, the two campaigns were also important because they cast a long shadow on the future of international-health activities.

Both WHO campaigns have been the subject of extensive retrospective analyses by program participants, other public-health experts, and, more recently, historians and political scientists. These analyses have tended to vilify malaria eradication and glorify the eradication of smallpox (though the most-recent historical scholarship on the smallpox campaign has taken a more critical stance).[2] In doing so, these reviews posited lessons about what worked and what did not in organizing health campaigns. These lessons shaped subsequent health strategies and continue to be used to advance particular approaches to global health.

This section compares the histories of these two campaigns. In chapter 7, I examine the conditions that led to the decisions to eradicate malaria and smallpox. How did the worldwide elimination of two major diseases come to be viewed as possible? I argue that politics, as much as scientific knowledge, shaped these decisions. Chapter 8 compares how the two campaigns were organized and implemented. In examining the histories of these two campaigns, I want to revisit past characterizations of them as having "failed" or "succeeded," and delve into the basis on which these characterizations were made. I argue that any comparison between the two campaigns needs to take account of fundamental differences in the biological characteristics of each pathogen and the instruments available to control and eliminate them, as well as differences in how the campaigns were implemented. This requires a biosocial analysis.

CHAPTER SEVEN

Uncertain Beginnings

PERHAPS THE MOST remarkable aspect of the malaria- and smallpox-eradication campaigns was that they happened at all. The decisions to launch the campaigns occurred despite uncertainty about the effectiveness of the strategies proposed for eliminating these two diseases; disagreements about the feasibility of the campaign goals; lack of financial resources to begin, let alone complete, the campaigns; and differences over who should lead the campaigns. In many ways, the decision to launch each campaign represented a leap of faith, stimulated less by a reasoned assessment of the chances of success than by the same factors that had led WHO and other postwar organizations to move away from broad-based approaches to health and development in the early 1950s. The discovery of new medical technologies; a growing confidence in the ability of health campaigns to eliminate major health threats in the absence of investments in broader development; a sense that the world faced a public-health crisis; the belief that rapid improvements in health could jump-start economic development; postwar political tensions; and competition for leadership in international health all contributed to the decisions that led to efforts to rid the world of malaria and smallpox. Tracing the history of these decisions reveals a great deal about the workings of WHO and the role of politics in shaping international-health agendas in the 1950s and 1960s.

Malaria Eradication: "No Other Logical Choice"

Marcolino Candau, who replaced Brock Chisholm as director-general of WHO in 1953, presented the case for global malaria eradication at the meeting of the World Health Assembly (WHA) in Mexico City in May 1955. In making the case for eradication, he noted the following facts. First, health workers operating in various parts of the globe had had tremendous success in employing residual pesticides, particularly DDT, to control malaria. Second, evidence from several countries, including Italy, Greece, Venezuela, Crete, Ceylon, and Mauritius, had demonstrated that malaria eradication by residual spraying was both economically and technically feasible. Third, there was growing evidence that vector resistance to DDT and other pesticides was increasing, threatening the ability of health authorities to keep malaria under control. Candau concluded that given these facts, "there is . . . no other logical choice: malaria eradication is clearly indicated, presents a unique opportunity, and should be implemented as rapidly as possible. Time is of the essence."[1]

Candau's argument for launching a global program for malaria eradication was based on faith in the power of pesticides to eradicate malaria and fear that vector resistance might soon eliminate the opportunity to do so. Other supporters of eradication added that although the cost of eradication would be high, it would be a one-time capital expenditure and, in the end, save millions of dollars in recurrent control costs. Eradication advocates further asserted that malaria eradication would remove one of the major blocks to economic development in tropical regions of the globe. Finally, supporters of the resolution argued that eradication efforts would contribute to the development of health services in countries where malaria existed.[2]

Yet to many public-health authorities at the time, the decision to make malaria eradication the centerpiece of WHO's efforts to promote world health was surprising. Skeptics pointed to the tremendous financial and logistical obstacles that lay in the path of eradication. They also warned that the health of millions of people living in malaria-endemic countries could be threatened if eradication efforts compromised their acquired immunity to malaria and then failed to achieve complete eradication, permitting malaria to return.

The WHA's passage of the resolution becomes even more perplexing in light of the following facts. The effectiveness of DDT for controlling, let alone eradicating malaria in many areas of the developing world was unknown. As late as February 1955, WHO's Executive Board, reviewing data on this topic, observed that it was not known whether the absorption of DDT by mud surfaces would compromise its residual effects. Early successes with DDT had, for the most part, occurred in locations where houses were not constructed with mud walls. One member of the board pointed out that this was a serious concern, since in most malaria-ridden areas, mud houses were the commonest form of dwelling. Advocates of malaria eradication recognized that in many parts of the developing world, malaria-control programs were poorly staffed and disorganized, leading to the misapplication of DDT. One of the most vocal and respected supporters of eradication, Paul Russell, had first-hand knowledge of the disarray that existed in many programs and the structural difficulties that stood in the way of eradication efforts.

At the time that the eradication resolution was passed, few of those who supported eradication believed that this could be achieved in sub-Saharan Africa in the foreseeable future. Logistical problems, limited public-health capacity, and the efficiency of *Anopheles gambiae* mosquitoes in transmitting malaria led Paul Russell and epidemiologist George MacDonald, who developed the mathematical models on which the eradication strategy was based, to conclude that malaria eradication in Africa would not take place for years. Several delegates to the WHA raised this issue. If Africa was not ready for eradication, how could the disease be eliminated from all parts of the world?

No one knew just how much eradication would cost, or how the funds would be raised, or how long it would take. Several delegates stated that the amount of funding proposed by the Executive Committee was inadequate, even though this was a preliminary amount and it was expected that additional funding would become available once eradication was under way.[3] The proposal called for the creation of a special malaria account into which countries could make "voluntary" contributions. But no one knew how much each nation would be willing to contribute. Moreover, the United States, Britain, and several other countries were opposed to the creation of such an account.

There was also disagreement about how malaria eradication should proceed. Frederick Soper, who was director of the Pan American Sanitary Bureau and a supporter of eradication strategies, believed that eradication required the application of pesticides until transmission not only had ceased, but until the last parasite had disappeared from the local population. Others, including Paul Russell and Emilio J. Pampana, chief of the Malaria Section of WHO, argued that spraying should be withdrawn once transmission had stopped, allowing the medical services to identify and treat the last cases of malaria. This approach would reduce program costs, as well as the risk that pesticide resistance would develop. In the end, the strategy supported by Russell and Pampana was chosen.

Finally, and most puzzling, the "Resolution for a Global Program for the Eradication of Malaria" was never reviewed by the Expert Committee on Malaria before it was debated and voted on by the WHA in May 1955.[4]

The Expert Committee on Malaria had met regularly since 1947 to discuss progress and prospects for malaria control and develop recommendations for improving control measures. At its last meeting in 1953, prior to the eighth World Health Assembly, the committee had expressed support for the goal of malaria eradication. It raised numerous concerns, however, about reliance on DDT and other insecticides to control malaria. It warned of a risk that "preoccupation with the power of residual insecticides could result in the derogation of other methods which have considerable utility." Instead, it advocated the use of a broad range of approaches, including "changes in agriculture and social conditions." The committee stressed that much was unknown about the application and effectiveness of pesticides under different epidemiological conditions and called for more research. It also noted that while anopheline resistance to pesticides had developed in a few places, it had not yet undermined control activities. Again, it recommended that more research be conducted on this problem. The delegates who voted to support malaria eradication largely ignored these issues.[5]

All of these concerns raise the question of why WHO would initiate such a program, given all that was unknown or uncertain about malaria eradication and the high costs of failure. There appear to have been several factors in play. To begin with, malaria control had been a central focus of WHO activities from the beginning. The US delegation to the fifth session of the IC in January 1948 presented a proposal urging WHO to direct "*a major share* of its energy and resources" [emphasis added] to applying new discoveries in malaria control, particularly to major food-producing areas.[6] The United States had been heavily invested in funding international antimalaria activities as part of its technical-assistance programs since the days of the Institute of Inter-American Affairs in Latin America. The proposal was accepted by the IC and adopted by the second session of the Expert Committee on Malaria in 1948.[7] The committee called for the creation of an "international programme of malaria control on a world-wide basis." When the full WHA met for the first time in May 1948, the delegates voted to make malaria control one of its top priorities. Funding for malaria-control projects represented roughly 25 percent of all WHO funding between 1948 and 1955.

US support for malaria control was an important factor in its adoption as an early focus of WHO activities. It is crucial to note, however, that successes achieved in controlling malaria with DDT had convinced health authorities across the globe that malaria should be a primary target of their national health programs. By the time that eradication was proposed, malaria-control programs using pesticides were under way in nearly every country in which the disease was a significant public-health problem.

Early successes, national acceptance and participation, and US support help explain why malaria came to dominate WHO's health activities during the 1940s and 1950s. But it does not explain the decision to move from control to eradication. WHO and UNICEF were involved in a number of disease control campaigns in the early 1950s (chapter 6), yet none of these diseases had been targeted for eradication. Reading the debate on eradication at the eighth WHA provides insights into the factors that shaped the decision to shift to eradication. Many delegates echoed the director-general's assertions regarding the potential of pesticides to rid the world of malaria and the strong concern that pesticide resistance would undermine this possibility if health authorities did not move quickly to eradicate the disease. While there was limited evidence of resistance to date, the fear that resistance would grow resonated with the delegates because, as Paul Russell noted at the 1955 Mexico City meeting, "there was not at present any satisfactory substitute method of attacking malaria." Despite the warnings of the fifth Expert Committee on Malaria about an overreliance on the use of pesticides to eliminate malaria, by 1955, the world of malaria control had become hooked on spraying.

Reliance on DDT and other pesticides to control malaria had grown exponentially following World War II, all but eliminating earlier methods for controlling the disease. The trend was reinforced by the flow of research funds from chemical corporations into universities to support research on new insecticide compounds after the war. The US Department of Agriculture estimated that 25 new pesticides were introduced between 1945 and 1953.[8] According to one observer:

> The resultant expansion of the pesticide industry was so rapid that it simply steam-rollered [sic] pest-control technology. Entomologists and other pest-control specialists were sucked into the vortex and for a couple of decades became so engrossed in developing, producing, and assessing new pesticides that they forgot that pest control is essentially an ecological matter. Thus, virtually an entire generation of researchers and teachers came to equate pest management with chemical control.[9]

Having become dependent on the use of pesticides, the realization that *Anopheles* species were developing resistance to pesticides was a major incentive to act fast to eliminate malaria before vector resistance became widespread, even if resistance had not yet seriously undermined control activities.

The growing reliance on the use of pesticides had another impact on the decision to move toward eradication. The vast majority of pesticide production after World War II occurred in the United States. Some of the biggest chemical companies in the nation were involved in the production and overseas sales of pesticides, including Montrose, Monsanto, Hercules, and DuPont. The winding down of the war, together with the dismantling of UNRRA, which had funded numerous disease campaigns using DDT, led to a huge drop in pesticide demand. As a result, production and prices had declined. Pesticide manufacturers lobbied public-health officials, including Paul Russell, to scale up malaria-control efforts. Manufacturers of spraying equipment, such as Hudson Manufacturers, and pharmaceutical companies, such as Merck (which manufactured antimalarial drugs, including chloroquine), also had much to gain from an escalation of the war on malaria.[10]

The delegates raised two more issues that played important roles in the WHA's final decision. First, other international organizations had already adopted an eradication strategy. The Pan American Sanitary Bureau and participants in the Malaria Conference for the Western Pacific and South-East Asia Regions (Second Asian Malaria Conference) had voted to shift from control to eradication in 1954. In addition, a National Malaria Conference, held in Pakistan at the end of January 1955, agreed that "the object of the Malaria Control Programme should be the eradication of malaria from Pakistan in five years."[11] These decisions highlight the extent to which WHO's decisions were influenced by national and regional interests and policies. Finally, in April 1955, UNICEF signed on to support the conversion of malaria-control programs to malaria-eradication programs.

In effect, by the time the WHA met in Mexico in May 1955, the eradication train was rolling down the tracks, having already been adopted by national programs, regional organizations, and, perhaps most importantly, by UNICEF, which was both a source of support for eradication and an emerging rival for international-health leadership (chapter 6). The delegates in Mexico had two choices: they could jump on the eradication train or watch it leave the station without them. To choose the latter would have been a serious blow to WHO's leadership role. In this respect, the most important intervention in the debate may have been Paul Russell's observation that, "the Organization already had a reputation as a leader in antimalaria campaigns and it would be tragic if it lost that leadership by failing to expand its facilities in the very modest way proposed by the Director-General."[12]

Second, while delegates were concerned about emerging pesticide resistance, they were equally worried about another kind of resistance, from political leaders who managed the flow of funds for malaria research and control. There was evidence that these leaders were losing interest in malaria and that complacency, driven by early successes, was setting in. As early as 1948, Paul Russell noted in his diaries that within the US National Institutes of Health (NIH), there was a move to wind down malaria research. At a meeting in January of that year, discussions of malaria research centered on the question, "what justification is there for continued research in malaria and what facts could be presented in support of such justification?"[13] In September, participants at a Malaria Section Meeting of NIH discussed the director's proposal to merge six sections into three and join malaria with tropical diseases. Russell noted, "The tendency is to consider malaria a finished story, but actually now is the time to give this disease a very great emphasis."[14]

Soper also recognized the threat that bureaucratic complacency represented for the war against malaria in the Western hemisphere. In his statement advocating a coordinated malaria-eradication program for the Americas, he noted: "The other resistance mentioned previously—that of those administering the funds—has already begun to appear in several countries in which malaria has lost the power to attract the attention it previously evoked when it produced such disastrous effects. Thus, it is also of the greatest importance to eradicate all sources of infection before this type of resistance has become deeply rooted."[15] Professor M. J. Ferreira from Brazil likewise referred to the problem of bureaucratic resistance at the 1955 WHA meeting in Mexico: "In certain countries, the malaria-control campaigns had been extremely successful and malaria had been reduced to a very low level. Governments were apt to be satisfied with that situation, and they might reduce their efforts if they failed to realize the danger of vectors developing resistance to insecticides and the urgent need for complete eradication of the disease."[16]

For Soper, Russell, and other leaders in malaria control, malaria eradication represented a means of reenergizing jaded bureaucrats and health officials with the promise that a one-time, time-limited investment in malaria control would result in a breakthrough for the health and development of their fellow citizens. These men feared that without the goal of eradication, support and resources for malaria control would dry up, and the gains that had already been achieved would be lost. As risky as eradication might be—given all that was known about existing problems of malaria control, and all that was unknown about eradication—the alternative of continuing on the same path of malaria control appeared untenable. Whether or not they believed that global eradication was possible, they believed that adopting eradication as a goal would reenergize malaria-control efforts. This same logic would play an important role in the renewed drive for malaria eradication in the twenty-first century.

Finally, one cannot overlook the role that Candau, as director-general, played in pushing the eradication agenda. Candau had a great deal of influence in shaping the agenda of the World Health Assembly meetings, and without his active support, it is unlikely that the eradication proposal would have passed. Candau's lukewarm embrace of the goal of smallpox eradication contributed to the campaign languishing for several years without substantial financial support. So why did Candau support malaria eradication? One answer can be found in his earlier career. Candau was born in Brazil and trained in public health and malariology at the Johns Hopkins School of Hygiene and Public Health, where a generation of international-health leaders learned how to improve health by eliminating diseases one at a time. From Johns Hopkins, Candau returned to Brazil and took up a number of positions in that country's health services. Perhaps most importantly, he served for several years under Frederick Soper in Brazil's malaria-control program. Through these activities, Candau gained an appreciation of the benefits of disease-control programs.

In 1950, Candau joined the staff of the World Health Organization in Geneva as director of the Division of Organization of Health Services. Within a year, he was appointed assistant director-general in charge of Advisory Services. In 1952, he moved to Washington, DC, rejoining Soper as assistant director of the Pan American Sanitary Bureau. When the post of director-general became vacant at WHO, Soper pushed Candau's candidacy.[17] Given his early work in malaria control and his long relationship with Soper, it is not surprising that Candau was an early supporter of eradication, and that he introduced the eradication proposal to the eighth WHA.

Thus, despite uncertainties about the feasibility of eradication, the availability of financial resources, and the benefits of eradication versus control, the WHA passed a resolution to move from malaria control to malaria eradication. The decision reflected a combination of hopes and fears regarding the power of pesticides, the threat of pesticide resistance, the future of malaria control, and the role of the World Health Organization in leading international-health efforts across the globe.

Launching Smallpox Eradication

The birth of the Smallpox Eradication Programme (SEP), beginning in 1958, also occurred in an environment filled with uncertainties. Perhaps even more than the adoption of malaria eradication, the decision to eradicate smallpox represented a leap of faith. At WHO, Director-General Chisholm had originally introduced a proposal for smallpox eradication in 1953. He argued that the program would demonstrate the importance that WHO had for every member state. Chisholm may have also heard Soper's footsteps. Soper, encouraged by the development of a freeze-dried vaccine, had urged ministers of health in the Americas to adopt a region-wide strategy to eradicate smallpox in 1950. Chisholm's subsequent proposal was rejected, as few believed that WHO should take on such a large project at that time.

The Soviet Union introduced a new proposal for the global eradication of smallpox at the WHA meeting in Minnesota in 1958. The Soviets had withdrawn from WHO in 1949 (chapter 6). Their proposal for smallpox eradication marked their return to the fold and their desire to take an active role in shaping WHO's agenda. The Soviets had successfully eliminated smallpox in the 1930s from Russia's Central Asian Republics, a region with little administrative infrastructure, using only a low-quality, unstable vaccine. They were now trying to prevent the reintroduction of the disease from neighboring Asian countries. Their proposal argued that the development of heat-resistant, freeze-dried vaccines made it possible to eradicate smallpox everywhere, eliminating the need for countries that had eliminated the disease to continue vaccinating their own populations and protecting their borders.[18] The Soviet representative also asserted that a coordinated global campaign would cost less than the indefinite continuation of national vaccination campaigns. The WHA was responsive to the proposal and called on the director-general to prepare a report on the administrative, technical, and financial implications of the proposal for the next WHA meeting. Candau did so, and his report and the proposal for a smallpox-eradication campaign were adopted in 1959.

Candau, however, was doubtful that smallpox eradication could be achieved. The Soviet proposal called for the mass vaccination of the world's entire population. How was this possible, when a number of countries were still not members of WHO, and large portions of the world's population lived in remote, difficult-to-access places? Moreover, vaccination in 1959 was a technologically challenging procedure, requiring the careful training of vaccinators to apply sufficient vaccine intradermally, with repeated needle pricks or with the painful rotary lancet, without causing bleeding. The quality of the vaccines varied greatly, so the chances of vaccination failure were significant. To make matters worse, UNICEF informed WHO that for financial reasons, it would be unable to offer the kind of material support it was providing for the Malaria Eradication Programme (MEP). Finally, both Candau and the US delegation were committed to malaria eradication, which was now consuming 33 percent of WHO's budget, and were opposed to allocating additional funds for a smallpox campaign. For these reasons, Candau resisted making smallpox eradication a priority.

The project was only allotted an annual budget of US$100,000, preventing any serious efforts to implement it. The program languished for the next six years. The disbursement for smallpox eradication amounted to only 0.6 percent of the total expenditure of funds placed at the disposal of WHO between 1960 and 1966.[19] Yet the landscape of both smallpox control and international health was changing. Advances in vaccine technology—in the form of a jet injector developed by the US Army and tested successfully in Jamaica, Brazil, Peru, and Tonga—allowed for the rapid vaccination of large populations with nearly 100 percent effectiveness. In addition, the voting membership of WHO was changing. In 1955, WHO had only 51 members, primarily representing nations in Europe and the Americas. European colonial authorities represented most of Africa and Asia, with the exception of China and Japan. As countries within these regions gained their independence, membership in WHO increased dramatically. By 1965, 121 countries were represented in the WHA. Many of the new members of the WHA viewed smallpox as a serious problem and called for increased WHO support for elimination efforts.

In May 1965, the WHA expressed dissatisfaction with the slow pace of the smallpox campaign, recommitted WHO to the goal of eradication, and called on the director-general "to seek anew the necessary financial and other resources required to achieve world-wide smallpox eradication."[20] The following year, Candau added US$2.4 million to WHO's general budget to support smallpox-eradication efforts. This budget was presented to the WHA, along with a detailed report on the state of smallpox and a plan for its eradication. The plan had been drawn up by Donald Henderson, from the US Communicable Diseases Center, which later changed its name to the Centers for Disease Control (CDC). Henderson would eventually lead the SEP. In presenting the proposed budget to the WHA, Candau endorsed eradication, but he warned that this goal should not be sought at the expense of the development of health services. Tellingly, he had not expressed a similar concern about health services when malaria eradication was being debated in 1955.

Debate over the proposed budget lasted three days. Most of the wealthier countries, which provided the bulk of WHO's funding, balked at the increased budget, which required them to make substantially larger contributions to the organization. The United States, along with several other nations, supported the eradication proposal but proposed an alternative budget that cut the proposed increase, including the smallpox budget, in half. Other alternative budgets were also tendered. A block of developing countries from Africa, Asia, and Latin America, however, supported Candau's budget and the smallpox proposal. In the end, the budget and Candau's proposal passed by just two votes over the two-thirds majority needed, the closest budget vote in the history of WHO. Lack of support from major funders nonetheless raised questions about the future of the eradication program. So, too, did Candau's lack of enthusiasm. Shortly after the vote was taken, Candau met privately with regional directors, expressed his doubts about the feasibility of eradication, and cautioned them about "imposing on countries a special programme like that for malaria eradication."[21]

More remarkably, the debate over smallpox eradication in 1966 focused on finances, not on the feasibility of the proposed program. The closest delegates came to questioning the feasibility of the campaign was to suggest that it might be wiser for the campaign to begin slowly (which would require a smaller commitment of funds), in order to test practices and avoid the problems that plagued the malaria-eradication campaign. Yet there was much that was not known about smallpox eradication, and what was known should have been unsettling to the WHA delegates. To begin with, there were many within WHO who remained opposed to the program. At the top of the list was Candau himself, who doubted that eradication could be achieved and worried that the declining fortunes of the MEP was already embarrassing the organization. A setback with smallpox might completely undermine WHO's credibility. WHO officials in the all-important regional offices, where much of the practical work of eradication had to be done, also remained skeptical of the program. The director of the Regional Office for South-East Asia (SEARO) wrote: "In our view, on account of the organizational and administrative weakness of health services and serious socioeconomic as well as financial difficulties, smallpox eradication is not likely to be achieved in the countries of this Region in the near future."[22]

Other regional directors saw smallpox eradication as either a lower priority or a distraction from the central goal of developing the capacity to deliver basic health services. Responding to these concerns, and perhaps attempting to deliberately undermine the eradication campaign, two of Candau's top deputies composed a memo to the regional offices two months after the vote, noting that "the establishment of permanent basic health services should be given the highest priority since it is a prerequisite for the success of the smallpox eradication programme in any area." They instructed the WHO staff to "be prepared to consider providing from the smallpox eradication programme resources, such assistance as may be required for developing and strengthening basic health services in the areas where the campaign is launched."[23] In essence, headquarters was giving regional offices the authority to divert funds from the campaign to support basic health services.

There were other reasons to be concerned. First, the appropriation of a special budget of US$2.4 million to conduct the campaign was thoroughly unrealistic. The director-general's report on smallpox eradication estimated the cost of the campaign to be US$180 million over ten years. But the budgeted US$2.4 million per year would hardly cover the annual cost of the vaccine, let alone the costs of transport, vaccination, and surveillance. In 1967, it was estimated that 250 million doses per year would be required for 2 or 3 years. Even at the estimated price of US$0.10 per dose, the cost of vaccinations alone would be US$25 million. To be successful, the campaign would need voluntary contributions by multiple countries, as well as bilateral aid. In the end, vaccine donations from the Soviet Union and the United States and reductions in the unit cost of the vaccine greatly lowered these costs, but these contributions and reductions had not yet occurred when the eradication proposal was voted on.

Second, the scope of the smallpox problem was unknown. WHO's Expert Committee on Smallpox, in its first report in 1964, noted that the reporting of smallpox morbidity and mortality was highly unreliable. The data indicated little more than the general trend of incidence in different regions. Donald Henderson observed that while 131,418 cases were reported worldwide to WHO in 1967, there may have been that many cases in northern Nigeria alone. He suggested that a more accurate figure might be between 10 and 15 million cases. But no one knew for sure.

Third, it was assumed that eradication would require the mass vaccination of populations across the globe. While progress had been made in many countries, a great deal more needed to be done, particularly in places where communication and transportation were poor, populations were mobile, and health-care services inadequate or nonexistent. Even if the needed vaccines were available, it was going to be a challenge to deliver them to people living in these areas.

Fourth, the administrative staffs needed to direct and carry out the campaign at both the central and regional levels were inadequate. Henderson noted in a letter to Karel Raska, director of the Division of Communicable Diseases at WHO: "The proposed staff both at headquarters and at regional levels is . . . far less than adequate to undertake the job. . . . The general scheme of the proposed programme is different from programmes of the past; surveillance techniques and their application are unknown in every one of the countries. . . . A substantial amount of training, guidance, and assistance will be required from well-trained full-time staff within the Organization."[24]

On top of this, smallpox eradication in the Americas, initiated by Frederick Soper and PAHO in 1950, was not going well. Numerous problems beleaguered the campaign. Difficulties in coordinating national programs, the absence of rural health services and surveillance systems, and problems with the production, delivery, and quality of vaccines all hampered progress toward eradication. In addition, the dominant virus in the Americas, *Variola minor*, produced relatively mild cases that were often difficult to detect through passive surveillance. Thus, by the early 1960s, despite a decade of efforts and some progress, smallpox remained endemic in Peru, Columbia, Argentina, Paraguay, and especially so in Brazil.[25] Given this history, the delegates at the WHA in 1966 had every reason to question whether a global campaign could be successful.

Finally, there was the experience of the ongoing Malaria Eradication Programme. By 1966, it was becoming clear to many at WHO that the campaign was not progressing as hoped. For a range of reasons, many national programs were struggling to achieve their goals. In the wake of growing evidence that the malaria campaign was in trouble, funding levels were beginning to drop.

Why did all of the challenges facing smallpox eradication not give pause to the delegates to the WHA? There is no easy answer to this question. What seems likely, however, is that the decisions to endorse eradication in 1959 and accelerate the campaign in 1966 had less to do with a change in attitude regarding the probability of achieving eradication and more to do with international politics. The desire to prevent cold war tensions from disrupting international cooperation on health issues encouraged many WHA delegates to support the Soviets' proposal for smallpox eradication. The delegates hoped that by endorsing it, they would encourage the Soviets to remain a part of the organization and cooperate within WHO on other projects.

Second, the 1960s saw a rethinking of the value of smallpox eradication on the part of both public-health officials and political leaders in the United States. Growing recognition that malaria eradication might not be possible led US foreign-aid officials to consider other opportunities for technical assistance in the health arena. For a number of reasons, smallpox eradication appeared to be a more attainable target than malaria eradication, even if it might not have the same bang for the buck in terms of stimulating economic growth. Smallpox had no animal reservoir; cases usually manifested themselves with clear, distinguishing symptoms; cases were only infective for a few weeks; and there was a vaccine that, when properly administered, was nearly 100 percent effective and provided 10 years of protection (chapter 8).

This rethinking encouraged President Lyndon B. Johnson's administration to initiate a smallpox-eradication program under CDC leadership in a contiguous block of sixteen West African countries in 1965. The decision to launch the program was stimulated by a combination of domestic and international concerns. Domestically, smallpox had been eliminated from the United States in the 1930s, as it had in the Soviet Union, but America faced the continued threat of reintroduction, requiring health authorities to spend roughly US$150 million a year on vaccinations.[26] In addition, there was growing concern among public-health officials that the risks of side effects from the vaccine were outweighing the likelihood of infection. For both reasons, US health authorities were becoming receptive to the idea of eradicating the disease globally.

Internationally, the Johnson administration saw the West African program as a way of gaining political support from countries in a continent that was increasingly being viewed as a cold war battleground. As one African country after another gained independence, the United States worked to prevent them from falling under Communist influence. Throughout the 1960s and 1970s, the US State Department, through the US Agency for International Development (USAID), invested millions of dollars in development schemes aimed at winning hearts and minds. America's Central Intelligence Agency also played a role, eliminating political leaders, such as Kwame Nkrumah and Patrice Lumumba, who had developed ties with Eastern bloc countries and were seen as threatening US regional interests. With malaria eradication faltering, and having not even begun in most of Africa, the United States saw smallpox eradication as a means of advancing its interests in this region of the world. For the United States, smallpox eradication in West Africa was a humanitarian mission driven by political calculations.

US support for a West African program in 1965 added fuel to efforts to put smallpox on WHO's agenda. In conjunction with the meeting of the WHA in May, the White House issued a statement in support of the goal of eradication. President Johnson asserted that the "technical problems" of global eradication were "minimal," while the "administrative problems," including assuring vaccine supplies, personnel, and coordination, could be solved through international cooperation. The statement concluded with the commitment that the United States was "ready to work with other interested countries to see that smallpox is a thing of the past by 1975."[27]

The United States accordingly supported the director-general's smallpox eradication proposal at the WHA in 1966, even though it tried to reduce the amount of funds allocated to the program. The US objection to a larger WHO budget for smallpox eradication reflected its view that WHO should be dedicated to providing technical assistance in the form of expert advice, and that other organizations should materially support operations on the ground. Specifically, the United States preferred funding bilateral programs, which would pay political dividends, rather than contributing to multilateral operations. The United States thus joined the Soviet Union and a block of developing countries in voting in favor of an intensification of the war on smallpox.

Like the Malaria Eradication Programme a decade earlier, the Smallpox Eradication Programme was launched despite many technical and administrative uncertainties and concerns, and with limited financial resources. The decisions to proceed with both programs were driven more by politics and a confidence in the power of biomedical technology than by careful research and planning. Lack of adequate preparation and resources would haunt the MEP for years, contributing to its inability to achieve its ultimate goal. The SEP nearly suffered the same fate, yet in the end was able to find ways to compensate for its lack of preparation. Why the two programs suffered different outcomes is the subject of the next chapter.

CHAPTER EIGHT

The Good and the Bad Campaigns

PROGRAM PARTICIPANTS AND SCHOLARS have chronicled the histories of both the malaria-eradication and smallpox-eradication campaigns. I do not intend to rehearse these accounts. Instead, I want to explore why the campaigns had different outcomes and question both the ways in which they have been characterized as having failed or succeeded and the reasons given for these opposing outcomes. Evaluations of the success or failure of the malaria-eradication and smallpox-eradication campaigns have focused on differences in the ways in which the two campaigns were organized and carried out. These factors were clearly important. They need to be combined, however, with an analysis of the biology of eradication. By this I mean an examination of the biological differences that existed between the two diseases and the technologies that were employed to eradicate them. A biosocial analysis raises questions about previous assessments of the two campaigns and the lessons that they have produced.[1]

The first point that needs to be made in comparing the two campaigns is that the MEP was not a complete failure. While it did not eradicate malaria from the face of the earth, and the disease remained a pressing problem over large areas of the globe, 26 countries, or just over half of the 50 nations that had initiated eradication programs, had succeeded in eliminating the disease by 1970. Many other countries had been able to dramatically drive down malaria morbidity and mortality. While malaria cases subsequently rose in these countries, they never reached the levels that had existed before eradication was initiated. This was a significant achievement.

In comparing malaria eradication with smallpox eradication, one needs to keep this qualification in mind. It is not that malaria eradication failed and smallpox succeeded, but that malaria eradication failed to achieve its goals in some places but did so in others, while smallpox eradication ultimately succeeded everywhere. This distinction is important. We need to identify the factors that led to program failure in some malaria-eradication programs, and why these factors did not disrupt smallpox-eradication programs. We also need to know why some malaria programs worked. Answering these questions will allow us to develop a better understanding of the various factors that contributed to success and failure and revisit the lessons that the two campaigns can teach us.

While the MEP and the SEP campaigns are thought of as being different in their organization and implementation, they had much in common. Both were designed to interrupt transmission through the application of specific technologies: pesticides for malaria, and vaccinations for smallpox. They both required surveillance systems that monitored progress and identified remaining cases to be treated or isolated. Both campaigns were also based on the assumption that the elimination of diseases could be achieved through the application of biomedical technologies alone, without any changes or improvements in the social or economic conditions that contributed to the transmission of the diseases, and in the absence of a functional, basic healthcare system.

The two campaigns were formulated in Geneva, but they had to be adopted by national malaria-control programs. This required WHO officials to engage in multiple negotiations in order to ensure global participation and coordination.[2] Country representatives to the World Health Assembly may have approved the resolutions creating the eradication campaigns, but WHO had to convince local malaria-control officers to follow the methods designed in Geneva. They also had to convince local political leaders. As was the case in India, where resistance in Madras derailed the WHO/UNICEF-led BCG campaign (chapter 6), this was not always easy.

Malaria officials sometime had to go to great lengths to gain the support of local leaders. For example, Dharmavadani K. Viswanathan, director of India's National Malaria Control Programme, needed to first convince Mahatma Gandhi that malaria control was essential to the health of the nation. Gandhi was a committed pacifist and protector of all living things, and thus was opposed, in principle, to the killing of mosquitoes. Viswanathan visited Gandhi at his ashram in Sevagram in an effort to get his blessing for the campaign. Joining them was the future minister of health who, according to Viswanathan, asked Gandhi, "Bapuji, how can you accept such a scheme with your creed of *ahimsa* (to not injure)?" Gandhi looked to Viswanathan and smiled, encouraging him to respond. Viswanathan replied:

> Sir, if I put a barbed wire fence around my house and if a thief scales over it in his attempt to rob me of my belongings and gets bleeding injuries all over his body, would you charge me with committing violence on his person? I do not propose to catch a mosquito, open its jaws wide, and put DDT in its mouth. I am only spraying the inside of the walls and roof of my house. The mosquito has the whole of the universe to pick from for its blood meal. Why should it come inside my house and seek my blood? If it does and in the process it gets killed surely it does not militate against *ahimsa*.[3]

Gandhi burst out laughing and subsequently blessed the National Malaria Control Programme. This may seem like a somewhat unusual example. Yet it was not uncommon for local religious and social beliefs to clash with the scientific certainties of eradication promoters.

The adoption of both the malaria- and smallpox-eradication programs was facilitated by the fact that the technologies employed by the campaigns were already being used extensively over wide areas of the globe. Antimalarial programs using DDT had been established in many countries in Europe, Asia, and the Americas and were achieving tremendous success by the early 1950s. The disease had been eradicated in the United States in 1951. Similarly, smallpox vaccination was widely practiced throughout most of the world by the 1950s. By the time the Smallpox Eradication Programme was launched, smallpox was limited to 31 countries, the disease being endemic in only Brazil, India, Pakistan, Afghanistan, Nepal, Indonesia, and most of sub-Saharan Africa. Both campaigns therefore represented an intensification of existing control methods in most areas of the globe. On the other hand, familiarity with the interventions could be an obstacle to the programs, in places where local health authorities had confidence in the methods they had been using to deploy these interventions and were reluctant to modify their practices to conform to the eradication guidelines developed in Geneva.

Even where there was cooperation, the application of eradication strategies frequently ran into technical and logistical problems. Pesticides did not arrive in time to carry out spraying when needed, spray teams failed to apply the right amount of pesticide to hut walls, vaccines shipments were delayed, or the quality of the vaccines was compromised. Surveillance systems failed to identify cases before they sparked new rounds of transmission. Local populations resisted efforts to spray their huts, or have themselves and their families vaccinated, for personal, religious, or political reasons. Population movements related to religious festivals, labor migration, pasture shifts for herds, or attempts to seek refuge from political upheavals and civil wars prevented populations from being protected, disrupted the application of eradication technologies, and contributed to the spread of infection.

Both programs were also constantly short of funds and needed to mobilize contributions from member states and other organizations in order to create and sustain their global programs. This meant appealing to the interests of donor nations, particularly those of the US government, which was by far the largest financial contributor to both programs. As a result, the MEP and the SEP found themselves enmeshed in the cold war politics that shaped technical-assistance programs throughout this period. External material support for eradication was linked to the political interests of donor countries and was subject to withdrawal. Malaria-eradication programs in India and Ceylon (known as Sri Lanka since 1972) were both seriously disrupted by the withdrawal of US funding, for political reasons, in the late 1960s and early 1970s.[4]

Finally, proof-of-concept for both campaigns was based on limited experiences with pilot programs conducted in a small number of settings. In the case of malaria, this was a particularly serious problem, because the disease varied dramatically from location to location in terms of its epidemiology. Yet even with smallpox, there were few models of success, other than the experiments carried out with jet injectors in a handful of countries in the 1950s, and the experiences of European countries and America, where eradication had been achieved slowly, over decades of vaccinations. The American and European experiences with smallpox vaccination had little relevance to the type of time-limited global campaign envisioned by the designers of the SEP. Moreover, smallpox also varied from region to region, though not as greatly as malaria.

At the end of the day, the SEP proved to be better able to overcome these obstacles and reach its goal than the MEP. The question we need to ask is, why? The reason most frequently given for this difference by participants and subsequent chroniclers of both campaigns was that the malaria campaign was too tightly scripted. The Expert Committee on Malaria established a set of protocols for eradication that left little room for variation. The training of eradicationists at various regional centers emphasized the importance of maintaining rigorous adherence to these protocols. Fieldworkers were discouraged from making innovations. As one SEP participant observed, "Malaria fieldworkers who found that the strategies were not working were discouraged from modifying the strategies and, in effect, were disempowered." The campaign, in other words, adhered to a top-down, cookie-cutter approach to malaria control. Efforts to deviate from the script, or to innovate, were discouraged and subject to criticism. In addition, once the campaign was initiated, very little attention was given to research, so when problems arose, there was little information available on which to base changes in the program.[5]

By contrast, the smallpox campaign is frequently praised for its flexibility, that is, for its ability to adjust and overcome new challenges. Participants and subsequent reviewers of the program cite many examples of innovation and stress that fieldworkers were encouraged to innovate. Local innovations were reviewed in a timely fashion and could be rapidly adopted by the national programs and, in some cases, the global program. Ongoing research was a critical component of the smallpox campaign. There is a good deal of truth to both characterizations.

The Bad Campaign Revisited

The sixth WHO Expert Committee on Malaria designed the malaria-eradication strategy in 1956. This strategy was based on a mathematical formula developed by British epidemiologist George MacDonald. At the heart of MacDonald's formula was the malaria reproduction rate, represented by the value R_0. The reproduction rate was the chance that one case of malaria would cause another case. In other words, R_0 equaled the likelihood that a person infected with malaria would be bitten by an *Anopheles* mosquito, which would survive long enough for the parasite to sexually reproduce within the mosquito, and that that mosquito would bite another human who would become infected with malaria. As long as R_0 was greater than one, transmission would continue. If it could be reduced to less than one, then transmission would cease.

A great number of factors determined whether $R_0 > 1$. These included the abundance of anophelines relative to the human population, the propensity of the vector to bite a human host, the proportion of bites that were infective, the duration of the reproduction cycle that occurred within the mosquito, the probability that a mosquito would survive a single day, and the rate of recovery of the human host. While all of these variables affected transmission, MacDonald concluded that the probability of the mosquito surviving a single day was the critical factor, suggesting that malaria transmission could be interrupted simply by shortening the lifespan of the mosquito vector, thereby preventing the parasite from developing fully. DDT would achieve this goal. MacDonald represented the disappearance of malaria with the following formula, where m=the man-biting rate, a=the abundance of the vector, b=the vector competence, and p=the expectation of infective mosquito life:

$$R_0 = \frac{ma^2bp^n}{-r \log_{10}p}$$

The problem with MacDonald's elegant formula was that it reduced a multitude of complexities associated with each element of the formula to a single value. For example, the man-biting rate was a product of the feeding habits of the malaria vector, the housing conditions and density of the human population, and the clothing and sleeping habits of the host populations. These, in turn, were shaped by local social, cultural, and economic conditions, which varied greatly over space and time. The formula projected a false sense of uniformity, which belied the complexity of malaria transmission. It made sense in the abstract, but the devil was in the details, as quickly became apparent.

It is unlikely that many of the WHA delegates in Mexico City in 1955 understood this reality. Paul Russell had noted that when the Expert Committee on Malaria met in Kampala in 1950 and engaged in a heated debate over the advisability of launching malaria-control programs in areas where malaria was hyperendemic, MacDonald had deployed his formulas to make the case for going ahead with control. Russell commented in his diaries that few people in the room understood MacDonald's formulas or were able to critique them. If this was true in a room full of malaria experts, it must have been even truer for a World Health Assembly composed of delegates from a wide range of professional backgrounds, most of whom were political appointees who knew very little about the technicalities of malaria transmission.[6]

The larger point here is that the use of mathematical formulas was an essential element of the culture of technical assistance. Formulas were widely deployed by economists, demographers, and public-health authorities to demonstrate the relationship between technical interventions, on the one hand, and positive development outcomes, on the other. Like MacDonald's formula for eradication, these formulas were designed to reduce the role of local variability and "stabilize uncertainty," in order to draw predictable lines of causation between independent and dependent variables, that is, between technological inputs and social and economic outcomes.[7] In doing so, they evoked a false sense of confidence in the power of technological interventions.

The Expert Committee on Malaria's strategy for achieving MacDonald's goal of reducing the lifespan of the malaria vectors involved four phases: preparation, attack, consolidation, and maintenance. Before eradication campaigns got under way, exploratory surveys were to be conducted to map malaria conditions, staff were to be recruited and trained, and pilot studies were to be undertaken to determine the feasibility of eradication methods. During the attack phase, all houses in malarious regions of a country were to be treated with residual insecticides. The committee determined the exact amount of insecticides to be sprayed per square inch on the walls of houses, as well as the frequency of spray applications, on the basis of earlier experience. The committee also noted that in some areas, local conditions would require the addition of larvicide treatments and drug therapy during the attack phase. But the central weapon was going to be pesticides. Moreover, WHO technical advisors actively opposed the use of pesticides for larvaciding in a number of countries, fearing it would hasten resistance and undermine the use of pesticides for household spraying.

Once spraying reduced transmission levels to near zero, spraying was supposed to cease and the consolidation phase begun. Surveillance systems were to be created to identify and treat any remaining cases, with the goal of eliminating all existing sources of infection. If no new cases appeared for a period of three years, national elimination would be certified, and the program would enter the maintenance phase, during which surveillance was to continue, in order to prevent the reintroduction of cases from outside a country's borders. Maintenance would last until global eradication had been achieved.

The Expert Committee acknowledged that local conditions might require national programs to adjust the ways in which they carried out each of these phases. Yet WHO technical advisors insisted that the guidelines be closely followed in all settings. There was very little effort on the part of WHO to identify potential problems before the campaign began, or to develop strategies for coping with these problems. Virtually no resources were invested in operational research. Soper was critical of those who invested time in research rather than in control. When the Mexican malaria-eradication commission conducted research on local vectors, Soper complained: "Mexicans have not yet gotten the idea regarding the eradication of malaria. . . . They are more interested in investigations and studies than they are in anti-mosquito measures."[8] The Expert Committee insisted that its basic guidelines needed to be followed and were effective, even when evidence from program after program indicated that these were not working. In the end, faith in MacDonald's formula, in the power of pesticides to kill mosquitoes, and in the ability of technical specialists to overcome obstacles led successive Expert Committees on Malaria to downplay the need for further research to address local problems and not question the fundamental soundness of the eradication strategy.

By 1961, however, it was becoming clear that many programs were running into problems, particularly in the attack and consolidation phases. Time and again, the procedures carefully worked out in Geneva came up against the social and economic realities of life in developing countries. The guidelines for the attack phase assumed that rural houses were all the same, yet experience showed that this was not the case. Different kinds of construction influenced the effectiveness of spraying. Mud walls absorbed DDT and reduced its killing power. This problem, referred to as sorption, had been flagged as a potential impediment to the use of indoor spraying by the fifth Expert Committee on Malaria in 1953, but it had been discounted in the actual planning of the eradication campaign. In a classic example of the refusal of eradicationists to acknowledge potential obstacles to their goal, Wilbur Downs, with the Rockefeller Foundation, reported having been physically assaulted by Soper at a meeting in Mexico City in the early 1950s for having raised the issue of DDT sorption on mud walls. Soper accused Downs of impeding the progress of eradication. In addition, some populations slept out of doors during hot seasons, making the use of indoor spraying ineffective.[9]

The eradication protocols also assumed that populations were stable. Yet in many places, households were mobile, or people resided in different places at different times of the year. In South Africa, spray teams working in the Eastern Transvaal Lowveld had to cope with the massive influx of displaced persons, expelled from the highlands, as part of the apartheid government's grand plan for racially segregating the country. This resulted in the overnight creation of large new settlements in areas that were thought to have been effectively protected. The sixth Expert Committee noted the possible problems that population mobility could create for eradication programs. It had few recommendations for how to deal with the problem, however. Moreover, these and other social or cultural practices were viewed as isolated problems that could be overcome. They were not seen as significant enough to undermine eradication campaigns. This turned out to be a serious miscalculation.

Spray teams also ran into resistance from house owners who found the activities of eradication teams disruptive, particularly after two or three rounds of spraying. Household members had to remove their furnishing to prepare for the spray teams, and the pesticides left a smelly residue on the

A malaria sprayer working in the Homs area of Syria, 1954. WHO/Paul Palmer.

walls. Repeated spraying sometimes altered local ecological balances, leading to an explosion of some household pests, such as fleas. Villagers also had to submit to periodic blood tests. Blood had powerful cultural meanings for many people, making routine screening a challenge in some locations.[10]

The Expert Committee guidelines made no provision for how to handle these situations, and WHO technical advisors provided little advice on how to develop effective communication strategies to ensure community cooperation. They simply assumed that the benefits of spraying would be self-evident. The lessons from community-based programs in the 1930s had been forgotten, and those from the BCG campaigns in India ignored. Instead of trying to understand and work with local populations, eradication teams tended to view resistance as a product of ignorance or irrational traditional beliefs, just as it had been by those running the yellow fever and hookworm campaigns at the beginning of the twentieth century. As often as not, the staff for national malaria programs shared these attitudes. In Mexico, as Marcos Cueto has shown, state authorities committed to a larger national project of integrating indigenous populations into the national mestizo culture were resistant to the very idea of adjusting spraying campaigns to accommodate local beliefs.[11] It is also worth noting that while similarly organized campaigns had been successful against yellow fever earlier in the twentieth century, these campaigns had been backed by strong colonial authorities who were able to impose health interventions and punish those who resisted. The MEP seldom benefited from such legal buttressing.

Programs often had difficulty carrying out spraying operations in the manner prescribed by the Expert Committee's guidelines. Spray teams failed to apply pesticides in a proper manner, missed dwellings, or were unable to carry out operations at the appropriate time. These problems

After clearing out her house, Sra. Concepcion Garcia of Altavista, Mexico, anxiously supervises the spraying of her house for malaria inside and outside, 1958. WHO/Eric Schwab.

were related to the inadequate training of spray teams, as well as to problems of procurement, transport, and communications. Trucks broke down, supplies were held up at ports or airports, spray teams were not paid on time. In northern Thailand in 1956, spraying operations carried out with the support of the US International Cooperation Administration (ICA), the precursor to USAID, were delayed for months because the Thai employee responsible for procuring pesticides and equipment refused to use the new procurement forms required by the Thai government in Bangkok. The planners of malaria eradication had not envisioned these kinds of technical and administrative problems; they were not part of the formula, and the planners had few solutions for them when they arose.[12]

The consolidation phase of malaria eradication depended on the creation of surveillance systems. Yet many countries lacked the resources to develop such systems. Poor surveillance left gaps in the program's knowledge of the existence of cases within a particular country. In some countries, this gap led programs to cease spraying before transmission had been interrupted, so they entered the consolidation phase too early. In others, it prevented programs that had entered the consolidation phase from identifying all the remaining cases, particularly asymptomatic cases and new cases introduced from outside the country. This led to fresh outbreaks and the need to resume spraying. Another problem during the consolidation phase was the failure of programs to adapt their surveillance methods to the changing epidemiology of malaria, once the disease began to disappear. Malaria surveys, which captured changing prevalence levels by screening sample populations for infection, allowed programs to evaluate the progress of the campaign and determine when spraying could stop. However, once consolidation began, it was essential to develop surveillance systems that captured *all* new cases. Many programs were slow to make this shift in surveillance.[13] The sixth Expert Committee provided very little guidance regarding the need for such changes and how they should be achieved.

To make matters worse, spraying campaigns often transformed the malaria ecology, eliminating vectors that were susceptible to pesticides as a result of their preference for indoor feeding and resting on sprayed surfaces, but leaving species that fed or rested out of doors relatively intact. This shifted the sites of transmission outdoors, requiring the introduction of new control methods. Spraying could also result in local *Anopheles* species developing resistance to one or more of the pesticides being used by the campaign. Without ongoing research, these changes often went unnoticed, and they undermined control efforts. The report on the successful Romanian eradication program noted: "It had been supposed that as the eradication programme neared completion there would be less need for research activities. Events have shown the reverse to be the case, since a number of unforeseen problems have arisen necessitating the introduction of new methods and techniques of a more effective and selective nature."[14] Many of these problems were acknowledged by the eighth Expert Committee when it met in Geneva in 1961. Yet the committee's report concluded: "On the whole, malaria eradication programmes are progressing toward their goal, *and the principles and technical bases of the strategy of malaria eradication stated in the sixth report of the Expert Committee on Malaria remain unchanged*" [emphasis added]. The report later noted that the concepts and processes needed "some clarification and some *minor changes* in definition of certain aspects of the campaign" [emphasis added].[15] It insisted that if there were problems with the program methods, it was in their implementation, not their conception. Thus the report stated that "the effectiveness of the methods advocated depends entirely on how efficiently and correctly they are applied." The failure of the MEP to reevaluate the feasibility of its original proposed methods, or to encourage investigations into ways to improve or modify program execution, was clearly an example of program inflexibility.

This does not mean that individual programs did not exhibit some flexibility in the ways they implemented WHO's guidelines. Many of the countries that succeeded in eliminating malaria successfully adjusted their programs, particularly in the consolidation phase, in order to achieve their goals. For example, Taiwan's malaria-control program owed much of its success to developing an innovative surveillance and response network during its consolidation phase. The network included basic health services, the military, the examination of schoolchildren, and both rural and urban Malaria Vigilance Groups. The first Rural Malaria Vigilance Groups were established in 1950. By 1963, a network of 241 such groups (made up of 1600 volunteers) had been set up in the former hyperendemic areas of the country. On average, every village with a population of 1000 had one vigilance group, composed of about ten volunteers who were trained in case detection and reporting, basic blood examination, and the distribution of antimalarial drugs. In 1963, a network of 307 Township Malaria Vigilance Units was established (with 2800 members). Each unit consisted of a malaria technician and several volunteers; they assisted in the preparation of blood slides and maintained close contact with physicians and teachers in that area. The program also offered small rewards to individuals who reported cases of malaria. Over time, as the program moved toward the maintenance phase, active case finding was reduced and greater reliance was placed on passive surveillance. This shift was possible, in part, because the country had developed an effective network of primary health centers, and because education programs had succeeded in teaching people with fevers to seek treatment in clinics, where cases of malaria could be identified and treated.[16]

Similar shifts occurred sooner in countries with well-developed health systems, and later (or not at all) in ones where the health systems were poorly developed. It was not until the late 1960s that the Expert Committee recognized that the successful completion of malaria eradication required the existence of a basic primary health-care system. The MEP, as a whole, was based on optimism and confidence in the methods developed to eradicate malaria in 1956, and it was slow to acknowledge that these methods might not work in all areas, or that they might need to be altered or supplemented by other methods.

The Good Campaign Revisited

The Smallpox Eradication Programme experienced similar technical and administrative problems in many countries. In addition, SEP workers had to cope with civil wars and refugee crises that threatened to disrupt the campaign in Nigeria, Sudan, the Horn of Africa, and South Asia during the early 1970s. In what would turn out to be the final days of the campaign, tracking down the last cases in Somalia, vaccinators had to navigate their way through a civil war. These problems had not been foreseen or planned for by the designers of the eradication campaign, any more than they had by malaria eradicationists. The SEP also had to achieve its goals with a much smaller budget than had been available to the MEP. The final cost of the SEP was just under US$300 million, while the MEP spent US$1.4 billion between 1957 and 1967. What was different was the ability of smallpox campaigners to develop work-arounds and overcome the administrative, financial, and technical problems that stood in the way of eradication.

The central example of this adaptability was the shift in strategy from mass vaccinations to a surveillance and containment strategy in which surveillance systems identified new cases, and teams were then sent to vaccinate all possible contacts in the surrounding areas to contain the outbreak. The change in strategy was tested in the context of an emerging civil war in southeastern Nigeria, where an eradication team led by the CDC's William Foege, together with A. Anezanwu, director of the Smallpox Programme for the Eastern Region, raced against time to eradicate smallpox before the war broke out. The new strategy ran counter to that of the Nigerian federal government, which was committed to mass vaccinations and was particularly sensitive to deviations initiated in a part of the country that was on the brink of rebellion. When the government cut off supplies for the campaign, the necessity of employing the new strategy increased. It was not until the war ended that the CDC team was able to confirm that their strategy had worked in ending transmission within the region.[17] Their success was quickly evaluated and adopted by other national programs in West Africa and subsequently played a central role in the success of the global program. While some national programs were resistant to this shift in strategy, particularly in India, it significantly reduced the time and resources required to achieve eradication.

Unlike the MEP, every part of the SEP was not defined by WHO headquarters in Geneva. From the beginning, the SEP was based on the assumption that the only common requirements guiding the various national programs was that people had to be vaccinated and surveillance systems had to be created to identify every case that occurred. How these two imperatives were achieved was left up to local program directors. WHO headquarters would coordinate and give advice. There were operation manuals. But WHO's technical advisors, who worked with national program heads, were given the freedom to develop techniques and approaches that fit local conditions. People at all levels within the programs were encouraged to make suggestions for improving program efficiency. The downside of this approach was that program success depended greatly on the abilities of the individuals put in charge of running the program in each country and district. The role of WHO headquarters was to negotiate agreements with national programs and regional offices, respond rapidly to requests for assistance by country teams, move the limited funds available to WHO to countries where they would have the most-immediate impact, and continually assess the progress of the campaign and investigate problems that threatened to delay success.

Ongoing assessment of program status and progress at the local level, along with the conduct of epidemiological research, was encouraged, and these efforts paid dividends in providing opportunities to improve program performance. Of particular importance were the early findings—in West and Central Africa and Madras State in India—that smallpox spread less rapidly and less easily than was thought, and that the prompt detection and immediate containment of outbreaks was the most cost-effective means of pursuing the goal of eradication. Studies also showed that cases seldom occurred among adults in endemic areas, and that few cases occurred among people who had previously been vaccinated. Vaccination campaigns therefore focused on children and on ensuring that all individuals had a vaccination mark.[18] In another example, in Wallega Province in Ethiopia, the number of new cases exceeded resources to contain the outbreaks using surveillance and containment methods. The local eradication team examined the epidemiological data and decided to transfer the limited resources they had to the borders of the area in which the outbreak was occurring. In effect, they focused on where the outbreak was heading, not where it was, vaccinating people in villages located around the area in which the outbreak was occurring. This resulted in the creation of a ring of vaccination around the outbreak that prevented its further spread, allowed the outbreak to burn itself out, and transmission to be interrupted.

In Uttar Pradesh, India, local vaccinators felt that the vaccination process was moving too slowly. They reviewed the process and concluded that the strategy of only vaccinating people who had not previously been vaccinated required lengthy discussions and examinations of the inhabitants in each village. This was consuming large amounts of time. So they decided to change strategies and vaccinate everyone in each village. This change in strategy significantly increased the speed of the local campaign. Eradication teams were also inventive in developing surveillance networks in the absence of either effective health systems or reliable communications. William Foege described how his team in Enugu, southwestern Nigeria, employed a network of shortwave radios that connected missionary stations and combined this with teams of runners who spread out from each mission to seek out cases in neighboring villages, in order to identify outbreaks.[19] These kinds of local innovations seldom occurred in the MEP, but were encouraged within the SEP.[20]

The SEP also made efforts to adapt their programs as best they could to local social and cultural conditions. In India, Pakistan, and Bangladesh, variolation, an alternative but more risky way of providing protection from smallpox, was widely practiced by traveling variolators. While variolation, in principle, reduced the number of people who needed to be protected with vaccinations, the practice could also spread the disease, because it involved transferring live *Variola* virus from person to person. SEP workers discovered that traveling variolators were causing small outbreaks of smallpox as they moved from village to village. The WHO teams studied their practices and found ways to recruit variolators to support vaccination and stop variolating people.

An additional problem facing vaccinators was the practice of women being kept in seclusion, or purdah, in Muslim communities. It was initially assumed that vaccinators would be unable to reach these women. The introduction of a team of US Peace Corps women, however, allowed the program to begin vaccinating women in purdah. Later discussions with local tribal heads and family members resulted in male vaccinators being permitted to vaccinate these women, though they had to change the site of the vaccination from the upper arm to the wrist or back of the hand, in order to observe customs of modesty.[21]

In another area, where children were traditionally tattooed as a protection against witchcraft, the scars of smallpox immunization came to be accepted for the same purpose. In one country approaching independence, a respected political leader convinced the people that a vaccination scar was a sign of their independence. In another country, vaccinations were combined with census taking, political meetings, and alphabet learning and reading programs, to save on transportation costs and time. Village midwives came to serve as advance motivators and helped persuade mothers to bring their children to be vaccinated. To increase coverage, smallpox teams awaited nomads at wells and water holes.[22]

Yet some forms of resistance could not be overcome with innovative strategies. When vaccinators were close to achieving their goals, refusals to be vaccinated for religious or other reasons could not be tolerated. People *had* to be vaccinated. Moreover, the methods used to ensure compliance were not always as culturally sensitive as program participants have suggested. Alan Schnur, who worked with the SEP in several countries in Africa and South Asia, described his experience searching out children to vaccinate among people living in a village in Nepal:

> It was found that many of the people could not be located and were reported as being outside the village. In the end, a good natured room to room search of houses with flashlights, accompanied by the village leader, would detect adults and children hiding under beds, in chests, large wicker baskets, and even in large clay storage vessels. However, the search was done as a game of hide and seek, with laughter after people were found and vaccinated, such as in one particular case where lifting the cover of a large wicker basket revealed two children hiding inside.[23]

It is hard to share Schnur's assurance that the hidden children and adults were engaged in a game of "hide and seek" with the vaccination team or that the children were not terrified by the experience. Paul Greenough's account of people being rounded up and held down to be vaccinated in Bangladesh in the last stages of the eradication campaign there provides a glimpse of what was sometimes necessary to achieve the program's goal of

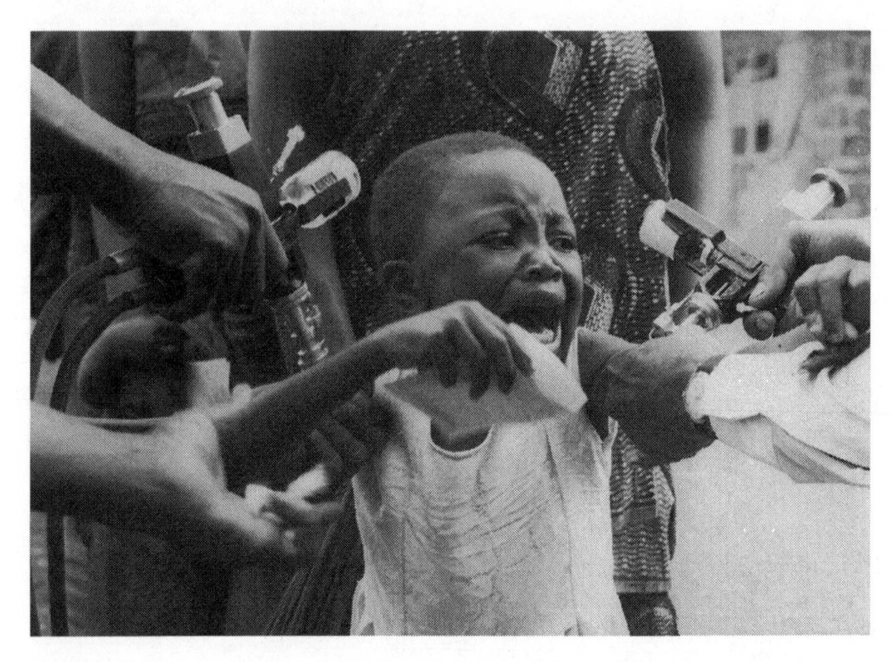

A 1968 photo of a young Cameroonian boy receiving his vaccinations during the African smallpox-eradication and measles-control program. The public-health technicians were administering the requisite vaccines using a "Ped-O-Jet" jet injector applied to each of the boy's upper arms. Centers for Disease Control and Prevention / J. D. Miller, MD.

eradication.[24] Greenough also described the cultural attitudes of expatriate fieldworkers, who viewed their local counterparts with the same disdain exhibited by many colonial officials or Rockefeller campaign workers in the 1910s and 1920s.[25]

The use of coercion does not appear to have been widespread. It was part of a larger pattern of behavior, however, in which expatriate SEP field teams disregarded local interests, and sometimes laws, in the belief that their mission justified any actions that were deemed necessary to the successful completion of the campaign. A close reading of William Foege's own account of his work in southeastern Nigeria reveals several moments where the ends appear to have been viewed as justifying the means. Early on in the campaign, he describes how his team "appropriated" unauthorized supplies from a governmental warehouse in order to carry out their vaccination campaign. He also describes having broken into a Nigerian official's desk to obtain a stamp to certify 50 vaccination booklets belonging to American missionary families who were trying to leave the country before the war broke out.[26] The tendency for aid workers to view themselves as somehow empowered to do whatever it takes to fulfill their mission, and to disregard local customs and laws in the process, was not limited to the SEP. These practices were, and are, more widespread than we would like to acknowledge. Then, as now, they reflect, in part, the near-absence of serious social, cultural, or ethical training in schools of public health.

Adaptability: A Biosocial Analysis

In assessing the argument that the MEP lacked flexibility and adaptability, while the SEP exhibited both traits, it is crucial to ask why this was so. Several reasons have been given. First, participants in the SEP have emphasized that its flexibility was built in from the beginning by the leadership of the program, particularly by Donald Henderson, who viewed a lack of flexibility as having been a major problem for the MEP. Another reason given for the flexibility of the SEP was that it lacked the resources to enforce uniformity in the eradication campaign. Henderson, from the beginning and throughout the program, had to operate with a very small administrative staff. He therefore viewed the role of WHO headquarters as one of advising and coordinating eradication activities, which were organized and run by the directors of the national programs. In his account of the campaign, Henderson repeatedly credited the progress of particular national programs to the leadership skills of that country's program director. But I would argue that disparities in the ability of the two campaigns to adapt to changes and overcome problems—in other words, to be flexible—were driven, in large measure, by differences in the pathogens each campaign was fighting and in the biological weapons available to them. A social analysis of the ways in which the two campaigns operated needs to be wedded to a biological analysis of these campaigns.[27]

Of all the diseases threatening the health of the world's populations in the 1960s, smallpox was the easiest to eradicate. It was low-hanging fruit. By contrast, malaria possessed features that made it one of the most difficult diseases to eradicate. In addition, the weapon available for attacking smallpox, the smallpox vaccine, was extremely effective in preventing smallpox and was easy to administer. The weapon of choice for attacking malaria in the 1950s and 1960s presented many more administrative and technical problems.

Each person who was infected with the smallpox virus normally developed a distinguishing rash that was easy to diagnose. There were no asymptomatic cases.[28] This fact greatly facilitated surveillance efforts. In addition, an infected person could only transmit the virus to others for a short period of time during the acute phase of the disease, when the severity of the symptoms greatly inhibited patient mobility. This restricted the speed with which the disease could spread, making it relatively easy to contain outbreaks. Finally, people infected with smallpox either died of the disease or survived and possessed a life-long immunity to it.

Malaria had very different characteristics. People infected with malaria exhibited a wide variety of symptoms, ranging from extreme sickness to being asymptomatic, depending on their level of prior exposure to the disease and degree of acquired immunity. This meant that it was difficult to identify cases clinically. In areas of intense transmission, where adult populations exhibited high levels of immunity, there were often large numbers of asymptomatic cases, which could only be identified through blood tests. During the MEP, the only means of doing this was by microscopic examination. But this method, even under the best of circumstances, was not very reliable. Moreover, many countries lacked adequate medical and laboratory services to conduct even this imperfect mode of case detection. Many cases went undetected. In addition, *Plasmodium vivax* malaria could reside in the human body for years after an initial infection, without causing symptoms, before remerging in the peripheral bloodstream, where it could infect mosquitoes.

Once cases were detected, they had to be treated, along with the infected individual's possible human contacts. The entire process—from clinically identifying possible cases, to taking blood slides, to having them read, and then organizing treatment followup—could takes days or, in some cases, weeks to complete. This lengthy process increased the possibility that one case could infect mosquitoes biting that person and restart transmission. The surveillance requirements of malaria eradication meant that, to a much greater degree than in the SEP, success was dependent on the existence of at least a rudimentary health system. Lastly, unlike a smallpox infection, malaria did not impart a life-long immunity to the disease. Individuals could be reinfected multiple times.

The smallpox vaccine was extremely effective in providing immunity to the disease for up to 10 years. Moreover, once a freeze-dried vaccine was developed that was heat stable, it could be administered anywhere in the world. The vaccine was also easy to administer, using teams of vaccinators, who needed very little training, to inoculate large numbers of people in a short period. The speed of vaccination was facilitated by the development of the bifurcated needle, which could hold the exact amount of vaccine needed to successfully vaccinate an individual and quickly administer it with a few skin pricks. In places where injection guns were impractical, either because of their heavy maintenance requirements or because vaccinations involved traveling from hut to hut, rather than assembling large numbers of people, the bifurcated needle, created in 1961, worked extremely well.

If administered correctly, the smallpox vaccine was nearly 100 percent effective, though early on in the campaign, there were problems with vaccine quality. Moreover, there were no examples of the virus developing resistance to the vaccine. Every successful vaccination resulted in a pustule and a distinctive scar, which remained for decades. Teams visiting an area could readily determine whether smallpox was present in the community and who had been vaccinated. This greatly facilitated the planning of vaccination campaigns and the speed with which the populations could be protected. In order to achieve eradication, populations only had to be vaccinated once. There was no need to return to communities on multiple occasions. This both facilitated eradication efforts and reduced the problem of local resistance to the campaign. Finally, the smallpox vaccine was only used for preventing smallpox. It had no wider application. This reduced possibilities for the virus to develop resistance to the vaccine.

Eradicating malaria through the application of residual insecticides or a combination of insecticides and antimalarial drugs represented a much more complex and demanding intervention. There were many moving parts, which had to be properly executed with precision, for the campaign to succeed. As Frederick Soper had stated, "perfection is the minimum permissible standard for a successful eradication campaign."[29] The proper application of the right amount of pesticide per square inch of wall surface, at the right time, over varied surfaces, required extensive training and supervision. It also necessitated a much more complicated distribution system. Enough smallpox vaccine to protect an entire province could be transported in a single vehicle, while it would take a small convoy to transport the pesticide and the spraying equipment needed to spray the same area in the fight against malaria.

Second, malaria eradication could not be achieved through a single application of pesticides to each household. It required repeated applications every six months to a year for several years. This increased the possibility of system breakdowns and an interruption of protection. It also increased the challenge of gaining local community support. Resistance to the repeated disruption caused by spraying was widespread. Finally, as was recognized at the beginning of the campaign, repeated applications of pesticides could

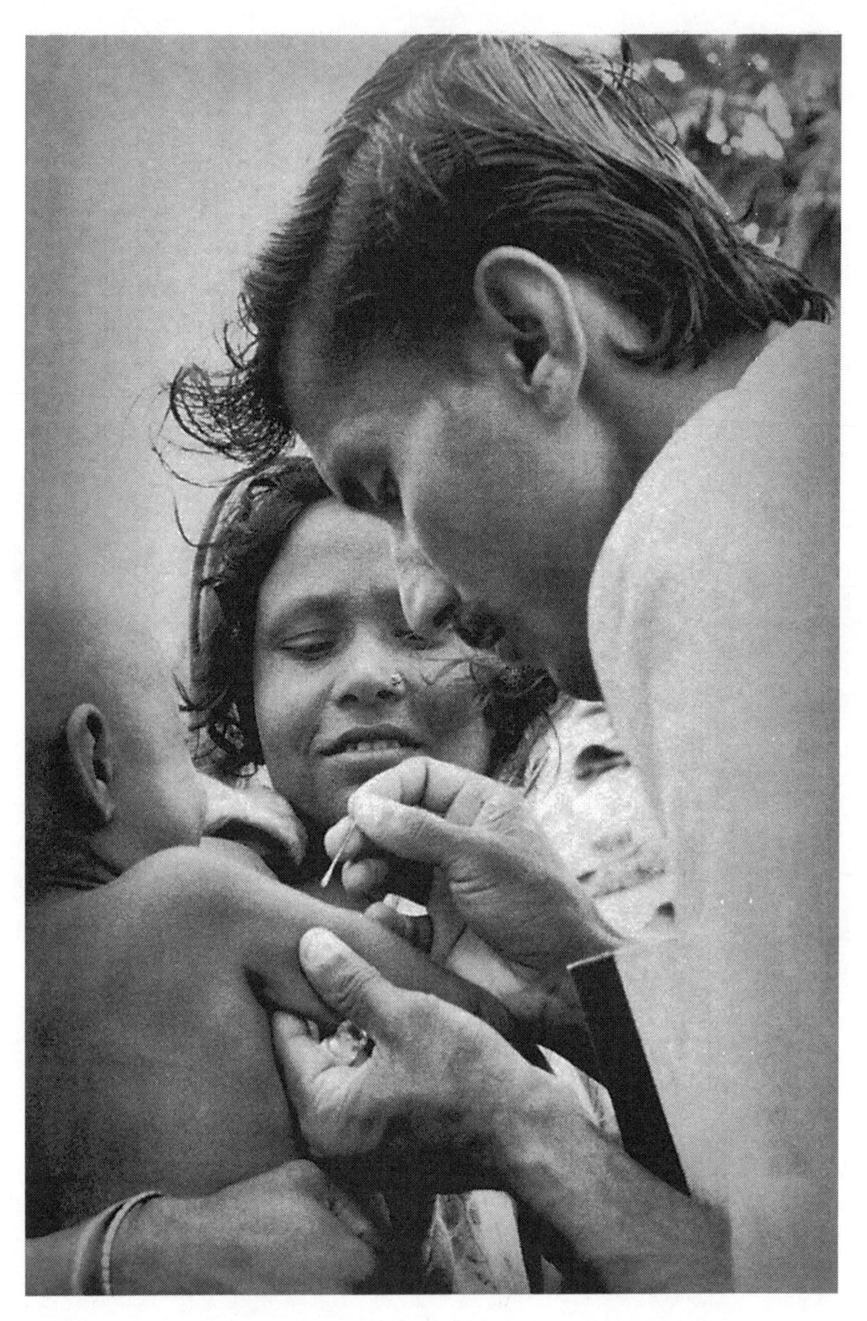

A smallpox-campaign worker in Bangladesh vaccinates a child, using a bifurcated needle. Centers for Disease Control and Prevention/WHO/Stanley O. Foster, MD, 1975.

lead to *Anopheles* resistance, undermining the effectiveness of the MEP's central weapon. The fact that, unlike the smallpox vaccine, DDT was used for multiple purposes, including protecting crops from insect damage, increased the risk of vector resistance. The concurrent use of pesticides for public-health and agricultural purposes undermined eradication efforts in Central America and India.[30]

These differences in the pathogens and the interventions made it much more difficult for malaria staff to achieve their goals, overcome obstacles, or be flexible. Too much spray could have toxic effects on humans and animals, but too little would be ineffective. Improper spraying could also contribute to pesticide resistance. It was for this reason that WHO's technical advisors created operation manuals and insisted on strict observation of program guidelines. Alterations to the guidelines, such as extending the period between spray visits or decreasing the amount of pesticide applied per square inch to wall surfaces, were actively discouraged, because they could undermine the campaign. Similarly, while it did not matter when smallpox vaccinators showed up, spraying had to be carried out at a precise time—at the beginning of the period of transmission—in order to be effective. In short, the effectiveness of the smallpox vaccine and its ease of delivery allowed SEP staff much more flexibility in how they approached their task. These characteristics meant that perfection was not essential.

SEP participants have also emphasized their ability to adapt to different cultural settings and overcome potential resistance to vaccination. Yet the fact is that they only had to do such things once at any particular location. Spray teams had to sustain the cooperation of the local population for years. This was a much more difficult task. I would argue that this difference in time scale, more than the ability and/or willingness of SEP fieldworkers to develop effective communication strategies for eliciting cooperation, accounted for their apparent success in this area of endeavor.

SEP participants frequently point to the fact that they were able to maintain their campaign even in the face of armed conflicts, such as had occurred in Biafra or the Sudan. This, no doubt, was an achievement. But it was an achievement that was possible because teams could move in and out quickly and still attain their goals. Malaria-eradication measures were much more difficult to sustain during wartime, because they required repeated visits over years and more-robust surveillance systems.

Other kinds of social and economic changes also had a greater impact on malaria-eradication efforts than on smallpox campaigns. Economic projects that led to population movements were much more disruptive to malaria-eradication efforts than to the smallpox-eradication campaign. This is because people infected with smallpox became acutely ill and were unlikely to travel. Even if active cases of smallpox entered into a region where the disease had been eliminated, they could easily be identified and isolated before infecting many people. People infected with malaria, however, could have minor symptoms or be asymptomatic. They could thus travel and reintroduce infection into a region where transmission had been interrupted. This happened in Swaziland in the 1960s. Workers were recruited from Mozambique to work on newly established sugar and citrus plantations that were located in the lowland region of the country, from which malaria had recently been eliminated. The workers came from areas of Mozambique where malaria remained endemic, and they restarted transmission in Swaziland.[31]

Finally, the fact that malaria eradication was, for multiple reasons, more susceptible to disruption meant that national malaria-eradication programs often lasted years longer than expected. By contrast, smallpox eradication was able to achieve its goals within four to five years in nearly every country. This difference played a critical role in the ability of each campaign to achieve its ultimate goal. The SEP was able to eradicate smallpox worldwide in just over ten years. The MEP's efforts, on the other hand, dragged on for years in many countries, and after a decade the goal of global eradication remained a distant vision.

Delays in reaching complete eradication led to growing disenchantment with the malaria campaign among administrators and policy makers at both the national and international levels. National health leaders, seeing the program extending years beyond what had been projected by campaign organizers, began to look askance at the viability of the strategy and to shift resources to other priorities. At the international level, major funders began questioning their continued investment in malaria eradication in the early 1960s. UNICEF, the principal provider of material support for the campaign in many countries, had, from the beginning, expressed reservations about spending large proportions of its available funding on a campaign that was not directly aimed at improving child health. For this reason, they had limited their commitment to providing material support for the attack phase of programs, which was expected to be largely completed by 1960. But by 1961, UNICEF recognized that the attack phase in many countries was extending beyond its projected time limits. In addition, the material requirements of the consolidation phase, which had originally been limited to the provision of drugs for treating the last cases of malaria, were increasingly being expanded to include sprays and equipment to stamp out outbreaks caused by the failure of programs to identify cases in a timely

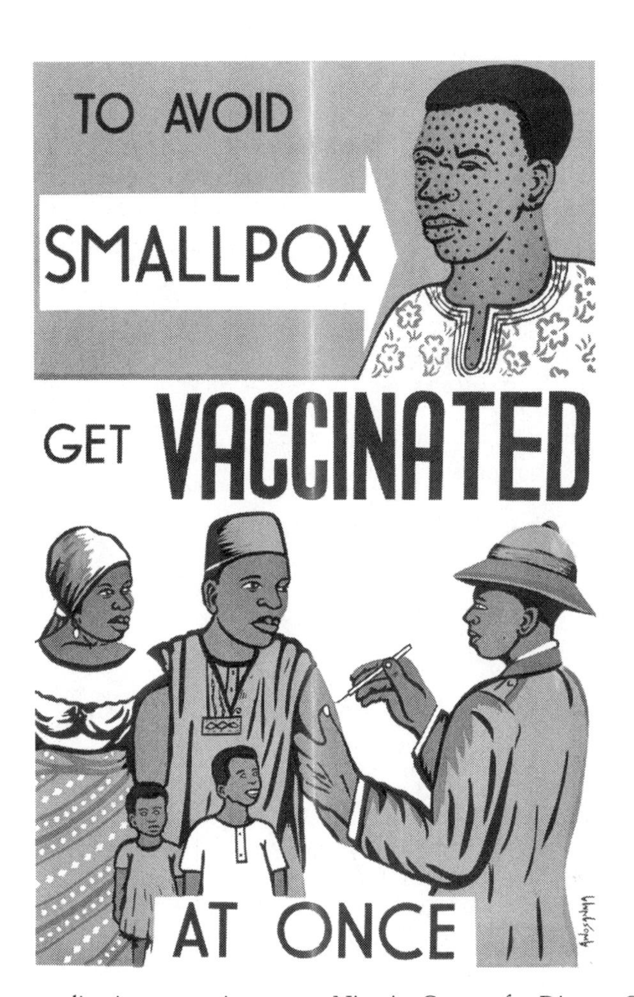

A smallpox-eradication poster in western Nigeria. Centers for Disease Control and Prevention.

manner. A number of programs were now requesting an extension of UNICEF support to cover these new consolidation-phase activities. Seeing both the slow progress of the attack phase and the emergence of new requests for material support for the consolidation phase, UNICEF placed restrictions on refunding programs in 1961 and, by 1964, had all but eliminated its support for the MEP.[32]

The other major funder of malaria-eradication programs was the US government. The United States continued its support for the campaign into the 1960s but increasingly shifted its contributions from WHO to bilateral US aid programs. This did not initially result in a decline in overall

US support for malaria activities. Over time, however, these program funds, disbursed by country-based USAID administrators, began to be reallocated from malaria programs to population-control programs (chapter 10), slowly draining national eradication programs of much-needed financial support.

The importance of the particular characteristics of the diseases and intervention measures in the success or failure of eradication campaigns was highlighted by Perez Yekutiel in a 1981 review of past eradication campaigns, including the MEP and the SEP. He identified six preconditions that needed to exist before launching an eradication campaign:

1. There should be a control measure that is completely effective in breaking transmission, simple in application, and relatively inexpensive.
2. The disease should have epidemiological features facilitating effective case detection and surveillance in the advanced stages of the eradication program.
3. The disease must be of recognized national or international socioeconomic importance.
4. There should be a specific reason for eradication, rather than control, of the disease.
5. There should be adequate financial, administrative, manpower, and health-service resources.
6. The necessary socioecological conditions should exist.

By socioecological conditions, Yekutiel was referring to the absence of such factors as major population movements, seasonal migrations, extreme dispersals of populations in remote areas, and cultural habits and beliefs that cause people to reject certain types of control. By these criteria, neither the MEP nor the SEP should have been initiated. Yekutiel acknowledged, however, that not all of these conditions needed to be met. Great strength in one area could offset weaknesses in others. Looking at the MEP in retrospect, he concluded that it had only met criteria 3 and 4 and should not have been attempted. The SEP only met criteria 1–4 and, like the MEP, had serious problems with criteria 5 and 6. The fact that the SEP was so strong in meeting criteria 1 and 2, however, allowed it to overcome weaknesses in criteria 5 and 6.[33]

None of this is intended to downplay the impressive energy and drive, as well as the inventiveness, of the people involved in the smallpox-eradication campaign, ranging from the very top in Geneva to the lowest fieldworker in Nepal or Somalia. There is no doubt that these men and women played a crucial role in the success of the campaign. Nonetheless, without the particular characteristics of smallpox and the smallpox vaccine, the disease would have been much more difficult to conquer, and that campaign would, in all likelihood, have failed. As it was, it only succeeded by the thinnest of margins. On the other hand, had malaria been a disease that was more like smallpox, and DDT as effective in preventing malaria as the smallpox vaccine was in preventing smallpox, the MEP would have appeared differently than it did in terms of how it was organized and managed. It could have been much more flexible and might even have been successful. Of course, all of this is speculative. Yet it is important to keep this perspective in mind when assessing the impact these two campaigns have had on the subsequent history of global health.

Revisiting the Legacies of the Good and Bad Campaigns

The eradication of smallpox resurrected the dream of disease eradication that had been tarnished by the failings of the MEP. It led many leaders in public health to call for WHO to launch eradication campaigns against other diseases. Many were considered, but only polio and dracunculiasis (guinea-worm disease) were subsequently targeted by WHO. Both were supposed to have been eradicated by 2000, but neither disease has yet disappeared.

The success of the smallpox campaign left a broader legacy. Even if no other disease could be eradicated, the smallpox victory convinced many in international health that the careful application of medical technologies, especially vaccines, could have a dramatic impact on mortality in resource-poor countries, and that this impact was possible without a well-developed health system or any significant improvements in the overall social and economic well-being of the targeted populations. A similar lesson had been drawn from earlier TB chemotherapy trials in South India in the 1950s, where the administration of antibiotics was shown to be equally effective in treating TB in impoverished slum areas as it was in well-managed institutional settings. WHO officials concluded from this result that intervention could be successful in the absence of social and economic resources.[34]

This was a powerful message that generated financial support for a range of health programs, from WHO's Expanded Programme on Immunization (EPI); to UNICEF's growth monitoring, oral rehydration, breastfeeding, and immunization (GOBI) program; to vitamin A supplementation programs; to the role of antiretroviral drugs to treat and prevent HIV/AIDS. It also inspired a generation of individuals who participated in the SEP to go on and lead organizations dedicated to this approach to global health. William Foege led later efforts to immunize the world's children in the 1980s, becoming director of the Task Force for Child Survival. He subsequently became director of the CDC and then the first global-health advisor to the Bill & Melinda Gates Foundation. Ralph Henderson, who had been director of the CDC's West African smallpox-eradication program, became the director of EPI. Donald Hopkins, who directed the smallpox-eradication program in Sierra Leone, later directed the guinea worm–eradication program at the Carter Center in Atlanta and became director of all of that center's health programs. The Carter Center set up an International Task Force for Disease Eradication to review possible targets for eradication. Isao Arita from Japan served on the Technical Advisory Board of the Expanded Programme for Immunization and Polio Eradication. Alfred Sommer, who participated in the smallpox-eradication campaign in Bangladesh, later demonstrated that the application of a single dose of vitamin A could reduce overall child mortality by 30–40 percent (chapter 16). He subsequently became dean of the Johns Hopkins Bloomberg School of Public Health, where he created the school's motto, "Saving Lives—Millions at a Time."

This legacy of the SEP was dependent on a particular reading of the histories of both the malaria and smallpox campaigns. It painted the malaria campaign as having been an administrative failure, compounded by the growing occurrence of vector resistance to pesticides, which might have been avoided with better program administration. By contrast, smallpox eradication was understood to have succeeded because it had better, more-flexible leadership and administration. These were the twin lessons that students of public health were and still are taught, to the extent that they are taught any history at all.

These linked readings ignored the biological challenges that malaria presented to eradicationists and the biological advantages that made smallpox an easy target. Leaving out biology allowed the success of smallpox eradication to be seen as a demonstration that good management could overcome social and economic disruption. A biosocial analysis allows us to see that the unique characteristics of smallpox and its vaccine trumped wars, famines, poor health systems, and poverty and, in doing so, once again empowered an approach to health-care delivery, centered on the application of narrowly defined technical interventions, that would dominate global health into the twenty-first century.

Viewing malaria eradication as an administrative failure, the product of bad planning and a lack of flexibility, without acknowledging the biological challenges it faced, further empowered a technical-intervention approach. The MEP should have taught policy makers that controlling or eradicating diseases that did not share the unique epidemiological qualities of smallpox, and for which there was not an exceptionally effective intervention, could be very difficult in the face of poverty, social and political instability, and the absence of effective surveillance and health services. The history of the MEP should have been a cautionary tale for policy makers. Labeling malaria eradication the bad campaign has allowed this tale to be ignored. Those currently struggling to eradicate polio are learning these lessons the hard way. Polio would seem to be a likely target for eradication. But the vaccine had weaknesses, which made overcoming local opposition much more difficult (chapter 16).

The MEP and the SEP had another legacy. International efforts to control or eliminate smallpox, malaria, tuberculosis, yaws, and leprosy, combined with the wider availability of both antibiotics to treat other infectious diseases and pesticides to attack vector-borne diseases, greatly reduced the global burden of disease and overall mortality. This reduction in mortality contributed to a rapid increase in population growth in many developing countries. While the causes of this growth were and are debated, many observers at the time pointed to the effectiveness of disease-control programs as the major driver of population growth after World War II.

In the immediate postwar period, population growth became a major concern for demographers, economic planners, and the makers of health policy, all of whom were concerned with global economic development (part V). These anxieties were translated into efforts to limit population growth. Just as they did with malaria and smallpox, world health leaders attempted to solve the overpopulation problem through the application of biomedical technologies—condoms, intrauterine devices (IUDs), sterilizations, and oral contraceptives—while continuing to ignore the social and economic forces that were driving population growth.

PART V

Controlling the World's Populations

In September 1944, economist Theodore W. Schultz convened a meeting at the University of Chicago to discuss population growth and the world's food supply. The meeting followed the international food conference that was held in Hot Springs, Georgia, the previous year, which had prepared the way for the creation of the United Nations' Food and Agriculture Organization. In organizing the meeting, Schultz was inspired by the interdisciplinary approach taken by the League of Nations Mixed Committee on the Relation of Nutrition to Health, Agriculture, and Economic Policy in the 1930s. He accordingly invited a diverse group of experts in agricultural economics, demography, nutrition, and international relations. Among the participants was Frank W. Notestein, director of the Office of Population Research at Princeton University.

Notestein presented a paper titled "Population—the Long View," which attempted to explain the rapid population growth that many postwar leaders in the United States and Europe believed was occurring in developing areas of the globe, blocking these regions' road to social and economic advancement. Attempts to address the "population problem" would dominate international-health and development efforts during the 1960s, 1970s, and 1980s.

Notestein's paper set forth a typology that divided the regions of the world into three types, based on their population-growth potential. He labeled these as areas of "high growth potential," "transitional growth," and "incipient decline." These three types existed on a continuum and were associated with different stages of industrialization. Notestein's stages of population growth recast an earlier set of prewar arguments about how the expansion of Western civilization would change fertility patterns in the rest of the world.[1] They were also a variant of modernization theory, which predicted the progressive transition of societies from a premodern, or traditional, society to a modern society. Modernization theory was made popular by American sociologists, particularly Talcott Parsons, during the 1950s and 1960s. For postwar planners in the United States and Europe, it became a blueprint for promoting economic growth in underdeveloped regions of the world. A particularly influential version of modernization theory was Walter Rostow's *The Stages of Economic Growth: A Non-Communist Manifesto*, which provided a road map for Third World development driven by increasing industrialization.[2]

Notestein's areas of high population-growth potential covered more than half of the world and included Egypt, Central Africa, much of the Near East, virtually all of Asia outside the Soviet Union and Japan, the islands of the Pacific and the Caribbean, and much of Central and South America. These areas were occupied by largely agrarian societies, which, according to Notestein, had been largely untouched by the forces of modernization. Death and birth rates remained high. The latter allowed populations to persist in the face of high levels of disease and warfare. At the same time, the economies of these preindustrial societies were labor intensive, requiring large families.

Areas of transitional growth had begun to move along the path toward industrialization. Both mortality and fertility rates had begun to fall, but the decline of mortality preceded that of fertility. This was because the new realities of industrial life were not apparent, and families continued to make reproductive decisions based on preindustrial modes of production. They continued to see large families as necessary to ensure sufficient numbers of offspring to maintain productivity and the survival of the family. The result of decreasing mortality and stable birth rates was rapid population growth. As agrarian populations transitioned to urban and industrial conditions, however, parents were faced with larger numbers of children, which they could not afford to support. Poverty and disease were the consequences. Notestein identified the populations of Eastern Europe as nearing the end of this stage; those of the Soviet Union and Japan, and of certain Latin American countries, as being in midcourse; while those of Turkey, Palestine, and parts of North Africa were just entering the transitional phase.

For Notestein, the rapid growth of populations in many transitional areas had been accelerated by a failure of colonial social and economic development. This was part of Notestein's model that received relatively little attention and would later become part of the structuralist critiques of population-control programs in the 1970s. He argued: "The modern nations of the West have imposed on the world's nonindustrial peoples that part of their culture [primarily medical and public-health interventions] which reduces mortality sufficiently to permit growth, while withholding, or at least failing to foster, those changes in the social setting out of which the reduction of fertility eventually developed in the West.

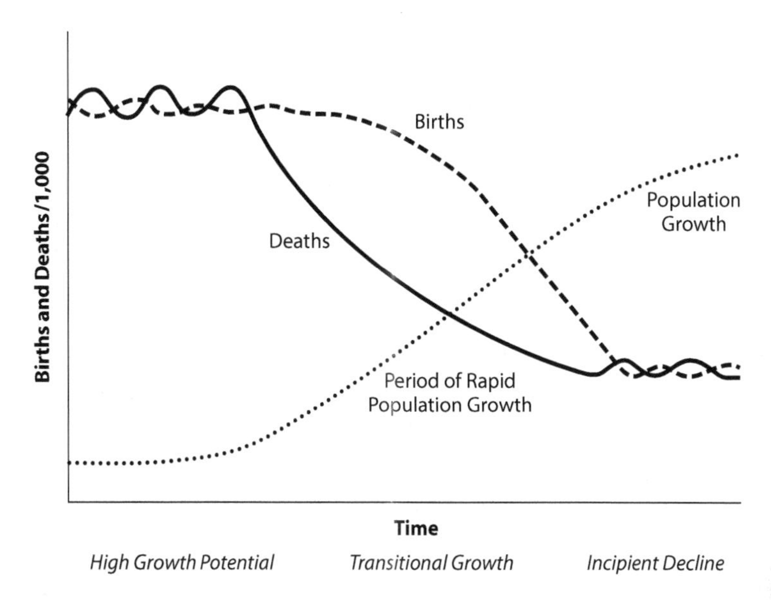

A demographic transition model

The result is large and congested populations living little above the margin of subsistence."[3]

Finally, there were areas of incipient decline, where population growth was beginning to slow. Notestein claimed that this decline resulted from a reduction in fertility caused by the increased use of contraceptive methods. He asserted, however, that the use of contraceptives was stimulated by broader social and economic changes associated with industrialization—the development of labor-saving technologies and new forms of production—that significantly reduced the need for family labor. In this changed environment, parents came to view large families as unnecessary for survival and economically disadvantageous. According to Notestein, "under the impact of urban life, the social aim of perpetuating the family gave way progressively to that of promoting the health, education, and material welfare of the individual child." The populations of northwestern, southern, and central Europe, as well as North America, Australia, and New Zealand, were all characterized as those experiencing incipient decline.[4] This adjustment would take decades, during which populations would continue to grow, but eventually populations in these areas would begin to decline.

Notestein's stages of population growth would become the basis of demographic transition theory, which he and his colleagues at Princeton University's Office of Population Research would present as a way of understanding and responding to global population growth. His colleague Kingsley Davis published an influential version of this theory in 1945.[5] Transition theory viewed social and economic transformations as an essential prerequisite, or determinant, of the decline in fertility, which many saw as essential to advancing health and development around the world. It was thus, in its original formulation, a model of change that was consistent with, and paralleled, the early, broad-based visions of health and development that shaped the missions of international aid organizations in the immediate postwar period (part III).

By the late 1950s, however, transition theorists like Notestein had shifted ground, promoting a more interventionist model of population change in which efforts to increase contraceptive use were viewed as essential to reducing population growth and promoting economic development. In short, they had reversed the causal relationship between social and economic development and population growth. Rather than social and economic development dictating family size, family size was now seen as dictating social and economic development. Population control had become a prerequisite for social and economic development.

Efforts to limit population growth during the 1950s, 1960s, and 1970s relied on technologies—spermicidal foams, condoms, IUDs, sterilization, and birth control pills—while failing to address the underlying social and economic determinants of reproductive decision making. They were thus part of the culture of technical assistance that had shaped efforts to control malaria and smallpox, only they represented an intrusive intervention into the intimate lives of millions of men and women across the globe.

Part V explores how efforts to actively limit the world's population became a central element of technical-assistance programs for three decades, beginning in the 1960s. In addition, it examines how population programs, or family-planning programs, transformed the landscape of international health. Unlike disease-eradication programs, efforts to modify patterns of reproduction in the developing world were not coordinated by international organizations. WHO, which might have been expected to take a lead in this area, played a limited role in population issues, for political reasons. Other UN agencies eventually assumed a larger role, organizing conferences and supporting local family-planning initiatives, but they were slow to take up these activities and did not do much in the way of coordinating family-planning activities. There were no unified campaigns on the scale of the malaria- or smallpox-eradication programs.

Much of the population work was carried out in a decentralized fashion by nongovernmental organizations (NGOs), including the Population Council, the International Planned Parenthood Federation, the Pathfinder Fund, Population Services International, and local family-planning associations, all of which were funded by private foundations, public donations, and, later, by bilateral and multilateral agencies. While the United States became by far the largest funder of family-planning programs around the globe, it funneled most of its financial assistance through the United Nations, NGOs, and universities rather than through bilateral agencies. It was within the field of population control that NGOs first became actively involved in international health. While their participation would increase in the late 1970s, due to financial constraints within USAID, and expand greatly in the 1980s and 1990s in response to neoliberal efforts to reduce governmental involvement in health care, NGOs emerged as important players in the field of international health through their role in promoting population-control activities during the 1960s and early 1970s.[6]

Family-planning activities also changed the landscape of international health by providing new opportunities for social scientists. Demography emerged as a policy science in the 1950s and played an important role in framing discussions about the need for population-control programs. Unlike earlier disease-eradication programs, family-planning programs engaged in extensive survey work, designed to assess local knowledge. Knowledge, attitudes, and practices (KAP) surveys were used for the first time in international health to measure local opinions regarding the use of birth control, motivations to limit fertility, and the acceptability of various forms of contraception. During the 1950s and 1960s, KAP surveys were conducted in developing countries in Africa, Asia, and Latin America. These surveys, which would become an essential part of international-health program planning and assessment, indicated widespread acceptance of the idea of birth limitation and were used to justify a supply-side approach to population-control efforts. This approach assumed that there was an existing demand for contraceptive technologies and that the central task of family-planning programs was to provide access to contraceptives to meet this demand.

Over time, the limitations of simply providing contraceptives resulted in more attention being given to convincing parents of the value of smaller families. This led to the collection of information on fertility patterns and reproductive decision making, and to the recruitment of demographers, anthropologists, and sociologists in ever-growing numbers. To a much greater degree than ever before, international-health interventions linked to family planning began to address the social and cultural attitudes and practices of peoples across the globe. In doing so, they created new opportunities for social scientists within international health. Their role would expand during the 1980s and 1990s as the need to design "culturally and socially sensitive" programs, obtain "local buy-in," and build "popular participation" became viewed as required elements in the design of new health interventions. Anthropologists would play an early and important, though not always positive, role in responding to the HIV/AIDS epidemic in the 1980s and 1990s (chapter 14).

Finally, and perhaps most importantly in terms of the longer history of global health, population-control and family-planning activities diverted attention and resources away from efforts to build and strengthen health-care systems or address the underlying determinants of ill health. Population-control programs were by far the largest recipient of international-health aid from the United States between the late 1960s and mid-1980s.

Chapter 9 traces early concerns about overpopulation prior to World War II. It then examines the conditions that gave rise to heightened fears among a diverse group of environmentalists, reproductive-rights advocates, and members of the emerging field of demography about population growth after the war. Finally, it explains why, despite these concerns, national governments and international-health organizations made only limited attempts to change reproductive behavior, leaving the field to foundations and nongovernmental organizations.

Chapter 10 describes how investments in population control and family planning grew dramatically in the 1960s, following the development of new birth control technologies and the United States' decision to begin actively supporting family-planning activities internationally. It examines the supply-side strategy employed by most organizations to make birth control technologies widely available and the weaknesses of this strategy.

Chapter 11 shows how the inability of family-planning programs to limit population growth by the early 1970s led family planners to begin rethinking existing approaches to the problem. Some countries chose to accelerate their family-planning programs by increasing the use of incentives and coercion. In other instances, some organizations and governments, including that of the United States, reconsidered their family-planning policies and, by the 1980s, partially withdrew financial support for them. By the 1990s, family planning had become less about limiting populations and more about protecting the health of women and children.

CHAPTER NINE

The Birth of the Population Crisis

THE FEARS ABOUT rapid population growth that emerged after World War II had a long history. There were, in fact, strong continuities between pre– and post–World War II population discussions. Thomas Malthus raised the problem of unrestricted population growth at the end of the eighteenth century. Malthus, like the subsequent post–World War II demographers, was concerned with the relationship between population growth and the availability of land and food. Populations, he argued, increased geometrically, while food supplies increased arithmetically. Therefore populations would always exceed available food supplies. This would result in either famine or wars over land, which would lead to a destructive reduction of populations—what later commentators referred to as a Malthusian crisis.

Concerns about population growth reemerged after World War I, spurred in part by the realization that population growth over the previous century had reduced the amount of land available for settlement. In northern Europe and the United States, these concerns were directed at colonial and other populations who were viewed as "less developed," including southern Europeans. This led to calls for immigration laws that would prevent these populations from overrunning more-developed countries, which had lower fertility rates. Immigration laws, by restricting people to limited areas, would have the added benefit of forcing people living in these areas to change their fertility patterns. This adaptation would be delayed, however, if these people were free to move to open spaces in Australia and North America.

For some, the ultimate solution to the imbalance between white and Asian fertility rates lay in promoting "white" fertility patterns around the globe. American sociologist Edward Alsworth Ross, in his influential 1927 book, *Standing Room Only?*, argued that Asians had to learn to adjust their fertility patterns and predicted that the spread of modernity and white civilization would cause this to happen. The idea that extending white civilization would transform the fertility patterns of the rest of the world represented an early version of Notestein's and Kingsley's transition theory.[1]

At the same time, nationalist and anticolonial commentators in South and East Asia pointed to immigration acts, along with earlier European colonial expansion in the nineteenth century, as having closed opportunities for Asian populations to expand and, thus, as being responsible for the growth of these populations. One of the most articulate promoters of this idea was Indian economist, ecologist, and sociologist Radhakamal Mukerjee. In his book *Migrant Asia*, Mukerjee argued for pro-emigration policies that would encourage Asians to occupy tropical regions of the globe, stating that Asians were more suited than Europeans to living in and developing these regions.[2]

Concerns about population growth, food supplies, and land also merged with the theories of eugenicists for improving populations. During the 1920s, many countries adopted eugenic policies aimed at improving the genetic stock of their nations. These policies often included both positive measures—such as encouraging individuals deemed particularly fit to reproduce to do so, as well as supporting maternal and child health programs—and negative measures, including laws prohibiting the unfit from marrying and having children. Many countries passed laws that permitted the forced sterilization of the unfit.[3]

As historian Alison Bashford has shown, the intersection of eugenics and population control had a long and complicated history. In the 1920s, any discussion of limiting population quantities naturally led to the questions of which populations to limit and how to improve population quality. While some, like US eugenics organizer Frederick Osborne, argued that universal birth control would ultimately improve the quality of all populations, others viewed population control as a way of limiting the fertility of particular populations who were viewed as inferior. This line of thinking was manifested in its most extreme form in Nazi Germany in the 1930s and 1940s. Yet reducing the fertility of less-developed peoples found broader support in a kind of global eugenics between the world wars. Even immigration laws, which were viewed as forcing the less-developed populations of southern Europe to change their fertility patterns in order to survive, were seen as a kind of eugenics.[4]

Many of those concerned with rapid population growth supported the use of contraceptives. In 1927, family-planning activist Margaret Sanger and biologist Clarence Little organized a World Population Conference in Geneva. It attracted an eclectic group of biologists, economists, ecologists, demographers, eugenicists, and people like Sanger, who viewed birth control in terms of the reproductive rights of women and the health of families. The participants expressed the belief that contraception was a necessary part of fertility change. Among those supporting the use of contraceptives were biologists Raymond Pearl and Julian Huxley, demographer Warren Thomas, and ecologist Radhakamal Mukerjee. All would play influential roles in post–World War II population discussions. Even British economist John Maynard Keynes, whose demand-side theories would dominate post–World War II economic planning, viewed birth control as an essential part of economic development.

Puerto Rico: A Colonial Laboratory for Family Planning

In the United States, governmental concerns about the impact of rapid population growth in developing countries focused initially on Puerto Rico, which would become a laboratory for the population-control programs that formed a central piece of US foreign assistance in the 1960s. US observers attributed Puerto Rico's continued poverty to population increases that had resulted from improvements in sanitation and the disease-control programs implemented by US authorities. These arguments deflected attention from the failings of US economic policies in Puerto Rico. They gained considerable purchase in federal policy discussions, and population control became a central piece of New Deal development policies in Puerto Rico under President Franklin D. Roosevelt.

In 1935, federal authorities, under the leadership of Ernest Gruening, chief administrator of the US Department of the Interior's Division of Territories and Island Possessions, established birth control clinics in Puerto Rico. Opposition from Puerto Rican nationalists and Catholic leaders, however, quickly forced Gruening to abandon the program. As a US senator in the 1960s, Gruening would play an important role in promoting US aid for family-planning programs abroad. This was another example of former colonial personnel shaping international-health efforts.[5]

Not all Puerto Ricans opposed birth control. A group of middle-class Puerto Rican nurses and social workers viewed family planning as a form of positive eugenics aimed at working-class women. They claimed that birth control would produce what they saw as the right sort of working-class family—small, nuclear, and legitimate. These professionals supported the creation of a family-planning-clinic movement in Puerto Rico. Thus family planning was not always imposed on unwilling Third World women by population planners from the global north.

Nonetheless, US-based philanthropic groups played a major role in promoting population-control programs in Puerto Rico during the 1930s. In doing so, these groups foreshadowed their later involvement in international family-planning efforts. Clarence Gamble, heir to the Proctor and Gamble soap fortune, established a Maternal and Child Health Association and created a network of clinics throughout the island between 1936 and 1939. Gamble insisted that these clinics provide cheap, easily usable birth control methods. He argued that poor families would not use diaphragms, which had to be medically fitted. He therefore promoted spermicidal jellies and creams, even though these methods were less effective than diaphragms. His goal was to change behavior and get women accustomed to using birth control. According to anthropologist Laura Briggs, Gamble reconstituted birth control in Puerto Rico on North American terms by making the island a laboratory for early contraceptive methods from the late 1930s to the early 1950s. Puerto Rico would later become a laboratory for testing Depo-Provera injections, IUDs, and birth control pills.

In 1937, the Puerto Rican legislature passed a law that made the distribution of birth control legal and assigned the task to the health department. At the same time, the legislature passed a eugenic sterilization law that included "poverty" as a legitimate reason for permitting sterilizations. In the absence of adequate supplies of alternative forms of contraception, sterilization became the method of choice for many Puerto Rican women seeking to control their fertility over the next decade.[6]

Population fears were not limited to leaders in Europe and the United States during the 1930s. They also found expression among economic and social leaders in Asia. China, Japan, and India contributed to contraceptive research and were leaders in legislating the use of contraceptives and encouraging parents to have fewer children. In Japan, Kyasaku Ogino pioneered methods to accurately chart ovulation, and Tenrei Ota was one of the inventors of the intrauterine device.[7] In 1935, the National Planning Committee of the Indian National Congress, headed by future prime minister Jawaharlal Nehru, adopted a resolution supporting state-sponsored birth control programs.[8] These efforts were stimulated, in part, by eugenic fears that the poor lower classes were reproducing faster than the upper classes, undermining the quality of the nation's racial stock. In addition, social and economic activists belonging to the All-Indian Women's Conference took up birth spacing as a strategy for improving the health of Indian women, particularly poor and working-class women.[9] Some Indian observers, including Nehru, viewed rapid population growth as a national economic problem. Asian countries would continue to play a leading role in promoting national birth control policies after World War II.

During the 1920s and 1930s, efforts were made to unite these various interest groups by convening national and international conferences, such as the 1927 Geneva conference, and supporting the global advocacy efforts of population activists like Sanger, who traveled across the globe in the 1930s to promote family planning. But attempts to widely legislate population-control policies, beyond targeting particular groups for eugenic sterilization, met with concerted opposition. Pronatalist groups, particularly in Europe, viewed underpopulation, not overpopulation, as a threat to national sovereignty, especially after World War I. Legal prohibitions against the dissemination of birth control devices, and even reproductive information, were widespread in these countries, as well as in the United States. In addition, the Catholic Church was strongly opposed to the use of contraceptive devices. The League of Nations Health Organization steered well clear of population control in the face of stiff opposition from the church and several Catholic countries.[10]

The Population Bomb

Population fears increased after World War II. There was growing evidence that advances in medicine and public health were rapidly reducing the global burden of disease. At the same time, birth rates remained stable or, in some cases, were increasing. As a result, populations rapidly increased. The island nation of Ceylon (now Sri Lanka) became a focus for efforts to measure the impact of disease control on demographic change. Death rates in Ceylon declined sharply after World War II. Writing in 1952, French demographer Henri Cullumbine argued that this drop in mortality was due in large measure to the impact of DDT spraying and a reduction in malaria mortality.[11] Within two years of the start of the campaign, the number of malaria cases had been cut by three-quarters, and six years later, life expectancy had increased from 46 to 60 years old, largely due to a decline in infant mortality. With every passing year, the lifespan expanded by a year or more, something that had never happened before in human history. This experience was repeated in Mauritius, Costa Rica, Mexico, and Barbados.[12]

At the same time, there was evidence that world food production was declining in relationship to population growth. The world was facing a potential Malthusian crisis. In 1949, the United Nations' Department of

Economic Affairs reported that worldwide food-production levels were 5–10 percent below prewar levels, while the world's population had grown by 10 percent. Concerns about the world's declining food supplies led to the creation of the Food and Agricultural Organization and to the 1944 University of Chicago meeting described at the beginning of this section. It had also led to Boyd Orr's proposals for the creation of a World Food Board (chapter 3). Yet many postwar observers viewed increasing world food production and distribution as unlikely to solve the food-supply problem as long as populations continued to grow. Following a neo-Malthusian logic, they argued that increases in the food supply would only contribute to a further population expansion that would eventually exceed food production. Population growth had to be slowed.

The postwar alarm over the dangers of population growth was ignited by the publication of two books in 1948: William Vogt's *The Road to Survival*, and Fairfield Osborne's *Our Plundered Planet*. Both books became bestsellers and influenced a generation of environmentalists and population activists. The authors warned that if current patterns of resource exploitation continued and populations continued to grow, the world would face a series of crises, ranging from mass starvation to political disruption and war. Vogt was particularly critical of Western efforts to improve the health and development of colonized regions of the globe, viewing these efforts as contributing to the population crisis. Writing about British rule in India, he argued:

> Before the imposition of Pax Britannica, India had an estimated population of less than 100 million people. It was in check by disease, famine, and fighting. Within a remarkably short period the British checked the fighting and contributed considerably to making famines ineffectual, by building irrigation works, providing means of food storage, and importing food during periods of starvation. Some industrialization and improved medicine and sanitation did the rest. While economic and sanitary conditions were being "improved," the Indians went to their accustomed way, breeding with the irresponsibility of codfish; as Chandrasekhar points out, sex play is the national sport. By 1850 the population had increased 50 percent; by 1950, according to State Department estimates, the population of India will be over 432,000,000.[13]

Elsewhere in the book, Vogt criticized a report by the FAO on the prospects of post–World War II development in Greece, because the report did not contain any "suggestion that a positive effort be made to reduce the breeding of the Greeks." He added, "How a group of scientists would justify such an omission on any rational grounds it would be interesting to know; such neglect would disqualify a wildlife manager in our most backward states!" These passages revealed both Vogt's concerns about the impact of development on population growth and his racist sentiments. Vogt explicitly viewed birth control as a necessary response to population growth and was in favor of research to develop a cheap and effective hormonal oral contraceptive.[14] Three years later, Vogt would become the national director of the Planned Parenthood Association of America.

Fairfield Osborne's book was a broad-based attack on humanity's mindless depletion—and ultimate destruction—of the planet's natural resources in the name of development. He lamented that most people still believed that the earth's living resources were without limits and could be drawn on endlessly, whereas, in reality, population growth was causing a worldwide depletion of these resources. As an example, he noted places like Mexico, where "the pressure of an increasing population, combined with the mounting injury to existing cultivable areas by erosion, is forcing people to use land that is totally unadapted to the growing of crops and at the same time is compelling the country to rely on imports for much of its basic food supply."[15]

The cause of population control was taken up by a number of influential individuals and groups who saw uncontrolled population growth as a social, economic, and political threat. Margaret Sanger continued her pre–World War II efforts to gain international support for contraceptive use as a way of improving the health of families. Sanger attempted to bring together all the various population-control constituencies, in order to create a united front after the war. This led to a conference on "Population and World Resources in Relation to the Family," held in Cheltenham, England, in 1948, and the emergence of the International Committee on Planned Parenthood, whose purpose was to support contraceptive research, publicize the problem of population growth and resource scarcity, and promote the creation of government-supported birth control programs.

American Hugh Moore, president of the Dixie Cup Company, became a convert to the cause of population control after reading Vogt's book. Unlike Sanger, who viewed birth control in terms of motherhood, reproductive choices, and healthy families, Moore, as a businessman, viewed the unfettered growth of Third World populations as leading to poverty and social unrest. This unrest, in turn, would open the door to Communist advances and disrupt the cause of global capitalist expansion. Moore sought to make the stakes involved in limiting population growth clear to business leaders, politicians, and the wider American public. To this end, in 1953 he published a short pamphlet titled "The Population Bomb," which laid out the political and economic dangers of uncontrolled population growth. He initially distributed 700 copies of the pamphlet to political and corporate leaders. He then added 10,000 names from *Who's Who* and had free copies sent to high schools around the country. Eventually, more than 200,000 copies were put into circulation. Moore helped make population growth a strategic issue. The booklet became required reading for the US State Department's Foreign Service School.[16] Paul Ehrlich later adopted the title of Moore's pamphlet (with permission) for his best-selling 1968 book on the population crisis. Like Moore, Vogt, and Osborne, Ehrlich warned against the dire consequences of unchecked population growth.[17]

While environmentalists and population activists worked hard to inform the public about the need for population control, their claims required scientific backing if they hoped to convince national and international funding bodies that family planning should be a central element in foreign-aid assistance. The discipline of demography, which was established before World War II, provided this backing. The field expanded dramatically in the postwar period, fueled by the financial support of private foundations concerned with population issues, particularly the Rockefeller Foundation. The Office of Population Research (OPR), which was established at Princeton University in 1936 with the support of the Milbank Fund, received large grants from the Rockefeller Foundation after the war. This funding allowed it to become both a major center for population research and a model for similar centers across the globe during the 1950s and 1960s.

The OPR and the field of demographic science was a special interest of John D. Rockefeller III, a longtime supporter of the eugenics movement. In 1948, he directed the Rockefeller Foundation to fund a fact-finding mission to the Far East to study population growth in Japan, Korea, Hong Kong, Taiwan, and Indonesia. Frank Notestein and demographer Irene Tauber were invited to participate in the mission. At the time, Notestein and his OPR colleagues believed that population growth was determined by social and economic conditions. They did not believe that reproductive behavior could be changed by direct interventions, but only as a result of changes associated with industrialization. The experience of visiting Japan, however, caused Notestein and Tauber to question these assumptions.[18]

Japan had experienced industrialization in the 1930s, but its population continued to grow. Tauber observed that industrialization had not improved living conditions, as had been projected by transition theory. Instead, gains from increases in productivity were reinvested in industrial growth. Japan's population grew further after World War II. While General Crawford

Sams, who headed the Public Health and Welfare Section of the Allied Command in Japan, believed in transition theory and urged General Douglas McArthur to promote industrial growth as a way of slowing Japan's population growth, the Japanese government proposed more-direct action.[19] The Japanese Diet passed legislation that rescinded measures limiting the dissemination of birth control information and the establishment of birth control clinics. It also passed a new Pharmaceutical Law, which permitted the manufacture and distribution of contraceptive devices.

Tauber and Notestein came away from their visit to Japan and other Far East countries convinced that transition theory would not work in this region of the globe, and that efforts to promote birth control both were needed and would reduce population growth. They interviewed farmers, who indicated that they would adopt birth control methods if these were made available. This led Tauber and Notestein, borrowing from Talcott Parsons, to conclude that it was not necessary to wait for economic change to transform a society's whole culture in order to alter reproductive behavior. There were always individuals, like the farmers they had interviewed, open to adopting new behaviors.[20] Education could increase the number of such individuals and eventually slow population growth, even though social and economic development had not yet achieved a level that would encourage lower fertility. This conclusion was presented in the mission's final report, published in 1950. The report was circulated widely among scholars, federal bureaucrats, foundation directors, and military leaders in the United States.[21]

Demographers at the OPR subsequently carried out a series of country studies in the early 1950s. These studies reinforced the arguments of environmentalists and population activists, who called for actions that would intervene directly in changing fertility patterns. Kingsley Davis of the OPR was particularly influential in making the case for population-control policies. He did so by linking the problem of unchecked population growth to the threat of Communist expansion. In his monograph on the populations of India and Pakistan in 1951, Davis pessimistically predicted the emergence of totalitarian governments and totally planned economies if these countries followed their current development pathway. He insisted that family-planning programs were essential in building a liberal or social-democratic path toward development.[22] He subsequently proposed that policies designed to control birth rates should be joined with efforts to lower death rates, through the provision of both technical assistance and economic aid. Such a combination of policies, if carried through effectively, would "strengthen the Free World in its constant fight against encroachment."[23] As elsewhere in the emerging postwar world of technical assistance and international health, cold war politics colored concerns and policy recommendations regarding population growth.[24]

Postwar concerns about the dangers of rapid population growth were not limited to US and European demographers and development planners. In addition to Japan, government officials in newly independent India and Egypt called for efforts to slow population growth by reforming the reproductive behavior of their citizenry. In India, an official inquiry into the 1943 Bengal famine pointed to the gap between food production and population growth and called for the establishment of population-control programs. The Bhore Committee's report (chapter 5) also called for the development of a population policy.[25] The Congress Party's National Planning Committee (NPC) ordered a report on population problems just before India's independence. The report reflected the fears of the upper classes that their own use of contraceptives, while the poorer classes continued to breed in an unrestricted manner, was leading to unequal patterns of population growth that would eventually result in an overall deterioration of India's racial fitness. Efforts needed to be made to encourage the poor to adopt birth control. The report also called for the wholesale sterilization of people who were insane or feeble minded. The NPC, headed by future prime minister Jawaharlal Nehru, considered the report and called for fertility limitation and cheaper contraceptives as part of a eugenics program. In 1951, Nehru chaired a planning commission that recommended fertility limitation for the sake of mothers' and children's health, as well as to stabilize the population. It also called for free sterilization and contraception, when recommended for medical or economic reasons.[26] Yet during the period of India's first five-year plan, the Indian government relied largely on economic growth and industrialization to stimulate reductions in fertility, following the prescriptions of demographic transition theory.

Egypt, after the Arab revolution that brought Abdul Nasser to power in 1952, also recognized the need to address its population problem. Egypt had been a key case for Princeton demographers tracking population growth in the 1940s. Egyptian officials were present at the International Planned Parenthood Federation conferences in Bombay in 1952 and Tokyo in 1955. Yet like India, Egypt initially put its faith in the power of economic development to transform the culture of the Egyptian family and reduce family size. As historian Laura Bier has noted, the management of reproduction, along with courses in personal hygiene in the public school system; subsidizing household appliances, such as refrigerators; and licensing village midwives—was an attempt by the state to create new sorts of families and a "modern" society, as well as a site for the articulation of new notions of citizenship. It was not until 1962 that the Nasser government indicated the state's intention to make family planning a part of its official development plan and established an extensive network of birth control clinics.[27]

Responding to the Population Crisis: The Role of International Organizations

Despite the growing chorus of voices calling for the limitation of population growth in the developing world, international organizations and Western donor countries were slow to respond. The United Nations created a Population Commission in 1946. The commission devoted most of its energies during the 1940s and early 1950s, however, to amassing population statistics. While these statistics heightened awareness of the problem of population growth, the commission did not propose polices to deal directly with it. The first effort by a UN agency to actively address the population issue came from Julian Huxley, director-general of the United Nations Educational, Scientific and Cultural Organization (UNESCO). Huxley, a biologist and eugenicist who had been a participant in population-control efforts before World War II, believed that UNESCO should address the pressing problems facing the world's populations, the most important of which was population control. Huxley viewed population growth in eugenic and racial terms. In a letter to the UN's director-general, Huxley noted that the world was already facing a Malthusian crisis, and that blacks in the United States and Africa were multiplying more rapidly than their white counterparts. He warned that the cumulative effect of such imbalances "on the human species would be disastrous." He called for a world population conference to prepare governments and public opinion for a world population policy. The UN's director-general, alarmed by the racial overtones of Huxley's memorandum, refused to circulate it to the UN Population Commission, and the proposed population conference did not take place.[28]

In 1951, the health minister of India, in response to the Indian Planning Commission's report, requested assistance from the World Health Organization in developing a national family-planning program. Brock Chisholm, first director-general of WHO, was an active eugenicist and promoter of voluntary sterilization. He also viewed population growth as a serious threat to global health. He was therefore receptive to the Indian health minister's request. Chisholm presented a proposal for assisting the Indian government in educating Indian women in the rhythm method. It made no mention of sterilization or other, more controversial methods of birth control. Moreover, the proposal would not involve a major commitment on the part of WHO. Nonetheless, the proposal drew the ire of the Catholic Church and the governments of Catholic countries. On the very day that Chisholm announced the Indian scheme, the Pope, speaking to the Catholic Union of Obstetricians, affirmed the concept of service in marriage and criticized any practice that would prevent either party in a marriage from restricting conjugal privileges. He did acknowledge, however, that there might be serious medical, eugenic, economic, and social reasons for such restrictions, leaving open the door for possible exceptions to this stance. But the Pope's injunction was broad enough to prevent Catholic representatives to WHO from supporting any effort to move the organization in the direction of supporting family-planning programs.[29]

The population issue came to a head again at the fifth World Health Assembly in 1952. Norway's Karl Evang moved that the WHA recognize the importance of the population problem by participating in the upcoming World Population Conference and establish an Expert Committee "to examine and report on the health aspects of the problem." The delegates from Belgium, Costa Rica, and Ireland responded sharply to the proposal, and the Irish delegate raised the specter of countries having to withdraw from the WHA if the proposal was adopted. The meeting cast a pall within WHO on the issue of population control, and for the next nine years the organization carefully avoided further discussion of the issue.

Family planning would seem to have been a natural focus for UNICEF, given its central concern regarding mothers and children. Yet the organization failed to take up the issue during the first decade and a half of its existence. While its various reports increasingly pointed to the economic costs of rapid population growth, UNICEF's Executive Board refused to propose any action in this area. In part, this was due to the organization's deference to WHO leadership on issues of health. UNICEF was also sensitive to the objections of Catholic countries. It was not until 1959 that the Swedish delegation raised the issue, arguing that UNICEF's recognition of the problem of rapid population growth obligated it to take action to support the use of birth control methods. This suggestion raised "a shock wave of disapproval and even disgust among some Board members."[30] It was not until the mid-1960s that UNICEF entered into the family-planning business.

Governments in Europe and the United States, which were providing the majority of postwar development aid, were equally slow to respond to the need for direct action on the population question. In the United States, President Dwight Eisenhower asked General William H. Draper to head a review of US military and economic assistance since the end of World War II. Draper's report highlighted the threat of uncontrolled population growth in the developing world. At a US Senate hearing in 1959, Draper stated that "the population problem . . . is the greatest bar to our whole economic aid program and to the progress of the world."[31] Eisenhower, however, viewed individual reproductive decisions as a private affair. He was also concerned about a potential domestic political backlash to US involvement in international family-planning programs.

The Kennedy administration also expressed concerns about the political and economic implications of rapid population growth in the developing world. Dean Rusk, secretary of state and former president of the Rockefeller Foundation, believed that fertility rates and population growth had social and political implications, and in August 1961, the US State Department set up a desk responsible for studying population questions. Rusk met representatives of more than 30 foundations concerned about population growth. Yet President John F. Kennedy was only slightly more willing than Eisenhower to engage the United States in supporting population-control programs overseas. Kennedy preferred foundations to take the lead, and in 1962 he supported a modest amendment to a foreign aid bill authorizing support for "research" on population programs.[32]

France and Britain were similarly slow to adopt family planning, either as part of their postwar colonial health and welfare programs or linked to their development assistance, once their former colonies gained independence. For Britain, this was largely due to fears of being criticized for trying to limit the growth of its colonial populations while maintaining a largely pronatalist stance at home. France's own pronatalist laws also made it difficult for French authorities to advocate population-control programs in its colonies. Even in Algeria, where high Muslim birth rates were viewed as a source of political unrest, the French failed to act.[33] Sweden and Norway, which would become major backers of family-planning programs in the developing world in the 1970s and 1980s, raised the question within UN agencies, but with few exceptions did not directly support population efforts until the 1960s.

Meanwhile, the Soviet Union largely rejected concerns about overpopulation as being exaggerated and rejected neo-Malthusian arguments. It pushed pronatalist policies in Eastern Europe and claimed that the world's resource crisis was one of overconsumption on the part of developed nations, more than overpopulation in developing nations. It also argued that socialist models of production would solve resource problems. Accordingly, Soviet delegates opposed UN resolutions regarding the need for nations to deal with their population problems.[34]

The failure of major donor nations and international bodies to take a more active stand in addressing population questions during the 1950s and early 1960s left the field of family planning in the hands of private organizations and foundations. In addition, a few developing countries took an early lead in creating their own population-control strategies.

The Rockefeller Foundation played a preeminent role in championing family-planning programs during this period. Frustrated by the lack of action taken by WHO in 1952, John D. Rockefeller III organized a Conference on Population Problems. The conference was held in Williamsburg, Virginia, and brought together an eclectic cast of academics from the fields of demography, embryology, physics, botany, and economics, along with activists like Fairfield Osborne and William Vogt, who represented the Planned Parenthood Association of America. It also included political, religious, and economic leaders. Rockefeller hoped to develop a consensus on how to effectively address population problems. Discussions at the conference were wide ranging. Yet again and again, speakers expressed a fear that uncontrolled population growth in the developing world, particularly India, was a threat to the developed world and, indeed, to the future of Western civilization. The image of centers of intellectual and industrial innovation in the global north being overrun by colored hordes from the global south was repeated by a number of speakers. Some, rehearsing the arguments of Hugh Moore and Kingsley Davis, raised the threat of Communist expansion, arguing that rapid population growth and poverty would encourage the leaders of developing countries to look to Communism as a solution to their economic problems. This, in turn, would threaten the economic and political interests of the West.[35]

The conference did not produce a blueprint for how to address these problems. It did, however, reinforce Rockefeller's belief that the foundation had to play an active role in advancing the cause of family planning around the world. His first action was to create the Population Council in 1952. The council organized a series of meetings that included demographers and representatives from the United Nations and the International Planned Parenthood Federation (IPPF). The IPPF had been formed following an international meeting of population activists in Bombay in November 1952, organized by Sanger and the International Committee on Planned Parenthood. Representatives of the Ford and Rockefeller Foundations, as well as major pharmaceutical companies, also attended these meetings. The meetings provided an ongoing forum for developing population policies, and the council quickly became the world's preeminent institution for population-policy research. It also became a conduit for the Rockefeller Foundation's funding for both research and population-control activities around the globe.

In addition, the Rockefeller Foundation poured money into the expansion of demography training, modeled on Princeton University's OPR, and provided fellowships for students to come to Princeton and other centers of demography. By 1957, the Population Council had spent roughly US$100,000 on training 32 fellows, 25 of whom came from foreign countries. The number of fellows continued to grow over the next decade.[36] The Ford Foundation joined the Rockefeller Foundation in supporting the expansion of demographic training, and it also contributed funding to the Population Council. These two foundations also established population-study centers in developing countries, including in Bombay (1954), Santiago (1957), and Cairo (1963).

The efforts of the Rockefeller and Ford Foundations to foster the field of demography around the world replayed the Rockefeller Foundation's efforts earlier in the century to advance the gospel of germs by building schools of public health and hygiene and providing training grants to foreign students. As before, the goal was to create a community of scholars and policy professionals who shared a common vision, in this case, a unified understanding of the causes of population growth and the need for population-control policies. Together, the Rockefeller and Ford Foundations ensured that an American vision of population studies became a global vision.[37]

The Rockefeller Foundation also funded studies on the effectiveness of birth control programs in limiting fertility. The earliest and most widely publicized of these was the Khanna study, conducted by Harvard researchers in the Punjab region of India between 1953 and 1960. The Khanna study was one of the first large-scale investigations to use knowledge, attitudes, and practices (KAP) surveys to collect quantifiable data on family-planning practices. The study's large data sets were processed on IBM computers.[38] The Khanna study was also the first controlled field trial of a contraceptive program in a poor, rural country. The study's field staff conducted monthly visits over a seven-year period, sampling homes in eleven villages containing over 16,000 people. It also collected masses of data on births, deaths, and migration. It failed, however, to demonstrate that the distribution of contraceptive devices could alter fertility rates.

More damaging to the family-planning cause, the study later became the focus of a scathing critique by a Harvard-trained anthropologist, Mahmood

Mamdani. Mamdani's *The Myth of Population Control: Family, Caste, and Class in an Indian Village*, voiced concerns that would haunt international family-planning efforts for decades. He had gone to Manapur, one of the Khanna study's villages, nearly a decade after the study was completed. He interviewed villagers, who told him that they had lied to the Khanna study researchers out of a sense of politeness, telling them what they wanted to hear about the villagers' use of the contraceptives they had received during the study. Many confessed that they had thrown the contraceptives away, because they believed that curtailing fertility was a recipe for financial disaster. Large families were essential to the labor needs of the villagers. Mamdani argued, moreover, that family-planning programs would not succeed in changing reproductive behavior in India in the absence of broader social and economic changes. In staking out this claim, he echoed the earlier conclusions of demographic transition theorists. Mamdani's book became a kind of bible for opponents of family-planning programs in the Third World and sparked a series of studies that highlighted the social and economic forces that worked to encourage high fertility rates, despite reductions in infant and child mortality.[39]

Although liberal academics would embrace Mamdani's critiques, foundations and smaller nongovernmental organizations, such as the International Planned Parenthood Federation, continued to work hard to promote the use of birth control in Asia and Latin America during the 1950s and 1960s. Clarence Gamble, who had supported family-planning clinics in Puerto Rico in the 1930s, renewed his efforts on a grander scale in Latin America in the 1950s and 1960s. Gamble established the Pathfinder Fund to finance birth control programs in Peru, Brazil, Honduras, Argentina, Chile, Uruguay, Bolivia, the Dominican Republic, Grenada, Haiti, Jamaica, and Turks and Caicos. While Gamble had used clinics to distribute spermicidal creams and jellies in Puerto Rico, the postwar programs focused on three rapid-intervention strategies aimed directly at target populations: (1) the dissemination of family-planning information via pamphlets and films, (2) the sale and distribution of spermicidal drugs, and (3) the promotion of intrauterine devices. Local physicians refused to participate in the program and did not examine women to determine whether an IUD was appropriate. Nor did they insert the devices or follow up to ensure that the women did not suffer from side effects. Instead, the Pathfinder Fund employed low-level fieldworkers affiliated with the National Evangelical Council, a small Protestant church organization, to distribute the information and contraceptive devices.[40]

Gamble's approach was based on the supply-side assumption that a pent-up demand for such technologies existed among women, and that simply making them available would result in their widespread adoption and subsequent declines in overall fertility. The Pathfinder Fund thus made no effort to understand local patterns of reproductive decision making. The fund was tone deaf to the needs of families. Supply-side approaches characterized early family-planning efforts in a number of countries in which nongovernmental organizations, like the Pathfinder Fund, or governmental agencies established family-planning clinics.

NGOs and foundations would continue to take the lead in supporting family-planning efforts in the developing world in the 1960s, 1970s, and 1980s. But their efforts received substantial political and financial support from both donor nations and international organizations, beginning in the 1960s. A change in US foreign policy with regard to population questions was crucial in this wider global participation in population-control efforts. This shift resulted in a massive increase in funding for family-planning programs across the globe.

CHAPTER TEN

Accelerating International Family-Planning Programs

UNTIL THE MID-1960S, supporters of family planning received limited financial assistance from major donor nations or international organizations. In particular, every US administration since World War II had acknowledged the problem of rapid global population growth but refrained, for political reasons, from actively supporting family-planning programs. This practice changed during the administration of President Lyndon B. Johnson, which placed birth control at the center of its foreign-assistance programs, beginning in 1965. Johnson's decision accelerated support internationally for population planning.

Johnson's stance was conditioned by a series of events that changed the landscape of family planning in the early 1960s. The most important was the development of new, cheap, and effective forms of birth control: the pill and IUDs. These technologies armed family-planning programmers with new tools that could be deployed to reduce the fertility of women across the globe, especially in poor, developing countries. The story of the development of the birth-control pill has been told by others but is worth reviewing, because it highlights again the ways that colonial contexts shaped international-health policies.[1]

In 1950, Katherine McCormick, heir to the International Harvester fortune, wrote to Margaret Sanger, asking how she could help with the problem of overpopulation. Sanger told McCormick, who had a long-time interest in birth control, that developing a low-cost, effective form of birth control to be used in "poverty stricken slums, jungles, and [by] the most ignorant peoples" was critical to the world and "our civilization."[2] McCormick, following Sanger's eugenic prescription, decided to bankroll a Harvard-based researcher, Harold Pincus, who had developed a new form of contraceptive, based on the consumption of estrogen and progestin in pill form. With McCormick's help, Pincus linked up with Searle Pharmaceuticals to organize clinical trials of the pill in Puerto Rico.

Knowledge of the effectiveness of an estrogen-progestin combination in controlling fertility had existed since the 1940s. But worries about the safety of the pill and its side effects had prevented drug companies, including Searle, from pursuing research on it until the 1950s. Anthropologist Laura Briggs argues that the rising concern among development planners in the United States and Europe about the economic and political consequences of unchecked population growth in the developing world made risky clinical trials morally acceptable. The health risks to a few hundred women enrolled in the trials was offset by the potential benefit for women around the globe—and the risk was especially tolerable if the women enrolled in the trials resided in an American colonial possession, which, as we have seen, had served as a laboratory for family planning since the 1930s.[3]

Searle organized two large trials of steroid contraceptive pills in Puerto Rico in 1956 and 1957. The trials had mixed results. Many women dropped out of the trials because of the side effects, which included headaches, nausea, vomiting, and midcycle bleeding. Others simply forgot to take their pills. The pills also proved to have a limited impact on preventing births. Among those taking the pills, there was a failure rate of 14 per 100 women-years of exposure. This figure was good enough, however, for Pincus and Searle to declare the trials a success, blaming the 14 percent pregnancy rate on the failure of women to take their pills. The fact that this failure raised concerns about how effective the pill would be in an uncontrolled setting was ignored.[4] Other pharmaceutical companies conducted further trials on the pill and IUDs in Puerto Rico in the following years. By the early 1960s, the efficacy of the pill had become accepted in family-planning circles, allowing population-control advocates to argue that family planning was not only necessary for promoting development, but now also feasible. International health had a new magic bullet for promoting health and development.

Yet birth control remained controversial in the early 1960s. Lyndon Johnson was hesitant to take on the issue early in his presidency, and he refused to meet with Rockefeller and Draper. Yet the early 1960s saw a growing drumbeat among public-health professionals, advocating for greater US commitment to population control. Within USAID, Leona Baumgartner and David Bell worked hard to push population programs. At the same time, the Population Council and researchers at population-study centers at Johns Hopkins University, the University of Michigan, Columbia University, the University of North Carolina, and the University of California system began publishing a series of articles, based on KAP surveys conducted in developing countries, to measure support for birth control. Beginning in July 1963, these researchers published a series of articles in the Population Council's journal, *Studies in Family Planning*, showing that women wanted birth-control interventions. The articles also showed that there was growing recognition among Third World administrators that rapid population growth was a threat to development.[5]

Equally important in raising the visibility of birth control was the women's movement in the 1960s, particularly in the United States. The movement focused attention on the needs of women, who were increasingly entering the workforce, to gain control over their reproductive lives. In 1965, the US Supreme Court, in the landmark case of *Griswold v. Connecticut*, overturned a Connecticut statue that prohibited the use of contraceptives. This decision made advocacy of family planning more politically acceptable. The women's movement itself, however, had little influence on President Johnson's decision regarding US funding of family-planning programs. He was no friend to women's rights and had to be pushed to extend the 1964 executive order on discrimination to include women.

More influential in shaping Johnson's view of population control was a report prepared for USAID by economist Stephen Enke. Enke laid out an argument that investments in family planning would provide direct payoffs in increased per capita gross domestic product (GDP) in developing countries.[6] Economists had paid little attention to population growth as an economic variable prior to the late 1950s and early 1960s. It was environmentalists, demographers, and population advocates who made nearly all of the arguments about the economic benefits of family planning. In the early 1960s, however, a group of economists illuminated the ways in which uncontrolled population growth could undermine economic development and highlighted the need for population-control programs.[7]

Enke's report went further, showing how investments in family planning could actually pay significant dividends in increased per capita income. To do so, he introduced the concept of the discount rate, which referred to the diminishing value of both costs and benefits over time. Enke used the concept to argue that in overpopulated countries, 15 years of investment in raising children from birth would outweigh the value of their future lifetime of adult employment. This calculation wrongly assumed that children under the age of 15 in agrarian societies were unproductive. But it was based on the application of a discount rate, which reduced the value of those future contributions when compared with the costs of current investments in raising children. Having discounted the value of future earnings, Enke was able to argue that the cost of a child's excess consumption, which he calculated using a discount rate of 15 percent, was US$279. Enke employed this calculation to argue for investments in interventions that prevented births. If a birth averted was valued at US$279, it made perfect economic sense to invest as much US$365 in sterilization or US$30 a year for the use of an IUD. Enke also argued that a US$4 vasectomy would have a positive impact on per capita GDP of US$1000.[8]

Many non-economists had, and continue to have, difficulty understanding arguments based on discount rates. Moreover, the logic of this kind of economic argument was difficult for many public-health professionals to accept, since it could be used to argue against any measures that save lives. Yet from the 1960s onward, these kinds of economic arguments gained increasing influence in development circles as economists began to play an ever-greater role in development planning. This trend would reach a peak in the 1980s and 1990s, as the World Bank became increasingly involved in global health (chapter 14). The complexity of the economic arguments was also their strength. Like MacDonald's mathematic model for eradication in the 1950s, they were difficult for nonspecialists to argue against.

Whether or not President Johnson fully understood the calculations behind Enke's arguments, he was impressed by their bottom line—preventing births would contribute to economic development. Johnson was critical of the limited achievements of US development efforts up to that time. He may well have been attracted by the idea that the problem lay in rapid population growth and that investments in birth control could pay direct dividends in increasing GDP per capita. Johnson put family planning on the US aid agenda in his State of the Union speech in January 1965. He proclaimed that the United States would "seek ways to use our knowledge to help deal with the explosion in world population." The president's speech was followed by an announcement from USAID that technical assistance in family planning would be made available to any country that requested it.[9]

Further support for US involvement in family planning was generated by a series of US congressional hearings, conducted by Senator Ernest Gruening (D-Alaska), who chaired the Senate Subcommittee on Foreign Affairs Expenditures. Gruening had directed efforts to implement a US-funded family-planning program in Puerto Rico when he had served as chief administrator of the Division of Territories and Island Possessions. The hearings were held intermittently for three years and included a steady stream of population experts, priests, and public-health and other professionals.

John D. Rockefeller III (*left*) meeting with President Lyndon B. Johnson (*right*) to discuss population issues, May 23, 1968. Courtesy of Rockefeller Archive Center.

The hearings made the problems of population growth more visible to the US Congress and more acceptable for public discussion.[10]

Johnson reaffirmed his commitment to family planning in his 1967 State of the Union speech, in which he noted that "next to the pursuit of peace, the really greatest challenge to the human family is the race between food supply and population increase" and concluded that "that race . . . is being lost." Shortly after this speech, the US Congress moved to earmark funds for the support of population-control efforts. This resulted in a massive infusion of dollars to US-funded international population programs. Johnson linked family planning to US foreign assistance, at one point telling Joseph Califano, who would become his secretary of health, education, and welfare (HEW), "I am not going to piss away foreign aid in nations where they refuse to deal with their own population problems."[11] US support for family planning would continue to grow over the next 15 years as the United States became the dominant funder of population programs in developing countries.

USAID missions were directed to encourage local governments to develop population programs and to support research to demonstrate the economic benefit of such programs. Despite these efforts, USAID was unable to spend the US$35 million that had been earmarked by the US Congress in 1968. President Richard Nixon continued the US commitment to population control, directing the Secretaries of State, the Treasury, and HEW, as well as the USAID administrator, the Peace Corps, and the United States Information Service, "to take steps to enlist the active support of all representatives of the United States abroad . . . to encourage and assist developing nations to recognize and take action to protect against hazards of unchecked population growth."[12]

Despite a growing commitment to supporting population programs abroad, US officials recognized that direct US involvement in these programs raised political concerns both at home and elsewhere. Domestically, there were many who believed that the United States should not be involved in promoting contraception. Catholic opposition, in particular, remained strong. But wider concerns existed. A law that prohibited the distribution of contraceptives under foreign-assistance programs in 1948 remained on the books in 1967. Internationally, US officials were sensitive to local concerns that the United States was trying to restrict the populations of developing countries. Some local governments were nervous about accepting population-control funds directly from the US government. These concerns were particularly intense in Latin American countries, especially Brazil and Mexico, for both political and religious reasons. To deflect these worries, the United States began funneling family-planning funds through intermediaries, including NGOs, local and international family-planning associations, universities, and international organizations. This strategy was also driven by the realization that USAID lacked the specialized personnel needed to run all of the overseas population programs it intended to fund.

One of the unforeseen consequences of channeling USAID money for population activities through organizations dedicated to family planning, rather than as bilateral aid to recipient governments, was that the USAID programs took on a functional rather than a geographical orientation. This was a critical development, for it encouraged the promotion of generic programs that were not shaped by local social and cultural conditions or needs. The funneling of USAID monies through organizations dedicated to family-planning activities had another consequence. USAID population funds were spent only on family planning and not on other kinds of development, which might have encouraged changes in fertility patterns through the expansion of general education or improvements in overall standards of living. USAID's Office of Population was, to a large degree, isolated from other development programs at USAID.[13]

Reimert Ravenholt, director of the Office of Population within USAID from 1966 to 1979, was the driving force behind the implementation of US-supported population programs abroad. Ravenholt, a strong believer in the ability of technology to solve development problems, was convinced that global population growth could be slowed by simply making contraceptive methods available to the world's populations. Like others at the time, he believed that there was an unmet demand for such technologies and that the central barrier to population control was access to effective birth control methods. He had no interest in developing broader programs aimed at encouraging behavioral change. Nor did he see the necessity of carefully tailoring family-planning programs to local conditions.

Because Ravenholt controlled access to the funds, he was able to implement this vision. Moreover, he reacted strongly to any efforts to question his approach. According to historian Peter Donaldson, Ravenholt was an intimidating figure, 6 feet 3 inches tall and weighing 205 pounds, and was known to react violently to those who opposed him. There were many who questioned his single-minded approach to population control, including USAID country administrators, who felt that more sensitivity was needed in promoting programs in their countries and in tailoring programs to local conditions. Others in Washington, DC, argued that alternative forms of development assistance should be used to encourage changes in reproductive behavior. Ravenholt listened to none of them. He made enemies. But he was able to build a population program at USAID that succeeded in its goal of making contraceptive technologies widely available in developing countries across the globe. Under Ravenholt, USAID flooded the world with condoms, oral contraceptives, and IUDs and made sterilization widely available in many countries. This success won him the support of private foundations and international organizations that shared his vision of family planning, as well as USAID higher-ups whose main objective was to see that the monies earmarked by the US Congress for family planning were spent.[14]

With the United States on board with funding family-planning programs, the way was cleared for international organizations, which had been reluctant to take on family planning, to become more actively involved. In 1965, the United Nations sponsored a large population conference in Belgrade. The following year, the organization's General Assembly agreed that the United Nations should help states that were developing population programs. This led to the creation of the United Nations Fund for Population Activities (UNFPA).

The UNFPA was a new kind of UN agency, which reported to the director-general and was independent of oversight by the UN Development

Programme's governing council. It was a semiautonomous organization that raised funds from voluntary contributions and was directed by a board composed of family-planning leaders, including John D. Rockefeller III. This arrangement not only provided the UNFPA with freedom of action, but also insulated the United Nations from its activities, which remained controversial in many quarters.

The new agency developed slowly, supporting only a few programs at first. By the 1970s, however, it had gained steam, linking its programs to the World Bank, which began funding population programs in 1967. UNFPA contributions to family planning increased from just under US$9 million in 1971 to US$78 million in 1979.[15] Most of the monies went to support national programs directly, but about one-third went to NGOs. Roughly half of the UNFPA's funds came from USAID.

WHO, for its part, continued to move slowly, as many member nations remained opposed to its active intervention in reproduction. In addition, Director-General Marcolino Candau viewed the rapid growth of family-planning funding as a threat to WHO's existing public-health priorities. Candau had reason to be concerned. By the late 1960s, USAID funds for malaria eradication were already being repurposed by USAID country directors to support population programs.[16] WHO simply supported the integration of family-planning activities into existing country-based health and development plans and was opposed to pushing programs on countries that did not have such plans. Its family-planning activities were thus limited to countries that already had programs. The reluctance of WHO to take a more aggressive approach to family planning meant that it remained a relatively small player in the global family-planning movement.

UNICEF also chose to move cautiously. It did not view population growth as a problem in terms of its impact on natural resources and economic development. The organization was concerned, however, about how repeated pregnancies and childbirths could adversely affect the health of mothers and children within households. In May 1966, UNICEF's executive director, Henry Labouisse, presented a proposal for the promotion of "responsible parenthood" to the UNICEF Board of Directors, gathered in Addis Ababa, Ethiopia. Labouisse was careful to emphasize that the plan would assist countries by expanding existing family-planning activities to ensure the overall health of children and mothers. UNICEF would not become actively involved in the distribution of contraceptives.

Nonetheless, the proposal was hotly contested. India and Pakistan, which had submitted proposals to UNICEF to support their existing family-planning efforts, were strongly supportive of the proposal. On the other side of the debate were Catholic countries, which were opposed to any policy that impeded procreation, and socialist countries, which viewed the population problem as a product of capitalist propaganda and argued that countries and international organizations should focus instead on advancing social and economic development. A number of African countries saw the population crisis as a racist invention and feared it would encourage female promiscuity.[17] As discussion of the proposal proceeded, it became heated and a number of countries threatened to withdraw from the organization, a tactic that had been successfully employed to defeat similar proposals before the World Health Assembly.

In the end, the WHO/UNICEF Joint Committee tabled the family-planning proposal for further discussion. Pakistan and India were granted assistance for programs that contributed to safe childbirth and the health of newborns. UNICEF also provided funds for educating women in the virtues of long birth intervals and small families, but without the remotest connection to the use of contraceptive devices. UNICEF continued to move cautiously toward meeting the family-planning needs of member countries by carefully defining them in medical terms, rather than in terms of economic development. This allowed the organization to slowly integrate family-planning initiatives into its efforts to improve medical care for mothers and children. But UNICEF never fully embraced family planning as part of its core mission, leaving the field to others—most importantly, to the UNFPA.[18]

As the leading international organizations were dragging their feet, USAID and the World Bank became the dominant funders of family-planning initiatives worldwide, supporting programs designed to actively promote birth control in Asia and Africa during the 1960s. India, Pakistan, Korea, Taiwan, Indonesia, Singapore, Ceylon, Thailand, Egypt, and Kenya all initiated supply-side programs that provided information extolling the value of smaller families and access to contraceptive technologies offered through family-planning clinics.

Communications strategies deployed by these programs included posters that contrasted the experiences and prospects of families with few children with those with many children. Large families were uniformly portrayed as unhappy and living in conditions of poverty, while small families were happy and prosperous. In Egypt, in a twist on this strategy, the promotion of small families was associated with a wider social-engineering effort designed to create the "modern Egyptian family," which was said to be essential to the creation of Nasser's postrevolutionary society. Having fewer children was a national obligation.[19] Yet the argument was the same. Small families were good, large ones were bad.

The ubiquitous use of these posters, with little modification, across the developing world speaks volumes about the ways that program organizers viewed the role of advertising. The posters were based on the assumption that young families, regardless of their social or economic situation, would immediately see the advantage of small families and adopt the birth control methods that were being offered. These advertising strategies also reflected the absence of any effort on the part of family planners to understand how parents might view children and how these opinions might vary across time and space. Even if there was evidence that some urban families were beginning to see children as a financial burden, the vast majority of developing-country populations lived in rural areas where, as Mamdani argued in his critiques of the Khanna study, agrarian families depended on adequate supplies of familial labor and children remained essential to the financial security of their parents.

Even urban families in many parts of the developing world continued to see the advantages of large families. As Cheryl Mwaria demonstrated for Kenya, patterns of land distribution and rural poverty, linked to colonial and postcolonial land policies, required Kamba families to support two households: one situated in an urban center like Nairobi, where some family members could earn wages or income from formal or informal economic activities; and a second, rural home, where other family members tended crops and maintained livestock. These arrangements were viewed as essential to a family's economic survival and demanded large families.[20] Similarly, anthropologist Dan Smith has shown that even in recent times, educated African parents living in Lagos, Nigeria, understood the burden of supporting children but continued to have large families. They did so because they recognized that large families were better positioned in modern Nigeria to participate in multiple patron-client networks, through which individuals gained access to jobs, education, or trading licenses. Having a large family was essential to their economic and social advancement and that of their children.[21]

None of these local realities were understood or viewed as barriers to a positive reception of the simplistic messages employed by family planners to encourage families to practice birth control. NGOs went about promoting family planning and conducting KAP surveys in remarkably uniform ways, making virtually no adjustments to local social, cultural, or economic conditions. Numerous KAP surveys conducted during the 1960s and 1970s focused almost exclusively on attitudes toward family-planning programs and how they might be affected by factors such as religion, socioeconomic status, place of residence, and ethnicity. Interviewers asked women about the ideal number of children they wished to have, but not *why* they wanted children. The surveys were time-saving substitutes for more in-depth inquiries into women's productive lives and decision making. They were a pervasive aspect of technical-assistance programs during this period. KAP surveys produced data that were quantifiable—and thus comprehensible—to project planners, turning complex social phenomena into percentages that could be compared across locations and used to draw generalizable conclusions about the demand for contraceptives. But they did little to advance family planning. Demographic surveys revealed that populations continued to grow at a rapid rate up through the early 1970s.

One of the few family-planning programs that seriously believed in the need to work with communities in reducing fertility was the Narangwal Project, organized by Carl E. Taylor in the Punjab region of India, near where Harvard researchers had conducted the Khanna study. The Narangwal study ran from 1965 to 1974 and included efforts to reduce child mortality and nutrition, along with family-planning activities. Control villages were established, in order to demonstrate the impact of the various components of the project. Integrating family planning with health and nutrition services paid dividends. As child survival improved, so too did acceptance of family planning. Most importantly, the Narangwal project included village leaders and residents in the planning and implementation of the program. Narangwal was one of the projects highlighted by WHO as an example of a successful primary health-care program in the late 1970s.[22] Narangwal, however, was an outlier, and by the early 1970s, there was growing dissatisfaction with the progress of population-control efforts.

CHAPTER ELEVEN

Rethinking Family Planning

THE FAILURE OF the supply-side family-planning strategies to bring about significant reductions in population growth in many parts of the developing world led to a reassessment of these programs at both national and international levels. This rethinking took several forms. In a number of countries, especially those with authoritarian governments, various incentives and coercive strategies were introduced to kick-start family-planning programs. These efforts to accelerate fertility decline had some success. At the same time, lack of progress led many supporters of family planning to rethink the supply-side strategies that had shaped family-planning programs. Finally, by the early 1980s, a number of major players, including the US government, were reconsidering their support for such programs.

Governmental Coercion

Lack of progress in limiting fertility led a number of governments to increase pressure on parents to limit the size of their families. In Singapore, the government launched a family-planning program soon after the island state had become independent in 1965. But the program did not take off until 1969, when the government introduced a comprehensive scheme of incentives and disincentives to encourage couples to give birth to only three or two children. For example, access to government-built housing was linked to the number of children one had.

In Indonesia, the Suharto government introduced coercive measures to accelerate the adoption of birth control in 1970. These included forcing rural women to attend lectures on the need for birth control. The massive population-control program, employing 30,000 workers and over 100,000 community leaders, held regular public meetings in villages in which male household heads were required to report on whether their families were using contraceptives and explain noncompliance. The results were compiled and represented on color-coded maps that were publicly displayed to promote community enforcement of compliance with contraceptive use. Some village leaders beat drums daily to remind women to take their pills.[1]

In India, Indira Gandhi came to power in 1966 and renamed the Ministry of Health the Ministry of Health and Family Planning. Under Gandhi, family-planning efforts were accelerated. Incentives were given to both doctors and women to encourage the use of IUDs or sterilization. States were directed to set targets for these methods. Women were counseled when they were giving birth, a time when they were believed to be most susceptible to family-planning messages. In addition, the government declared that employees having three or more children would lose their benefits. Yet population numbers continued to grow. In 1976 the Indian government declared the impracticality of "wait[ing] for education and economic development to bring about a drop in fertility" and resorted to what it described as a "frontal attack on the problem of population." Mass-education campaigns were stepped up, promoting the use of the pill and IUDs and the sterilization of men and women. That same year, after Prime Minister Indira Gandhi declared a state of emergency, family-planning officials began a program to promote male sterilization. Incentive payments for sterilization were raised to 150 rupees for those with two children and 100 for those with three. Men were pressured to undergo sterilization in order to secure their jobs, and governmental workers were under heavy pressure to meet sterilization quotas. State and local authorities were also rewarded for promoting vasectomies. Campaigns to clear Delhi of slums, which were viewed as a product of the population boom, were tied to sterilization efforts, as slum dwellers with three or more children were required to produce a certificate of sterilization in order to be eligible for new housing. In some provinces, forced sterilization was undertaken, and sterilization camps popped up across the country.[2]

Closer to home, the use of sterilization in Puerto Rico began in the 1930s and continued into the 1960s. According to sociologist Harriet Presser, 34 percent of mothers aged 20 to 49 had been sterilized by 1965. This was by far the largest percentage of women sterilized in any country.[3] Sterilizations, known locally as *la operación*, were often performed on women who were giving birth for the third or fourth time. It is unclear to what degree these operations involved coercion. Sterilization was clearly promoted by health authorities and was part of an island-wide population-control program. Yet many women may have sought out the operation to control their reproductive lives. Laura Briggs suggests that the operations had been so common for so long that they had become an accepted part of the culture of the island.[4] Sociologist Iris Lopez, on the other hand, claimed that adverse social and economic conditions compelled many women to accept sterilization. She also noted that misinformation influenced their decisions. Many wrongly thought the operation could be reversed and later regretted their decision to be sterilized.[5]

The most dramatic example of the use of incentives and coercive measures to stem population growth in a developing country occurred in China. Before the 1960s, China had followed a pronatalist policy that encouraged large families. By 1970, however, in the wake of the economic disruption caused by the Cultural Revolution, Prime Minister Mao Zedong and the Politburo concluded that population control was essential to the country's economic growth. The production and distribution of contraceptives was increased, and information promoting birth control was widely disseminated. The government also urged couples to marry later and to space out their children, allowing four to five years between their first and second child. These measures resulted in a reported 50 percent decline in fertility. Following Mao's death and the ascendency of Deng Xiaoping, China accelerated its efforts to curtail population growth, launching its famous one-child policy in 1979. The policy was implemented in alternative waves of coercive measures, incentives, and disincentives. Compulsory mass meetings were held to indoctrinate parents. Pregnant women were made to participate in repeated group discussions and, at the height of the campaign, forced to undergo abortions and sterilizations. The number of abortions in China reached 5 million a year in the early 1970s, and more than 120 million between 1980 and 1990. IUD insertions and tubal ligations also numbered in the millions. Vasectomies, however, played a minor role, as China's one-child policy was directed toward women.[6]

International funders, while not condoning such measures, did not intervene to stop them. Moreover, several of the programs were funded by either bilateral US assistance or international organizations. In India, the UNFPA and the Ford Foundation continued to finance family-planning programs during that country's state of emergency. The UNFPA and the World Bank funded Indonesia's program, and the UNFPA and the IPPF provided support for China's emerging population programs in the 1970s.

USAID supported the Pathfinder Fund's sterilization programs in Latin America in the 1970s and funded a program to train thousand of physicians from more than 70 countries in a new technique—called minilaparotomy, or minilap—for female surgical sterilizations (tubal ligations). The technique was developed by two gynecologists at the Johns Hopkins School of Medicine. It allowed surgeons to perform sterilizations without hospitalization. Using this technique, a physician could perform up to 90 operations a day. The Johns Hopkins physicians formed the Johns Hopkins Program for International Education in Gynecology and Obstetrics (Jhpiego), which has become a major global-health NGO in the fields of maternal care and reproductive health, as well as in combating AIDS and malaria. Jhpiego not only promoted the use of the surgical technique but also, with the support of USAID, provided some 5000 laparoscopes to programs and medical schools around the globe; 90 percent of medical schools worldwide were recipients of this equipment.[7] USAID financed the use of this technique through a grant to the Association for Voluntary Sterilization. According to historian Matthew Connelly, the association's subgrantees were supposed to recognize national laws regarding sterilization. They were told, however, that "in those localities where sterilization may not be 'acceptable' they may be supplied under the euphemism 'to detect and treat abdominal disease.' "[8]

Wavering Support

Lack of evidence that family-planning programs were reducing population growth led a number of major supporters to rethink the supply-side strategies that had dominated family-planning programs in most developing countries up to the early 1970s. Kingsley Davis, one of the early advocates of demographic transition theory and a major promoter of population-control policies, published a blistering critique of family-planning programs in *Science* in 1967.[9] In it he questioned the basic assumptions guiding existing programs. He argued that family planners failed to ask why women desired so many children. They instead assumed, on the basis of demographic surveys, that there was a desire on the part of women to limit family size and that the problem was lack of access to the means to do so. The problem with this assumption was that a desire for birth control technologies was compatible with *high* fertility. The goal of family-planning programs was to bring women around to accepting a limitation of the number of children they wanted, but not to accepting a *specific* number that would actually reduce population growth. Programs failed to set such goals because free choice was a basic premise of family-planning programs.

While family planners assumed that the individual decisions of women to reduce the number of children they produced would collectively lead to population declines, there was a gap between the number of children women desired and the number that would actually lead to population reduction. In country after country, Davis noted, surveys showed that while women wanted access to birth control, they also wanted three to four or more children, and that many of those seeking birth control were at the end of their reproductive lives, having already produced several children. Family planning might reduce the birth rate, but as a voluntary program, it would not substantially curtail overall population growth as long as forces existed that encouraged women to give birth to many children. The conditions that caused births to be wanted or unwanted were beyond the control of family planners and, hence, beyond the control of any nation that relied on family-planning alone as its population policy. Davis argued instead for the societal regulation of reproduction through a combination of incentives and rules that would discourage men and women from marrying early and having large families. They would, in effect, counter the social forces that encouraged large families. For example, in developed countries, providing economic incentives in the form of tax breaks for people who remained single and eliminating tax credits for having or caring for children would limit population growth.

John D. Rockefeller III, another major supporter of family-planning programs and the founder of the Population Council, was also becoming disillusioned by the lack of progress. He hired Joan Dunlop, who had worked for the Ford Foundation and for New York City's Mayor John V. Lindsey on urban policy issues. Rockefeller told her that he felt something was wrong with family planning and asked her to attend family-planning meetings and report back on what she heard. What she heard, and saw, was that the family-planning movement was being run almost exclusively by men, with little to no input from women, despite the fact that nearly all family-planning programs focused their activities on women. Moreover, it was not just men who headed the programs, but *white* men, a number of whom had gotten their start serving as colonial administrators.[10] Ernest Gruening, the US Senate's most active promoter of family planning, was a former director of the US Division of Territories and Island Possessions. The first secretary-general of the International Planned Parenthood Federation was a 30-year veteran of the British colonial service. Many of the leaders of the family-planning movement had little understanding of or empathy with the concerns of women of color, on whom the success of family-planning programs depended. As Kingsley Davis had argued, these men viewed the problem simply as a matter of technological innovation and distribution: making the right technology accessible to the target populations. They had little interest in understanding why women wanted children, or in the complex social and economic forces that shaped these decisions.

Concerns about family-planning programs came to a head at the 1974 World Population Conference in Bucharest. The United States hoped that the conference would endorse population-reduction targets for both developed and developing countries. But no agreement could be reached on this issue. Worse, the conference raised serious questions about the direction of family-planning programs and their importance for development. To begin with, the conference clearly revealed the absence of women in the leadership of the family-planning movement: 80 percent of the official representatives were males, and men headed 127 of the 130 delegations. In addition, the conference was asked to ratify a World Population Plan of Action that contained only a single paragraph on women. This lack of attention to women led to protests by feminist leaders Betty Friedan, Germaine Greer, and Margaret Mead, all of whom attended the meeting. The role of women in controlling their own fertility played a central role in the pro-choice movement in the United States and was an important principle of the women's movement more generally during the 1970s.

John D. Rockefeller III noted the neglect of women in family planning in his speech to the conference. He also shocked the family-planning movement by saying that he had changed his mind about the conduct of family-planning programs and had come to Bucharest to call for a deep, probing reappraisal of the population-control movement. Family planning, he argued, could no longer be just about fertility reduction, but had to be made part of a broader development program aimed at meeting basic human needs. Such a program, moreover, had to give urgent attention to the role of women and recognize that women themselves should determine what their role should be.[11]

In citing "basic needs," Rockefeller was channeling language that had become a central element of the World Bank's revised development agenda under the bank's new president, Robert McNamara (chapter 12). Rockefeller's call for linking family planning to broader development efforts not only replayed the arguments of early transition theorists, but resonated with the concerns of the leaders of a number of developing nations, who came to the conference demanding a refocusing of attention from population control to the wider development needs of Third World countries. For these leaders, narrowly designed family-planning programs were ineffective, as well as a distraction that was preventing serious efforts to be undertaken to address the economic problems of the developing world. Focusing on population growth as a factor in underdevelopment, moreover, allowed developed nations to ignore its real causes, which lay in the prevailing economic order. Part VI looks more closely at these arguments and the call for a new international economic order, as they also played an important role in the rise of the primary health-care movement.

These arguments would also become part of a growing structural critique of population-control programs and the assumptions underlying them. Soon after the Bucharest Conference, and in the face of serious drought and famine in West Africa, which many linked to overpopulation and resource depletion, political scientist Susan George published her scathing critique of the world food and population crises, titled *How the Other Half Dies*. In it, she claimed that arguments about the impact of rapid population growth in developing countries on the world's food supply and other natural resources were seriously flawed, as they conveniently ignored that it was the developed countries that consumed the lion's share of the earth's natural resources at an ever-growing rate, an argument the Soviet Union had made for some time. According to the World Bank, the United States, with 6 percent of the world's population at that time, consumed 35 percent of the world's natural resources. One billion people, living in countries with a per capital income of US$200, used 1 percent as much energy as people living in the United States. If population-control programs were needed, should they not be aimed at the populations consuming the most resources? Second, she argued that the threat of population growth seriously affecting the world's food supply ignored the fact that so much of the available agricultural land was being employed inefficiently, to raise cattle and other livestock to meet the dietary preferences of the rich. If the same land were used to grow grains, there would be no food crisis.[12]

Similar lines of argument appeared in *Seeds of Famine*, anthropologist Richard Franke and sociologist Barbara Chasin's 1980 analysis of the causes of the Sahelian famine, and geographer Michael Watts's monumental analysis of the structural causes of malnutrition and famine in northern Nigeria, *Silent Violence*.[13] It was also the focus of political scientist Timothy Mitchell's insightful analysis of how development agencies in the global north promoted population-control programs in Egypt by framing that country's underdevelopment in terms of its geography and dependence on the floodwaters of the Nile. According to Mitchell, development documents from USAID and the World Bank claimed that Egypt suffered from overpopulation and malnutrition because its people were crowded into small areas of land bordering the river. These analyses ignored massive inequalities in land ownership, as well as the use of large areas of farmland to support the production of beef to meet the food tastes of Egyptian elites and tourists.[14]

The 1974 World Population Conference was, in many ways, the turning point in the history of family-planning programs. Shortly afterward, with the new US presidential administration under Jimmy Carter in 1977, USAID's population programs were decentralized, giving much more control over planning to the regional offices and country missions. The revised arrangement was intended to ensure that local issues would be addressed in the design and implementation of programs. The change was also driven by the concerns of USAID's new leadership about the population program's heavy reliance on supply-side approaches to family planning and its failure to address the determinants of fertility. These points had been expressed as early as 1970 but had been systematically rejected by Ravenholt, who also fought the decentralization proposals. Ravenholt was dismissed from his position as director of the Population Office in 1979.

A similar rethinking of family planning after the Bucharest Conference occurred in the Ford Foundation, where a 1977 strategy paper asked, "To what extent is it appropriate for us as a Western foundation to support activities that profoundly affect traditional mores and value structures?" At the Rockefeller Foundation, concerns were raised about the Population Council's activities. A number of staffers, led by demographer Paul Demney, were critical of population control, particularly the sterilization programs that the Council had supported, and felt that the council should walk away from the whole business. Between 1974 and 1977, the number of Population Council staff fell from 275 to 174.[15]

At the World Bank, McNamara continued to support family-planning programs but, at the same time, argued that expanding other activities—such as providing increased educational opportunities for women, reducing infant mortality, and improving nutrition—would also reduce fertility rates. It was not just about providing contraceptives. A number of developing countries, including India, Pakistan, the Philippines, Nicaragua, and Iran, also were backing away from family planning, at least in its most aggressive forms. In 1979, efforts by UN officials to solicit information about family-planning activities resulted in more than half of the countries queried not responding.

More importantly, the administrations of Presidents Jimmy Carter and Ronald Reagan were influenced by the right-to-life movement in the United States and chose to ratchet down support for family-planning initiatives, cutting off any organization that advocated or supported abortion. The US delegation announced its new policy at the 1984 Population Conference in

Mexico City.[16] It opened with the following assertion: "Population growth is, of itself, a neutral phenomenon. It is not necessarily good or ill. It becomes an asset or a problem only in conjunction with other factors, such as economic policy, social constraints, need for manpower, and so forth. The relationship between population growth and economic development is not necessarily a negative one. More people do not necessarily mean less growth. Indeed, in the economic history of many nations, population growth has been an essential element in economic progress."[17] With these words, the United States rejected a set of assumptions that has supported family-planning programs from the 1950s on, eliminating the primary economic justification for population control.

The statement went on to assert that there had been an overreaction to the rapid population growth that had followed reductions in mortality. The population boom was a challenge; it need not have been a crisis. Seen in its broader context, it required a measured, modulated response. It provoked an overreaction by some, largely because it coincided with factors that, together, hindered families and nations in adapting to their changing circumstances. The most critical of these factors, according to the US delegation, was governmental regulation of the economy, which had stifled economic growth and prevented the kinds of material improvements that would have led families to want fewer children: "Too many governments pursued population control measures without sound economic policies that create the rise in living standards historically associated with decline in fertility rates." Thus, in an odd evocation of demographic transition theory, the US delegation promoted market-driven economic policies and a reduction in the role of governments, not population control, as the key to unlocking Third World development. In terms of specific policies, the delegation's statement asserted:

> The United States will no longer contribute to separate nongovernmental organizations which perform or actively promote abortion as a method of family planning in other nations. With regard to the United Nations Fund for Population Activities (UNFPA), the U.S. will insist that no part of its contribution be used for abortion. The U.S. will also call for concrete assurances that the UNFPA is not engaged in, or does not provide funding for, abortion or coercive family planning programs; if such assurances are not forthcoming, the U.S. will redirect the amount of its contribution to other, non-UNFPA, family planning programs."

Yet family-planning programs would continue to operate around the globe. The UNFPA and the World Bank still provided resources to support programs that encouraged family planning, often integrated into broader reproductive-health and maternal and child health programs. The United States reentered the family-planning field during President Bill Clinton's years in office, only to reduce its commitments again under George W. Bush's presidency. The Cairo Conference in 1994 reaffirmed the importance of reproductive health and established the right of women to control their reproductive lives, while rejecting all forms of coercion. With this shift, population programs in many schools of public health morphed into maternal and child health programs. The Bill & Melinda Gates Foundation became a major funder of these programs (part VII).

NGOs continued to play a key role in supporting reproductive-health programs. In the 1980s and 1990s, social scientists began to produce more finely detailed studies of reproductive decision making and the social, economic, and cultural conditions that shaped fertility decisions in various parts of the globe.[18] Yet, to a remarkably large degree, family-planning programs remained separated from broader health efforts and from social and economic development. While there were moments when various actors invoked the language of demographic transition theory and the role of social and economic transformation in shaping reproductive behavior, they were fleeting and, as seen in the United States' Mexico City declaration, self serving.

Worldwide fertility declined by over 50 percent from the 1960s to the 1980s. The global rate of population increase peaked at 2.2 percent in 1963 and declined to 1.1 percent by 2012. Most of the decline occurred after 1970.[19] The causes of this decline are debated. Family-planning programs no doubt contributed, as did the coercive policies and forced sterilizations in some countries, particularly China. But fertility declines in Latin American countries, where less-intensive campaigns occurred, were just as great as in countries in Asia, where programs were most intensive.[20] There is evidence, moreover, that social and economic development, and, most importantly, the education of women, played a critical role in changing reproductive decision making and contributed to the decline in populations that occurred in many parts of Asia and Latin America.[21] Conversely, the absence of economic advancement and education limited population decline in the poorest countries, especially in sub-Saharan Africa. At the end of the day, the transition theorists may have had it right after all. Transforming reproductive behavior requires broader social and economic changes.

Like the disease-eradication programs (part IV), family-planning programs during the 1960s and 1970s were a product of the culture of technical assistance that dominated efforts by development planners in the global north to transform the lives of people living in the global south. They were directed from above and delivered with little effort to promote community participation in their design and application. They also ignored the underlying social and economic determinants of reproductive decision making. At the same time, they drew attention and funding away from the need to build up basic health services. Most family-planning programs operated as vertical entities, with little connection to governmental health services. This approach to international health would be challenged once again, at the end of the 1970s, with the rise of the primary health-care movement, but it would not be abandoned. In fact, it would grow in importance with the rise of the concept of global health at the beginning of the twenty-first century.

PART VI

The Rise and Fall of Primary Health Care

In September 1978, representatives to the World Health Organization met in the remote city of Alma-Ata, located in a mountainous region of what is now Kazakhstan. The purpose of the meeting was to promote the development of primary health care as an essential building block for health around the globe. The meeting, originally proposed by representatives of the Soviet Union in 1974, took place after a long set of discussions, the preparation of planning documents, and a series of preparatory meetings. Three thousand delegates from 134 governments and 67 international organizations from all over the world attended the conference, along with a number of NGOs. It lasted seven days, at the end of which consensus was reached on a document that would become known as the Alma-Ata Declaration.

The document put forth a set of principles that were intended to guide member nations in restructuring their health systems around the provision of primary health care for all of their citizens. In doing so, it reaffirmed WHO's original commitment to a broad definition of health "as a state of complete physical, mental, and social well-being, and not merely the absence of disease or infirmity."[1] It also stated that health was a basic human right, and that national governments had an obligation to ensure the health of all their citizens. While noting that health was essential to economic advancement, it stressed that social and economic development were crucial in the attainment of health. The declaration also asserted that primary health-care services should meet the needs of local communities and be designed and built with community participation. Finally, the declaration reaffirmed the goal defined by WHO in the previous year: achieving "Health for All by the Year 2000."

The Alma-Ata Declaration was, in many ways, a revolutionary document that set forth a new pathway to health. In its call for broad-based, integrated approaches to health, its commitment to equity and community participation, and its acknowledgment of the need to address the underlying social and economic determinants of health, it represented a radical departure from the culture of technical assistance—with its reliance on

WHO Director-General Halfdan Mahler (*right*) and US Senator Ted Kennedy (*left*) at the International Conference on Primary Health Care, Alma-Ata, Soviet Union, September 6–12, 1978. WHO.

top-down programs based on technological solutions and curative medical care provided by physicians—that had dominated international-health efforts since the 1950s. It also implied a major shift in the allocation of health and financial resources toward peripheral populations and away from large urban medical centers.

Yet placed within the longer history of international health, the Alma-Ata Declaration represented a return to the earlier vision of health and development that had characterized the work of the League of Nations Health Organization and the Rockefeller Foundation's China program during the 1930s. The same vision had framed the original goals of the health and development organizations that emerged immediately after World War II. As others have noted, the text of the Alma-Ata Declaration resonated with the principles articulated at the Intergovernmental Conference of Far Eastern Countries on Rural Hygiene, held in Bandoeng, Java, in 1937.

So why, in the late 1970s, did WHO and other international organizations present at Alma-Ata change course and veer back toward this earlier vision of health? One can point to the role of Halfdan Mahler, who became director-general of WHO in 1973, in shaping the agenda of WHO, or to the rising demand of developing countries for better health services, or to the lobbying efforts of the Soviet Union. All played a role. But to really understand what happened at Alma-Ata, one needs to look at the bigger picture. The Alma-Ata Declaration was the result of a fundamental rethinking of how international development, including efforts to improve the health of the world's populations, had been organized since the end of World War II. This rethinking, in turn, resulted from the convergence of a constellation of events, personalities, and movements, beginning in the late 1960s.

While the goals of primary health care (PHC) remained a gold standard for international, or global, health for years to come, the commitment to putting these goals into practice was remarkably short lived. Within less than five years, the pendulum shifted back again toward a set of approaches that contained certain elements of the Alma-Ata vision and much of the rhetoric of PHC but, in their design and execution, represented a return to the culture of technical assistance. The demise of primary health care resulted in part from its own failings. Yet the PHC movement was also a casualty of broader political and economic shifts. Specifically, the worldwide recession of the 1980s, combined with the rise of neoliberal economic policies and the subsequent emergence of new organizational actors and approaches to international development, turned the tide of international health back toward more-narrowly constructed, technological solutions to health problems.

Chapter 12 describes discussions among international-health leaders in the late 1960s and early 1970s regarding the need to focus health interventions on building basic health services, rather than on large-scale disease-elimination campaigns. It then examines the broader context in which these discussions occurred, arguing that the reassessment of international-health priorities was part of a broader critique of development. The chapter ends by tracing the negotiations and discussions that led up to the Alma-Ata conference on primary health care in 1978.

Chapter 13 examines the subsequent fate of the PHC movement. It identifies the obstacles that limited efforts to build primary health-care systems and the broader economic and political forces that contributed to a shift away from primary health care and toward more-selective approaches to health care.

CHAPTER TWELVE

Rethinking Health 2.0

The Rise of Primary Health Care

THE PRIMARY HEALTH-CARE MOVEMENT did not emerge overnight. It was the result of a long process of intellectual and ideological change, both within and outside of WHO, that began a decade before Alma-Ata. If we want to understand why this movement arose in the late 1970s we must begin in 1969, with what many people saw as WHO's greatest failure, the global Malaria Eradication Programme (MEP).

The MEP was launched in 1955, with the hope of eliminating malaria from the globe (part IV). The MEP was approved, despite the opposition of many within the World Health Assembly who felt that WHO should be concentrating its energies and resources toward building up health-care services around the world, not in organizing single-disease campaigns. As the MEP began to falter in the mid-1960s, and it became increasingly apparent that the program was unlikely to achieve its goals in many of the countries in which it had begun operations, these same opponents called for a reappraisal of the MEP and a redirection of WHO activities toward the development of basic health services.

The Soviet Union, which had not been a member of the WHA when the MEP proposal was passed, led the effort to shift WHO's priorities. In 1969, its representative called for a reassessment of the eradication strategy. This reassessment acknowledged the limitations of eradication and urged programs that had not achieved eradication to return to control strategies, with the ultimate goal of eradication at some distant time. The reassessment also requested that malaria-control programs be integrated into basic health services. Yet in many places it was clear that basic health services did not exist or were insufficiently robust to support the inclusion of malaria control. The integration proposal highlighted what many of those who were involved in malaria-eradication efforts had come to realize: eradication had faltered in many countries because the basic health services were underdeveloped and unable to manage the surveillance and treatment activities that the eradication strategy required. Disease control required basic health services.[1]

WHO had not ignored health-care strengthening. Despite the dominance of disease-eradication campaigns during the 1950s and 1960s, considerable effort had gone into creating plans for improving health services in developing countries. While disease-control and family-planning campaigns attracted the lion's share of funding, advocates of health-services development had been active, holding conferences and meetings designed to create a consensus around how to improve the delivery of health care in the developing world.

Primary health care, or the idea that health-services development should focus on providing essential health care at the village or neighborhood level, was not a new idea. WHO's Expert Committee for Public Health Administration had set out a model for basic health services in 1953. It contained many of the elements that would be part of the later PHC model, including maternal and child care; communicable-disease control; environmental sanitation; maintenance of health records; health education; public-health nursing and medical care; and, where needed, access to larger hospitals.[2]

Yet up through the 1960s, there had been little progress in implementing this model. Neither the governments of developing countries nor WHO and its partner organizations were able or willing to invest in building basic health services. Moreover, within the cold war environment, which shaped much of development thinking during the 1950s and 1960s, building clinics was much less appealing to bilateral donors than large-scale disease-eradication campaigns. Clinics might do more to improve the health of populations over the long run, but they did not have the dramatic impact of disease-eradication programs.

To the extent that bilateral donors were investing in health-care services during this period, they were directing this support primarily to building hospitals—monuments to technological advances in medicine. Hospitals were also attractive to the leaders of developing countries, because they were symbols of modernity and met the needs of their country's elites. Even the Soviet Union and East Germany, which had invested heavily in building primary health-care systems in their own countries, supported the construction of large, modern medical centers in Africa as reflections of the scientific achievements of socialism. For donor countries, constructing hospitals in developing countries also created markets for medical equipment and the products of pharmaceutical corporations. It could be argued that

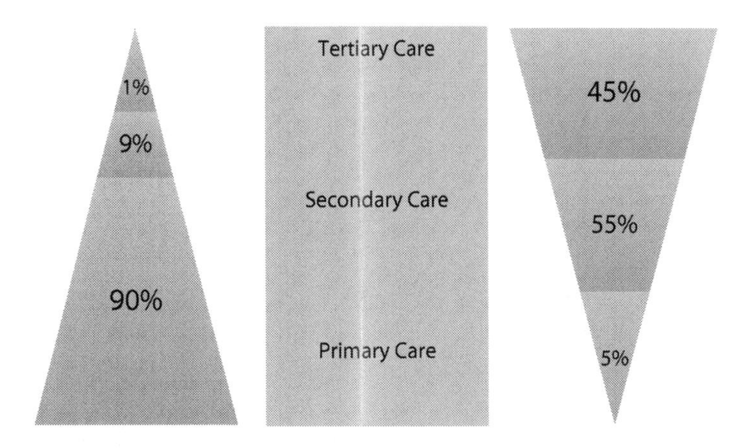

The health dilemma in Ghana: hospital-centered health systems. Ministry of Health, Republic of Ghana, *A Primary Health Care Strategy for Ghana* (Accra, Ghana: National Health Planning Unit, 1978).

these financial interests discouraged donor governments from investing in primary health care when there was more to be made from selling medicines and medical equipment.

Funneling limited development dollars into building tertiary-care centers produced an inverse relationship between the amount of money spent on different kinds of health care and the numbers of people served by each level. This relationship was captured in the inverted-pyramid model produced by Richard Morrow, a professor at Johns Hopkins, to illustrate health inequalities in Ghana in 1970. A working group established by WHO's Executive Board in 1972 reported that the basic health-services approach had failed to achieve its goals. Neither countries nor international donors had invested in this effort to reorient health services to support efforts to build effective health services.

In 1973, Halfdan Mahler became the third director-general of WHO. Mahler had joined the organization in 1951 and spent 10 years as a senior medical officer attached to the National Tuberculosis Programme of India. Mahler's field experience in India had convinced him that basic health services were critical for the elimination of diseases like TB. More importantly, the experience had made him realize that health services had to serve the needs of local communities and be established with community participation. He was critical of top-down planning and the imposition of health-care models from above. He also felt that the medical establishment was a barrier to the development of effective primary health care. He is reported to have told the leader of a developing nation, who asked what he could do to improve health care in his country: "I think the first step is to close the medical schools for two years. Then we can discuss what the medical schools were supposed to do, because they really constitute the main focus of resistance to change."[3]

Mahler believed that the absence of community participation was a major stumbling block to the creation of effective basic health services, and he came to the director-general's office committed to developing a new participatory model for the provision of such services. In his first annual report on the work of WHO during 1973, Mahler stated that "the most signal failure of WHO as well as of Member States has undoubtedly been their inability to promote the development of basic health services and to improve their coverage and utilization."[4] He called for a recommitment on the part of member states to create cost-effective, scientifically grounded, community-based health systems. At the same time, he acknowledged that there were few successful models for countries to emulate and encouraged member states to design systems that could provide them.

Over the next two years, a series of reports and meetings called for the creation of a new model of basic health services that would be integrated into the life of local communities, maximize local resources, and involve community participation. These health services would be provided by new kinds of health workers, who were trained specifically to meet the needs of local communities. There was also agreement that primary health services should be integrated with systems of secondary and tertiary care, as well as with other sectors involved in community development. Repeated resolutions called on WHO to work with member nations to build such systems, and in 1974, the Soviet Union requested an international conference to explore ways of implementing new models of primary health care. It would be another four years before the meeting was held, and there would be many disagreements about the ideal form of this new model of health services, but by 1975, promoting primary health care had become a major part of WHO's efforts to improve the health of the world's populations.

Critiquing Development

Before examining the final push to Alma-Ata and the subsequent history of the primary health-care movement, it is important to step back and examine the wider context in which this shift in thinking within WHO and other international organizations occurred. The change was part of a broader critique of existing development strategies that arose in the early 1970s from both the centers of power in the United States and Europe and from developing nations.

In organizations such as the World Bank, the IMF, and development agencies within European and American governments in the 1950s and 1960s, economic development was understood in terms of raising the GNP of countries. In 1961, the United Nations, following the lead of US President John F. Kennedy, declared the 1960s to be a Development Decade and set targets for developing countries to increase their GNP by 5 percent by 1970. Strategies for achieving development goals focused on increasing capital accumulation and foreign-exchange earnings and building large infrastructure projects, such as dams and power plants. These projects were designed to fuel industrial growth, which was viewed as the driving engine of development, as laid out in Walter Rostow's highly influential *The Stages of Economic Growth*.[5] Efforts were also made to spur the export of commodities through the creation of large-scale farming projects, and the World Bank provided millions of dollars in concessionary loans to fund them.

By the 1970s, many countries had achieved the goal of a 5 percent increase in their GNP. Yet there was widespread concern about the results. The ILO director-general, in his 1970 report, *Poverty and Minimum Living Standards: The Role of the ILO*, captured this discontent, stating: "The reason for my concern is basically that the immense—and in global economic terms, not altogether unsuccessful—efforts for development during the past two decades have not so far resulted in many perceptible improvements in the living standards of the majority of the world's population."[6] The development decade had, in effect, produced growth without development. GNP had risen, but poverty persisted. Dams, highways, hospitals, and factories were being built, but the basic needs of the people were not being met.

The failures of previous policies were thrown into stark relief by the global recession that began in the early 1970s. This recession was marked by high unemployment and inflation, caused by a sharp spike in oil prices, which began in 1973 with the decision of a number of major oil-producing countries to cut down production. The recession significantly curtailed a long period of economic growth that had begun following World War II and had fueled the growth of GNP in countries across the globe. Export revenues and terms of trade for developing countries plunged. Like all recessions, this one took its greatest toll on those who already were impoverished. Finally, the massive famines that hit the Sahel region of West Africa in the early 1970s highlighted the fragility of subsistence for millions of Africans.

The ILO conducted a series of country studies during the early 1970s and presented its results both at the World Employment Conference and in a 1976 ILO report titled *Employment Growth and Basic Needs: A One-World Problem*. The report highlighted the failure of current development strategies to reduce poverty around the globe and proposed that national development strategies be redirected, with an explicit goal of meeting the basic needs of entire populations. "Basic needs" was defined as ensuring that every family possessed sufficient income for the purchase of food, shelter, clothing, and other essential requirements, together with the provision of vital services to ensure basic education, health, safe drinking water, and sanitation. Basic needs also required "the participation of people in making decisions that affect them." In short, the definition of basic needs mirrored the emerging vision of primary health care within WHO.[7]

Since its inception in the 1930s, the ILO had been an advocate for policies that addressed the economic and social needs of the working poor. It led the way in highlighting the impact of the Depression on working families (chapter 4). But the ILO was not alone in calling for a new approach to development, focused on basic needs. Other international actors joined in, the most important being the World Bank. The World Bank had been a major funder of growth-led strategies during the Development Decade. Beginning in 1973, however, it began to rethink its policies. Leading this reappraisal was the bank's new president, Robert S. McNamara.

McNamara, a graduate of the Harvard Business School, had worked his way up in the Ford Motor Company, becoming its president, before being recruited by President John F. Kennedy to serve as his secretary of defense. McNamara continued in that position under President Lyndon Johnson and was charged with overseeing America's growing military involvement in Vietnam. The Vietnam War convinced McNamara that one of the greatest threats to the well-being of United States and other developed nations was the poverty of populations living in the developing world. In a speech delivered just before leaving the Defense Department to become president of the World Bank, McNamara observed, "I am increasingly hopeful that at some point soon . . . the wealthy and secure nations of the world will realize that they cannot possibly remain either wealthy or secure if they continue to close their eyes to the pestilence of poverty that covers the whole southern half of the globe."[8] Wars of insurgency were the result of the frustrations born of poverty.

Under McNamara's leadership, the World Bank shifted gears, moving from supporting large-scale infrastructure projects designed to boost GNP to poverty-alleviation programs that focused on providing for the basic needs of populations. In his first speech to the bank's Board of Directors, he put poverty on the table and spoke not only about GNP, but also about directing the bank's lending toward improving the living conditions of poor people. Most fundamentally, unlike his predecessors, he spoke about the poor as individual human beings, rather than as countries. He called for increases in support for all levels of education and basic nutritional requirements for each person's health and well-being. In subsequent speeches, McNamara stated that he would require borrowing countries to enact social and economic policies that would permit a more equitable distribution of wealth.

To achieve these goals, McNamara doubled the World Bank's lending and made sure that much of it was funneled through the International Development Administration to programs that would benefit the poorest countries. This required the bank to borrow more money, which made many on the governing board and the bank's general staff uncomfortable, especially in the context of the economic recession and rising inflation. McNamara was able to raise the needed capital by borrowing millions from members of the Organization of the Petroleum Exporting Countries (OPEC), which were awash in dollars earned from the sale of oil. McNamara also reorganized how the World Bank operated, making it more hierarchical, which allowed him to control bank activities more closely. In addition, he strove to change the bank's working culture, which had been characterized by what one staffer described as a "leisurely perfectionism." The bank's staff had been focusing on a relatively small number of large-scale projects, which they would follow in minute detail, and take months to make decisions. This style was not compatible with McNamara's desire to fund many smaller projects and to get projects moving rapidly. Under McNamara, staffers were required to make decisions in short timeframes with less information—something that made them very nervous.

While McNamara was committed to poverty-alleviation strategies, he was uncertain about which strategies would be most effective. He and his staff considered programs focused on slowing population growth that McNamara, along with many others, viewed as a drag on economic development. They also looked into programs aimed at improving education, health, and nutrition. In the end, they concluded that these areas deserved support but did not fit well with the bank's expertise and were being addressed by other UN organizations. The bank therefore chose to focus most of its resources on integrated rural-development schemes and aid to small subsistence farmers. Under McNamara, the World Bank's lending policies, like WHO's approach to health, shifted from large-scale projects to programs focused on grassroots development and individual needs.

This reorientation of international development efforts also occurred in the United Kingdom. In 1975, the newly elected Labour Government repurposed British overseas-aid efforts, calling for a basic-needs approach that directed future foreign assistance toward the world's poorest populations. In a white paper titled *The Changing Emphasis of British Aid Policies: More Help for the Poorest*, the government laid out new principles for aid, which included giving increased emphasis to bilateral aid to the poorest countries, especially those most seriously affected by the rise in the price of oil and other commodities, and emphasizing programs oriented toward the poorest groups within these countries, especially rural development.[9]

The Birth of the New International Economic Order

The critique of international development policies was by no means limited to policy makers centered in the global north. An even-stronger attack emanated from developing countries in the global south and, in the early 1970s, led to a call for the creation of a new international economic order. One of the earliest centers for critiques of development in the global south emerged in Latin America. In 1948, the United Nations set up the Economic Commission for Latin America (ECLA) in Santiago, Chile, under the leadership of Argentine economist Raúl Prebisch. Prebisch was charged with preparing a report, presented at the ECLA conference in Havana in 1949, on the economic situation in Latin America. The report was a critique of the international division of labor, in which industrial nations in Europe and America were producers of manufactured goods exported to the world, and developing countries in Latin America were limited to producing agricultural commodities and raw materials while importing industrial goods from Europe and the United States.

Prebisch argued that this system, and reduced terms of trade, where the purchasing power of Latin American exports was declining, were the major causes of underdevelopment in Latin America. His solution was a strategy of import substitution industrialization, or ISI, by which Latin American countries would begin building industries that would produce goods currently being imported from Europe and the United States. Only through industrialization would Latin American countries obtain their share of the benefits of technological progress.[10] ECLA's prescription for change was greeted enthusiastically by a number of countries in Latin American and the Caribbean.[11]

The ECLA declaration influenced academics and policy makers in other parts of the developing world and became associated with what became known as structuralist economic thought during the 1960s and 1970s. Structuralism—also known as dependency theory, or underdevelopment theory—viewed poverty in the developing world as a product of the structural relationship that existed between the industrial economies of Europe and the United States and the largely commodity-producing economies of the developing world. One of the most influential promoters of these ideas was Jamaican historian Walter Rodney, whose 1972 book, *How Europe Underdeveloped Africa*, described the long history of European policies that had led to Africa's underdevelopment.[12]

In Africa, the adoption of dependency theory led to several efforts at ISI, most spectacularly in Ghana under President Kwame Nkrumah. Nkrumah attempted to rapidly industrialize Ghana, in order to reduce its dependence on foreign capital, technology, and manufactured imports. He viewed industrialization as necessary for the country to achieve true independence. He invested heavily in industrial infrastructure, including the massive Volta Dam project, which was supposed to provide electricity for much of southern Ghana and support industrial growth. He also built hotels, airports, and convention centers. While industries grew, the country's foreign debt rose faster, and by 1968, the country had amassed US$1 billion in debt. To pay off the debt, Nkrumah raised taxes on the country's cocoa farmers. Dissatisfaction over the tax increase contributed to Nkrumah's overthrow in 1968.[13]

The critique of development, which began in Latin America, led the leaders of a number of developing countries to call for the creation of a new international economic order in the early 1970s. Emboldened in part by the successful oil embargo initiated by OPEC, other countries dependent on commodity exports sought to transform the economic relationship between developed and developing countries by changing a wide range of trade, commodity, and debt-related policies that were viewed as detrimental to the economic growth of poorer, commodity-producing countries. Among the changes called for were lowering of import tariffs in developed countries and improving the terms of trade that currently favored industrial nations. They also sought trade mechanisms that would ensure stable and ruminative prices for raw materials and reduce the cyclical fluctuations in commodity prices that were detrimental to the economic interests of commodity-producing nations.

These proposals were put forth in 1964 at a meeting of the UN Conference on Trade and Development. The elected secretary-general of the meeting was Raúl Prebisch.[14] The proposals gained purchase among the leaders of a large number of developing countries and were imbedded in the Declaration for the Establishment of a New International Economic Order, adopted by the United Nations General Assembly in 1974. There was thus a broad-based, worldwide consensus that efforts at global economic development had to be transformed, even if there was disagreement over the blueprints for this change.

Critiquing Health Care

International development policies and structures were not the only areas subject to a critical rethinking during this period. The wider field of health and health care, not just international-health policies, was the target of a set of strong critiques. The early 1970s was a time when modern biomedicine was placed under growing scrutiny, and earlier assumptions about the role of physicians as responsible for the health needs of populations were questioned. In the early 1970s, biologist and public-health worker David Werner was working as an advisor and facilitator for Project Piaxtla, a village-run program dedicated to assisting disabled children in the remote town of San Ignacio in west central Mexico. A group of health workers associated with the project had been stapling together hand-copied pages of health tips that they then distributed to local villagers who had no access to doctors. In 1973, Werner turned the packets into a book, which was initially published in Spanish and titled *Donde No Hay Doctor* (*Where There Is No Doctor*).[15] The book arguably became the most widely used public-health manual in the world, selling over 3 million copies.

At one level the book was a how-to manual. It provided useful and potentially life-saving information on how to treat a wide range of medical problems with simple, readily available interventions, without the care of a medically trained physician. It also contained guidance for village health workers on how to be effective educators as well as health-care providers. In doing so, it laid out the principles of community participation, which would become a central piece of the primary health-care movement. At the same time, it rejected top-down, elitist approaches to health education and delivery. Werner wrote, "A good teacher is not someone who puts ideas into other people's heads; he or she is someone who helps others build on their own ideas, to make new discoveries for themselves." The book was a manifesto for community participation and empowerment. Its underlying message was that you could have effective health care without trained physicians if you invested local people with the knowledge they needed to take care of their own health needs. Not surprisingly, Werner's book was not initially well received by the medical profession.

Where There Is No Doctor went beyond describing how to provide medical care, preventative measures, and health education. It also addressed the need to improve a population's economic condition, supplying information on how to increase agricultural outputs, and the need to create a balance between people and land. On this topic, Werner questioned the logic of family planning, noting that children were a financial necessity for the poor, and that in many parts of the world, the problem was not too many people for the land, but the unequal distribution of land. In all of this, Werner provided perspectives that would become embodied in the principles of the primary health-care movement, as well as in critiques of family planning.

If Werner's book suggested that people could take care of their health needs without the presence of physicians, others launched a more general criticism of the role of physicians in society. In 1975, Ivan Illich published his blistering critique of medical practice, titled *Medical Nemesis*. Illich denounced what he saw as the medicalization of life, especially birth and death. He decried both the wastefulness of devoting excessive resources to extending the end of life and the impact of such extraordinary measures on the dignity of the individual. Illich also pointed to the negative consequences of medicalization, introducing the notion of iatrogenic disease—physical harm caused by medical practices. Finally, like Werner, Illich argued that medical care and technology should be put in the hands of lay people.[16]

Illich's critique appeared at nearly the same moment as Thomas McKeown's *The Modern Rise of Population*, which questioned the role of medicine and public health in causing declines in mortality and population growth in the industrial world since the late 1700s.[17] McKeown argued that improvements in overall standards of living, especially diet and nutritional status, that resulted from better economic conditions were responsible for these demographic changes. McKeown presented his data in a series of articles in the journal *Population Studies*, beginning the 1950s. But they did not gain purchase until his book appeared in 1976. While McKeown's data and arguments were criticized by other historians and demographers, the book remained influential, for it resonated with a growing skepticism about the role of medicine and physicians in society during the early 1970s. It also supported the arguments of those who believed that efforts to improve the health of populations needed to address the social and economic determinants of health.

While the early 1970s were marked by a growing critique of established medical and public-health practices, there was also an increased awareness of alternative models of health care that focused on providing care to the poorest of the poor. In 1971, Victor Sidel, from the Department of Social Medicine at the Montefiore Hospital and Medical Center in New York City, was invited by the Chinese Medical Association to visit the People's Republic of China to observe that country's medical system and its use of village-level health workers, popularly known as barefoot doctors. At that time, little was known in the West about the Chinese health-care experiments. The Cold War, and adversarial relationship between China and the United States, had made travel to China and cultural exchanges rare occurrences. Praising anything Chinese opened one up to political scrutiny.

Sidel used the visit and his description of the selection, training, and practice of barefoot doctors as an opportunity to both inform his colleagues in the United States about what was happening with primary care in China, and to critique various aspects of the American health system. For example, he noted that "many physicians in the United States performed many tasks that could be done as competently, if not more competently, by personnel who are less broadly trained than physicians."[18] Similarly, he praised the Chinese for shortening the medical training period: "They have made a conscious effort to shorten the training time by eliminating irrelevancy and redundancy and by stressing the practical in relation to the theoretical." Such criticisms were politically risky, but Sidel was committed to publicizing what he saw as a new model for primary health care. Interest and information about the barefoot doctors increased dramatically following President Richard Nixon's trip to China in February 1972. Nixon's visit made it possible for more US academics to travel to China and talk about its experiments in primary health care and the use of barefoot doctors without fear of being labeled a Communist sympathizer. A slew of articles appeared in the early 1970s, describing China's health system and the role of barefoot doctors. Most of them praised the program.

Researchers also visited other countries in which experiments in primary health care using village health workers appeared to be having a positive impact on reducing infant mortality and controlling diseases. Cuba, Tanzania, Costa Rica, and Mozambique provided examples of countries that had been successful in driving down infant- and child-mortality rates and increasing the expected lifespan of the general population by using village health workers and minimal medical resources.

Efforts to rethink international health and to direct health resources toward the development of more-effective health services, based on the needs and participation of local communities, was thus part of a larger reassessment of international development and the wider field of public health and health care in the early 1970s. In a similar way, a second fundamental rethinking of development in the early 1980s would contribute to the demise of primary health care.

Alma-Ata and the Difficult Beginnings of Primary Health Care

By 1976, planning was well under way for an international conference on primary health care (PHC). In 1975, the director-general's report to WHO's Executive Board on "The Promotion of Health Services" declared that "primary health care services at the community level is seen as the only way in which the health services can develop rapidly and effectively." It laid out seven principles that should guide the development of primary health-care systems. These stressed the need for

1. the formation of primary health care "around the life patterns of the population";
2. involvement of the local population;
3. "maximum reliance on available community resources" while remaining within cost limitations;
4. an "integrated approach of preventive, curative, and promotive services for both the community and the individual";
5. interventions to all be undertaken "at the most peripheral practicable level of the health services by the worker most simply trained for this activity";
6. the design of other echelons of services in support of the needs at the peripheral level; and
7. PHC services to be "fully integrated with the services of the other sectors involved in community development."[19]

The report was prepared by Kenneth Newell, whom Mahler had tapped as his point man to advance his PHC agenda within WHO. It built on a study that had been commissioned by the Joint UNICEF/WHO Committee on Health Policy, *Alternative Approaches to Meeting Basic Health Needs in Developing Countries*.[20]

That report, submitted to the Executive Board in 1975, had reviewed existing models of primary health care in several countries, including China, Cuba, Tanzania, Venezuela, and Yugoslavia. The report recommended that "WHO and UNICEF should adopt an action programme aimed at extending primary health care to populations in developing countries, particularly to those that are now inadequately provided with such care, such as rural and remote populations, slum dwellers, and nomads." It further stated that PHC systems should be based on a set of principles that essentially mirrored those included in the director-general's report. Finally, the report urged WHO and UNICEF to continue studying existing examples of PHC and to encourage changes in the training of health-care personnel "to enable them to discharge their duties as envisaged in a health service system oriented toward primary health care."

Newell subsequently published a book titled *Health by the People*, which revisited a number of the examples included in *Alternative Approaches* and presented the narratives of those who had played a key role in the development of those cases.[21] It was intended as a motivational book, presenting positive examples of what could be achieved and how it could be done.[22] Carl Taylor's Narangwal project (chapter 10) also served as an inspiration for the primary health-care movement. Taylor was a friend of Mahler and played a key role in preparing background materials for the Alma-Ata meeting.

Meanwhile, Soviet representatives to WHO pushed hard for the convening of an international conference on primary health care to be held in the Soviet Union, offering to provide funds to cover the travel expenses of representatives from developing countries. They hoped to use the meeting to showcase their own highly developed health system. Other countries proposed conference sites, but in the end the Soviets won out, and in 1976 it was decided that the conference would be held in Alma-Ata in September 1978.

Mahler was not happy. As much as he believed that primary health care needed to be at the center of WHO's efforts to improve world health, he had major reservations about holding a conference so soon, as well as with holding the conference in the Soviet Union. Mahler felt strongly that despite the reports of the Joint Committee and Newell's *Health by the People*, there was not enough knowledge or experience with organizing primary health-care systems that were truly participatory and met the needs of local communities. He feared that the proposed conference would be a premature launching pad for primary health care. How could a program be started effectively without knowing what the best practices were or what the obstacles to creating those practices might be?

As Socrates Litsios has noted, there was no consensus about exactly what primary health care should be, despite multiple reports and studies. The Russian health-care system, while extensively developed and achieving high levels of coverage, lacked key elements that Mahler believed were critical to the success of the movement. It was a centralized, top-down system that was highly medicalized and made little effort to address the needs of communities or to draw on communities to mobilize their own health resources or participate in the design of their health care. The Russian system was antithetical to Mahler's vision of primary health care, and Mahler had no interest in highlighting this alternative model of PHC. Despite his reservations and the lack of consensus about the definition of primary health care, in 1976, the Executive Board voted to move ahead with the meeting and to hold it in Alma-Ata in 1978.[23]

During the following two years, background papers were prepared to provide guidance to the conference attendees. The drafts captured many of the ideas that had appeared in the reports that had been presented to the Executive Board and in Newell's *Health by the People*. Newell, in fact, prepared two initial drafts of the background papers. The background papers, however, suffered from two major weaknesses. First, they provided little discussion of the constraints and challenges facing the implementation of PHC. These had been highlighted in Newell's drafts but were dropped as being too negative. Second, little was said about the need to develop successful models of PHC, which could help countries create their own programs. This was in stark contrast to the Executive Board's position in 1972, which called for the utilization of trial areas to test methods and the means to ensure that these were suited to local conditions and attainable with available resources. In addition, the background papers were overly optimistic about the current state of knowledge and the readiness of countries to embark on building effective PHC systems.

These weaknesses were also apparent in the final documents emanating from the conference in 1978. The report of the conference laid out a set of recommendations on what countries needed to do in order to build effective health systems centered on the provision of primary health care. Yet the recommendations were general guidelines and provided no information about how they should or could be translated into local settings. Without workable examples that provided detailed information on program implementation and its subsequent impact, local governments had to proceed in a more-or-less trial-and-error manner. Moreover, without examples of successes, it would become difficult to convince donors of the value of PHC programs. In addition, the documents produced by the conference were equally devoid of any discussion of the challenges facing efforts to implement PHC programs or ways to overcome various forms of resistance. In effect, the conference documents presented an idealized vision of what should occur, a vision that failed to acknowledge the challenges that lay ahead and would plague efforts to build primary health-care systems around the globe (part VII). As it turned out, the devil was indeed in the details.

This is not to say that there were no positive examples of building primary health-care systems that reflected the values of the Alma-Ata Declaration. One of the most successful was started before Alma-Ata, in Jamkhed, India, by physicians Raj and Mabelle Arole, who had earned their Master of Public Health degrees at Johns Hopkins under the guidance of Carl Taylor.[24] The Jamkhed Comprehensive Rural Health Project (CRHP) was intended to bring both preventive and curative health care to some of India's most vulnerable people in a rural, poor, drought-prone part of Maharashtra State. It was designed to be community based and was implemented in cooperation with local populations. Funds for initiating the project were arranged by the Christian Medical Commission, which also played an important role in the discussions of primary health care leading up to Alma-Ata.[25]

The CRHP began in 1970, with a small hospital serving 30 villages and a population of 40,000 people. Central to the project was the recruitment of local women to serve as health workers. These women collaborated with local communities to identify problems affecting their health and develop solutions to them. These not only included issues related to nutrition and sanitation, but also extended to the topics of gender inequality and women's needs to find a means to generate income. A basic principle to the CRHP was that improvements in health could neither occur nor be sustained without improvements in the overall economic and social development of the communities served by the project.

The CRHP displayed remarkable longevity and an ability to grow in a world where many such projects experienced a relatively short existence and found it difficult to "go to scale," that is, to expand beyond local communities to encompass entire regions or countries. By 2015, the project had expanded to provide training for community health workers who delivered health care to over a half a million people in Maharashtra State. The CRHP also recorded significant improvements in the health of the communities it served.[26] These statistics, produced by the CRHP, were exceptional, not only for what they showed in terms of program impact, but also because few primary health-care programs were able to document this kind of success. Lack of statistical evidence limited the ability of programs to convince donors that primary health care was more cost-effective than selective interventions.

The CRHP, which was an early inspiration for the primary health-care movement, became a model for community-based health care and trained

Health outcomes in Jamkhed Comprehensive Rural Health Project villages, 1971–2011

	Year							
	1971	*1976*	*1986*	*1993*	*1996*	*2004*	*2011*	*India* 2004
IMR (infant mortality rate = n/1000 live births)	176	52	49	19	26	24	8	62
CBR (crude birth rate = n/1000 live births)	40	34	28	20	20	19	24	24
Antenatal care coverage	0.5%	80%	82%	82%	96%	99%	99%	64%
Safe-delivery coverage	<0.5%	74%	83%	83%	98%	99%	99.4%	43%
Family planning	<1%	38%	60%	60%	60%	68%	*	41%
Immunization coverage	0.5%	81%	91%	91%	92%	99%	*	70%
Malnutrition in children under 5 years of age	40%	30%	30%	5%	5%	<5%	*	47%
Leprosy (cases per 1000 individuals)	4	2	1	0.1	0.1	<0.1	*	0.24
Tuberculosis (cases per 1000 individuals)	18	15	11	6	6	2	*	4.1

Source: Comprehensive Rural Health Project, Jamkhed, "Impact," www.jamkhed.org/impact/impact/.

*First eight months

thousands—local people as well as members of international NGOs—in the program the Aroles had developed. Yet in many ways the CRHP was an exception, and the broader primary health-care movement quickly ran into multiple obstacles. Some of these were of its own making. Others were a product of developments elsewhere within international health—particularly the success of the smallpox-eradication campaign. Still others were raised by a changing political and economic environment, beginning in the early 1980s. These obstacles will be examined in the next chapter.

CHAPTER THIRTEEN

Challenges to Primary Health Care

IN SOME WAYS, plans to build the kind of primary health-care system embodied in the Alma-Ata Declaration were seriously compromised before the conference even occurred. Mahler had been correct when he argued that moving to a planning conference was premature. There were few examples of successful programs to learn from. Those that existed, like China's barefoot doctors program, were the product of specific historical circumstances that would be difficult to replicate in other parts of the world. Many small-scale programs designed and run with the help of nongovernmental organizations showed promise, but there was no experience with taking such experiments to scale. Creating a lasting and significant impact was particularly problematic in instances where early experiments enjoyed extensive resources provided by external donors, resources that would be difficult to replicate at a national or even a regional level.

Second, the commitment to primary health care required a radical redirection of investments in health care in most countries. Building PHC demanded that resources be shifted away from existing centers of secondary and tertiary care, located most often in urban centers, and toward the building and staffing of clinics in rural areas, in collaboration with local communities. Maintaining medical centers and private medical services devoured large portions of a country's available resources for health care. But existing medical authorities and the private medical sector had a vested interest in sustaining established patterns of health-care allocations and were often supported by local, urban-based elites, who wished to maintain access to advanced forms of medical care.

Mahler recognized that such jolting changes would meet with resistance, but he did not foresee how great the opposition would be. Redirecting resources to rural areas was like trying to turn the *Titanic*. Something as basic as changing medical-school curricula to provide doctors with the skills and perspectives needed to work in rural communities met with a serious pushback from medical students, who feared that tailoring their medical education to the needs of PHC would make it difficult for them to seek employment elsewhere, particularly in Europe or the United States. Medical students in Mozambique in the mid-1970s went on strike to protest efforts to create a national medical curriculum.[1] Efforts to build PHC also involved the devolution of decision making to the community level, something that was often strongly resisted by members of the existing medical hierarchy in each country.

At the heart of PHC was the commitment to popular participation. Communities needed to be actively involved in planning for their own health needs. These requirements could not be dictated from above. Yet the goal of popular participation often reflected ideological commitments on the part of the advocates of PHC more than a well-developed understanding of how rural communities functioned. These commitments were also based on an idealized view of rural societies as homogenous, without any disparities in power and in the ability to participate in decision making. The reality was often quite different. Inequalities in wealth and authority meant that decisions were seldom made in a democratic fashion.[2]

The time constraints of daily labor often prevented some members of communities from participating in planning discussions. Women, in particular, who often played a major role in meeting the health needs of communities, commonly experienced high demands on their labor, which limited their participation in community planning activities. Working all day in the fields and then coming home to prepare family meals left little time or energy to attend community meetings. In addition, in many settings, patterns of patriarchal authority prevented the participation of women in any kind of decision making. On the other hand, efforts to democratize village decision-making processes from above often met with resistance and undermined primary health-care initiatives. This was the case in Benjamin Paul and William Demarest's classic study of the unsuccessful efforts of an outside American community organizer to build a participatory rural health program in a village in Mexico. Rather than working through existing village hierarchies, the organizer insisted that the community develop democratic processes and, in the end, failed to make any progress. The study suggested that the imposition of idealized visions of popular participation might have undermined efforts to build community-based PHC programs in a number of locations.[3]

Efforts to build community-based health systems also assumed the existence of clearly defined communities. In many places, this kind of social organization did not exist. Serving as a Peace Corps volunteer in eastern Uganda in the late 1960s and early 1970s, it became clear to me that extant social networks seldom corresponded to the official administrative boundaries that divided the population into distinct villages. It made little sense to try to develop community participation within these official boundaries. In addition, the health units to which people traveled for care were often far from their homes, even though other health units were close by and, in a sense, more a part of the community. This was because the medical staff that ran the closer units belonged to a different ethnic group. In short, real-time conditions were seldom as simple as the idealized community models envisioned by PHC planners.

It should not be surprising that those who envisioned PHC as being built on the basis of popular participation had little understanding of the realities of village life. Few of those doing the planning had much experience working in rural communities. As Judith Justice showed in her pathbreaking ethnography of PHC programs in Nepal in the late 1970s, expatriate health workers seldom left the capital and instead relied on the knowledge of Nepalese counterparts, who themselves seldom visited villages outside the Kathmandu Valley. When expatriate aid workers did travel to the rural areas, they were frequently shown model villages located near major highways, seldom penetrating into the more remote areas where the majority of rural populations lived.[4]

More generally, international-health consultants rarely spent long periods of time in any one country. They were trained to do specific tasks, rather than to work in particular settings, and often moved from project to project, following the trail of contract funding. They did not have time to develop a deep understanding of the countries in which they worked, the languages spoken there, or the actual needs of the people they hoped to help. Ignorance of local conditions was a systemic problem, created by the structures of international assistance funding. It was also a product of a system of public-health education that provided little opportunity for students to learn about the languages and cultures of the places in which they would work. Students were trained to be specialists in particular disciplines or fields and expected to learn about local realities on the ground, while carrying out their missions.[5] Sadly, most of these educational conditions continue today.

Efforts to integrate indigenous healers and birth attendants into PHC programs also met with very limited success, in large measure because those planning rural health services failed to appreciate the roles that indigenous health workers played within rural communities. The motivation behind their incorporation in PHC was partly to solve manpower needs, and partly to adapt health services to local social and cultural settings. Yet the two motives were inherently at odds. Bringing indigenous medical practitioners and birth attendants into a biomedical system required a redefinition of the roles they played in local communities. As Stacey Pigg has shown, again for Nepal, the process of incorporation separated those elements that PHC organizers viewed as functional and complementary to biomedicine—the indigenous healers' knowledge of local medicines and their role in addressing mental aspects of illness—from religious and cultural practices that were, from the perspective of local villagers, a central part of a healer's identity and medical authority.[6] In short, traditional healers and midwives were enlisted to be medical auxiliaries within a biomedical system of health care, and in the process became something that had not existed before. This may have satisfied some of the manpower needs of PHC programs, but it did little to make them culturally acceptable to local communities.

In another Nepalese study, Linda Stone suggested that efforts to mobilize traditional healers largely disregarded the interests and needs of the community. Healers were trained as health educators within the PHC program, whereas local villagers expected them to provide curative services and were not interested in health education.[7] In addition, program organizers assumed that villagers would respect their traditional healers and follow their advice. This view failed to appreciate the interactive relationship that existed between healers and patients, in which both sides contributed to therapeutic decision making.

Finally, despite the commitment of those at Alma-Ata to address the broader social and economic determinants of health, including improvements in sanitation, in access to clean water, and in the overall economic well-being of populations, few countries were able or willing to address these broader issues. In many developing countries, bettering the social and economic status of the poor required addressing serious inequalities in access to land, education, and power. Correcting these inequalities required governments to attack the social and economic interests of wealthier classes, who were often the base of governmental support. Few were willing to do so. In this regard, the challenges facing PHC organizers in the 1970s and 1980s replayed problems encountered by the Rockefeller Foundation's IHD in China and Java in the 1930s.

The Impact of Changing Economic and Political Conditions on Primary Health Care

The difficulty of implementing the PHC model envisioned at Alma-Ata was only one of the challenges facing primary health care. A set of events and political and economic changes following the Alma-Ata conference conspired to undermine the PHC movement even before the meeting ended.[8]

In October 1977, Ali Maow Maslin, a cook working at a local hospital in Somalia, was diagnosed with smallpox. There was nothing unusual about his case. Smallpox had been endemic in the Horn of Africa for centuries. Yet in retrospect, his illness was special, in that it became the last case of naturally transmitted smallpox in the world. His cure marked the successful completion of the campaign to eradicate smallpox.[9] By December 1979, world health authorities determined that eradication had been achieved, and in 1980, they announced to the world that this was so.

The eradication of smallpox was arguably WHO's greatest victory (part IV). It was also a much-needed vindication for an approach to public health that was based on the mass application of simple medical technologies through vertical programs. The failures of the Malaria Eradication Programme had discredited this approach in many international-health circles and opened the door for a reassessment of international-health priorities in the early 1970s. This rethinking helped set the stage for the PHC movement. The success of the smallpox-eradication campaign threw a monkey wrench into that reassessment effort.

Even before smallpox eradication was declared, efforts were afoot to build on its success and develop vertical programs for the expanded use of vaccines and other biomedical technologies to address specific health problems. In April 1979, Rockefeller Foundation President John Knowles organized a small meeting on "Health and Population in Developing Countries." The meeting was held at the foundation's conference center in Bellagio, on the shores of Lake Como in northern Italy. It was the first of a series of meetings that would take place at Bellagio and other resort towns that would question the feasibility of primary health care, as defined in Alma-Ata, and give rise to a new approach to international health known as selective primary health care. The heads of a number of key international and bilateral organizations attended the 1979 meeting in Bellagio, including Robert McNamara from the World Bank; Maurice Strong from the Canadian International Development and Research Center; David Bell, vice president of the Ford Foundation; and John Gillian, the administrator for USAID.

The centerpiece of the 1979 meeting was a paper that had been published earlier that year in the *New England Journal of Medicine* by Judith Walsh and Kenneth Warren. Warren was the head of the Health Sciences Division of the Rockefeller Foundation. The paper, titled "Selective Primary Health Care: An Interim Strategy for Disease Control in Developing Countries," is often cited as the first to argue for the adoption of selective primary health-care programs as an initial step toward the development of PHC.[10] It is worth looking at in some detail. Walsh and Warren began by lauding the goals of PHC, but quickly noted that these goals were unattainable, because of their prohibitive costs and manpower requirements.[11] They asked, "How then, in an age of diminishing resources, can we best attempt to secure the health and well-being of those trapped at the bottom of the scale long before the year 2000 arrives?" To answer this question, they provided a metric for deciding which diseases were of the highest priority, based on a combination of their impact on mortality and the feasibility and cost of controlling them. The authors then compared different strategies for attacking these diseases, including total primary health care; basic primary health care; vector disease-control and nutrition programs; and categorical disease control.

They concluded that categorical disease control, what they called selective primary health care (SPHC), was the most promising and least wasteful for most parts of the world. As an example, they described a program that would include measles and DPT (diphtheria, pertussis, and tetanus) vaccination for children under six months old; tetanus toxoid injections for pregnant women; the encouragement of breastfeeding; chloroquine for children under three years old in malarious areas, to ingest during febrile episodes; and oral-rehydration packets for and instructions on the treatment of diarrhea. Fixed or mobile units, visiting once every four to six months, could provide these services. They estimated that the cost of avoiding one child's death though this categorical, or selective, approach would be US$200–US$250. This was much less than the cost of building PHC programs on the scale proposed by the Alma-Ata conference. But Walsh and Warren acknowledged that they had no actual data to support this conclusion, because their approach had never been tried. Nonetheless, the authors concluded that "until comprehensive primary health care can be made available to all, services targeted to the few most important diseases may be the most effective means of improving the health of the greatest number of people."

Their conclusions were weakened by two points, which they acknowledged. First, they noted that effectiveness was calculated solely in terms of cost-effectiveness, measured in terms of cost per death averted. This metric did not account for other benefits provided by alternative approaches, including illnesses and disabilities prevented, as well as the possible secondary benefits of various interventions. For example, vector control to reduce malaria might also decrease the transmission of filariasis and leishmaniasis. Nutritional supplements might be distributed to the entire family, increasing their overall well-being; water supplies located close by might provide release time for women that could be devoted to other productive tasks having health benefits. Those who jumped on the bandwagon of SPHC often overlooked this caveat. Second, by focusing only on the costs of deaths averted, the authors ignored the many other health benefits that would be provided by more-comprehensive health services or by the provision of clean water and sanitation. The paper represented a narrow, econometric view of health care, a view that would gain tremendous purchase under the changed political and economic conditions of the 1980s.

Yet for supporters of PHC, as envisioned by the Alma-Ata Declaration, arguments based on cost-effectiveness and mortality reduction failed to recognize that PHC was about more than averting deaths. It was also about community mobilization and participation, redirecting decision making and resources to the people. PHC was about community empowerment. For many, this was a political issue. SPHC programs continued the vertical model of public health, in which health interventions were packaged and implemented from above, with no effort to have communities participate in their design, let alone in broader discussions of how to improve local health conditions.

In addition, SPHC services, delivered though episodic, or campaign, modes of organization, did little to provide for the development of a permanent infrastructure for the delivery of health services. SPHC was also found wanting by the supporters of PHC because it totally ignored the wider social determinants of health. For PHC supporters, SPHC represented a Band-Aid approach, treating symptoms rather than causes. Oral rehydration of children with diarrheal diseases would save lives. But it would not address the causes of the diseases: lack of clean water and sanitation.[12]

Other participants at the Bellagio meeting made a number of these points. In addition, two papers questioned the statistical basis on which claims for the cost-effectiveness of specific medical interventions were based.[13] By no means was there a consensus among the presenters at the conference as to the relative value of SPHC over other approaches to health. Nonetheless, the Bellagio meeting initiated a set of discussions among a number of key actors in international health that led to their advocacy of SPHC strategies. Central to these discussions and to the advancement of the SPHC movement was Jim Grant, who became the executive director of UNICEF in 1980.

Grant's advocacy of SPHC was somewhat surprising, given his personal history and early career. Grant's father and grandfather had been missionaries in China. His father, John Grant, was a Rockefeller Foundation medical officer who had been centrally involved in the foundation's North China Rural Reconstruction Program (chapter 4). He was also a pioneer in primary health care in China, establishing rural health posts with simple laboratories and a few trained health workers. Jim's father was committed to a system of health for all that would cost no more than US$1 per capita per year, the maximum that the peasants at that time could afford. Hence the emphasis was on health promotion, hygiene, home gardens for improved nutrition, and immunization. As a boy, Jim Grant traveled with his father to rural villages in China and came to understand that poor health was not so much a medical problem as a social and economic one. Perhaps for this reason, he chose to study economics and law instead of medicine and devote his early career to international development. He worked his way through college, earning a degree in economics from the University of California–Berkeley in 1943, and then returned to China, where he worked for UNRRA until the Communist revolution in 1949.[14]

Returning to the United States, Grant went back to school, earning a law degree from Harvard in 1951. He worked for the International Cooperation Administration and then for its successor, USAID. As USAID director in Ceylon, and later in Turkey, he nurtured the Green Revolution as a means of assuring self-reliance and empowering small farmers. He was fascinated by the application of technology and its potential to change peoples' lives. Leaving USAID, Grant founded the Overseas Development Council, an NGO dedicated to guiding US policies toward international development. While there, he developed new ways to measure social progress, as opposed to purely economic growth. He promoted the Physical Quality of Life Index (PQLI) and the Disparity Reduction Rate (DRR) as measures of social achievement, based on a combination of infant-mortality, life-expectancy, and literacy rates. He was thus influential in the move to rethink development in the early 1970s.[15]

Grant's background made him a natural choice to become the executive director of UNICEF in 1979. At that time, UNICEF was a highly decentralized organization supporting a wide range of largely small-scale community programs aimed at providing basic needs and appropriate technology. It was involved in education, water supply and sanitation, housing, and help for street children. It was about integrated development and community participation, approaches that Grant had supported. UNICEF had also been a partner in the planning of the Alma-Ata conference and a supporter of the primary health-care movement. No one at the time envisioned that within the first two years of becoming executive director, Grant would lead UNICEF in a radically different direction. But this is what he did.

According to Peter Adamson, who, for 15 years, authored UNICEF's *State of the World's Children* reports, Grant's vision for UNICEF was highly influenced by a paper that had been presented by Jon Eliot Rhode at a conference in Birmingham, England. Rhode was a physician and public-health worker who had served in Indonesia and Bangladesh with the Rockefeller Foundation, and in Haiti as a director with Management Sciences for Health. His paper made a simple but startling point—more than half of all the deaths and diseases among the children of the developing world were simply unnecessary, because they were now relatively easily and cheaply preventable. Vaccines were available to protect children from measles, whooping cough, tetanus, and polio, which were killing 4 or 5 million children every year. But only 15 percent of the children living in resource-poor countries received these immunizations. In addition, cheap oral-rehydration therapy could save over 3 million children who were dying from diarrheal diseases each year. The methods of prevention or treatment were available, tested, and affordable, but they were not getting to those most in need of them. Rhode's paper, in effect, reinforced the arguments that Warren and Walsh had made at the Bellagio meeting in 1979.[16]

In September 1982, Grant called a meeting of UNICEF staff at the organization's New York City headquarters and announced that he wished to redirect UNICEF's activities toward ensuring that these simple technologies were made available to all the world's children. He wanted to launch a worldwide Child-Survival Revolution. At that time, UNICEF projects in the developing world were reaching a few hundred, and very occasionally thousands of children in villages and urban neighborhoods. Grant wanted to reach 400 or 500 million children in the developing world, and the 100 million that were being born into it each year, providing them with immunizations, oral-rehydration kits, and other cost-effective and appropriate interventions that could save millions of lives.[17]

Grant was not opposed to primary health care, even though he had not been involved in the planning for the Alma-Ata conference. Much of his early life and career had taught him the value of the principles on which PHC was built. But Grant was impatient. He refused to sit by and watch millions of children die from preventable deaths while the world built effective primary health-care systems. He was also concerned, as were Warren and McNamara, that in the economic environment of the 1980s, it would be difficult, if not impossible, to raise the financial resources needed to implement the Alma-Ata vision of PHC. Grant asked, "How can human progress be maintained in the absence of increased economic resources?"[18]

In autumn 1982, UNICEF launched the Child-Survival Revolution, aimed at reaching millions of children. The program focused initially on four interventions: growth monitoring, oral rehydration, breastfeeding, and immunizations. These were packaged together under the acronym GOBI. There were many in UNICEF who questioned Grant's shift in policy. They worried not only about the feasibility of reaching so many children, but also about how it radically changed the culture of UNICEF, shifting from an organization supporting innovative, community-based projects aimed at improving the lives of children on multiple fronts to a highly centralized, vertical program for distributing a few technologies. There were many more outside of UNICEF who would also object to Grant's proposed Child-Survival Revolution.

Grant pushed ahead, as was his way, and sought out partners to help finance the program. He turned to Robert McNamara, president of the World Bank, who was sympathetic to programs based on appropriate technologies for meeting basic needs, and the Rockefeller Foundation's Ken Warren, who supported the types of targeted interventions Grant was proposing. In addition, Jonas Salk was brought on board because of his experience in developing vaccines, including the polio vaccine. The four men met in New York City the following May, along with Robert Clausen, who had replaced McNamara as president of the World Bank. McNamara argued that instead of mounting all four interventions at once, they should focus on immunizations, which were measurable and promised to save the most lives in the shortest time. WHO had created the Expanded Programme on Immunizations nine years earlier. It was delivering high-quality immunization services, but it was only reaching 10–15 percent of the world's susceptible children and was expanding very slowly. They needed to move faster. This was agreed on, and McNamara suggested that they should convene a major meeting in Bellagio, Italy, in 1984 to plan the immunization campaign. Representatives from the various agencies that would administer the vaccines, including WHO, and those who would pay for them would be invited to devise strategies that would ensure immunization of all newly born children in developing countries.[19]

Meanwhile, word of UNICEF's policy shift had reached Mahler in Geneva. Far from being supportive of the new endeavor, Mahler was outraged. In a speech to the WHA at the same time that Grant was meeting with his Child-Survival Revolution allies in New York, Mahler warned the assembly against programs that purported to support primary health care but actually threatened to subvert its basic principles. He stated:

> I am all for impatience if it leads to better and speedier action along collectively agreed lines. But I am all against it if it imposes fragmented action from above. I am referring to such initiatives as the selection by people outside the developing countries of a few isolated elements of primary health care for implementation in these countries; or the parachuting of foreign agents into these countries to immunize them from above; or the concentration on only one aspect of diarrheal disease control without thought for the others. Initiatives such as these are red herrings that can only divert us from the track that will lead us to our goal.[20]

With this speech, the battle lines were drawn between two distinct visions of global health—one focused on the rapid deployment of cost-effective interventions from above; the other on building a broad base of health services from below that addressed community needs as well as the social determinants of health. It continued the struggle between two alternative approaches that dominated international health for most of the twentieth century.

Mahler would eventually agree to participate in the Bellagio meeting, recognizing that WHO needed to work with UNICEF and its powerful institutional allies: the Rockefeller Foundation and the World Bank. But throughout his tenure as director-general, Mahler remained opposed to the principles on which GOBI was founded and continued to promote the original version of PHC that had motivated the Alma-Ata Declaration. It was a battle he was destined to lose, as a result of both the failings of early primary-health efforts to achieve their goals and the changed economic and political environment of the 1980s, which greatly favored Grant's vision of global health. We need to examine these altered conditions before returning to efforts to implement the Child-Survival Revolution and selective primary health care.

Primary Health Care, Global Recession, and the Challenge of Neoliberalism

The ability of governments to address economic inequalities or invest more broadly in development was seriously limited by the major economic recession of the 1980s. This recession, which had begun in the early 1970s with inflation and the shock of rapidly rising oil prices, deepened in the early 1980s as industrial production declined in the major manufacturing countries in the global north. This reduced the demand for agricultural commodities and other raw materials, which provided the bulk of foreign exchange in many developing countries in Africa, Asia, and Latin America.

The impact of this decline was compounded by the fact that many developing countries, especially in Latin America, had borrowed heavily from foreign governments and international banks in order to fuel their development efforts. Between 1975 and 1982, Latin American debt to commercial banks increased at a cumulative annual rate of 20.4 percent. In 1983, Latin America debt quadrupled, from US$75 billion in 1975 to more than US$315 billion, or 50 percent of the region's GNP. International banks were encouraged to lend money by the previous pattern of economic growth in the region, with the expectation that these countries would be able to repay their loans. The flood of deposits the banks were receiving from OPEC countries also increased the willingness of these banks to loan money to resource-poor countries. The banks needed to pay interest on the OPEC deposits, so they had to earn interest on the funds received from OPEC countries. One way to accomplish this was to loan the money to cash-starved developing countries. In some cases these banks pushed loans on countries, even when the possibility of their repayment was questionable. The banks cared less about recovering the principle that had been loaned to a country than in continuing to receive interest on the loan.[21]

As the inflation of the 1970s began to take its toll on industrial economies, lowering the demand for developing-country exports, the ability of countries that had borrowed heavily during the 1970s to repay their loans also declined. In 1982, the government of Mexico declared a unilateral moratorium on its debt payments and, in effect, defaulted on their loans. This sent a shock wave through the international banking system. Banks were faced with the possibility that other countries would default on their loans, forcing the banks to count loans as losses on their balance sheet. They therefore cut back on this form of lending, initiating an international debt crisis. Countries with high levels of outstanding debts found it difficult to refinance their loans in order to keep making payments. The cost of money soared as banks charged higher interest rates on loans to hedge against possible losses. Debt payments rose sharply in many countries, taking a growing bite out of governmental revenues. Total debt, as percentage of export goods and services, rose from 10.9 percent in 1980 to 20.5 percent in 1990 in sub-Saharan Africa. The immediate result of these events was that low-income countries had fewer dollars to invest in development and found it increasingly difficult to obtain international funding to support development projects. Building health infrastructures and addressing the social and economic determinants of health, even where there was the will to do so, became increasingly impossible during the 1980s.

The debt crises and the recession in the 1980s had more long-term consequences on efforts to improve the health of the world's poor and on the development of basic health services. As more and more countries found it difficult to meet their debt payments and faced the prospect of default, the International Monetary Fund, which was charged with securing international financial stability, together with the World Bank, stepped in to provide funds in the form of short-term loans to ensure that heavily indebted countries would not default and cause a global financial crisis. The goal of these interventions was first and foremost to secure the loans made by banks and governments. The IMF therefore required that borrowing countries adopt a series of financial policies designed to stabilize their economies and ensure their ability to repay the loans. These were known as structural adjustment policies (SAPs). We should be clear here. Structural adjustment was first and foremost about securing the international financial system and catering to the interests of loaning countries and banks. Second, it was about transforming the economies of developing countries in ways that would support free-market economies that were beneficial to the economic interests of donor countries. Lastly, and most cynically, it was supposed to advance economic development in debtor countries.

SAPs reflected the economic thinking of a group of neoliberal economists who ascended to positions of authority in the World Bank and the IMF in the wake of the election victories of Margaret Thatcher in Great Britain in 1979 and Ronald Reagan in the United States in 1980. These economists espoused a set of principles, developed by University of Chicago economists Milton Friedman and Friedrich Hayek, that became known as neoliberalism.

Much of the economic thinking regarding international development after World War II had been influenced by the work of John Maynard Keynes and focused on the role of governments in stimulating economic growth through governmental spending. Friedman and Hayek argued to the contrary: only free markets distributed societies' resources in an optimal way. The neoliberal economists who took over the reins of the World Bank and the IMF were opposed to governmental spending and to policies that interfered with free-market forces. This included such things as the imposition of tariffs to protect local industries and price supports to assist farmers (policies that had existed in the United States for decades). All of these policies were viewed a distorting market forces and disrupting free trade. Neoliberals were also opposed to state spending on social services, such as food subsidies or health services for the poor. These expenditures increased governmental spending, which, in turn, required increases in state revenue through taxes. Taxes restricted private-sector investments and, thus, economic growth.

With the IMF and the World Bank firmly in the grasp of neoliberal economists, and with developing countries facing economic collapse, the door was open for the creation of a neoliberal world order that would end the hopes of developing countries for a new international economic order that would better serve their interests. The IMF and the World Bank would dictate that new international economic order. In a cruel irony, the success of OPEC, which had inspired efforts to create a new international economic order in the 1970s, ultimately increased the economic dependency and subordination of commodity-producing countries in the global south to the economic interests of wealthy industrial countries in the global north.

IMF officials required governments facing default to enact a series of economic measures designed to free up markets and reduce governmental spending. These included currency devaluation, aimed at making exports less expensive and more competitive, while increasing the cost of imports to reduce spending and the loss of foreign exchange. Tariffs and subsidies were also eliminated, to encourage free trade. Finally, governments were required to reduce their overall spending. This meant reducing salaries for governmental officials and cutting government-funded programs. Just which programs were cut was decided in closed-door negotiations between the IMF and officials from governments receiving loans. In most cases the bulk of the cuts occurred in social services, including in health care. This choice may have been made because governmental officials wished to protect other sectors, such as the military, which kept them in power. But IMF officials were opposed to government-supported social programs, in particular programs related to health, which neoliberal economists viewed as a commodity that should be delivered within a market context, instead of as a right, as proclaimed by the Alma-Ata Declaration. The effects of SAPs were severe on local populations. Farmers saw the prices they received for crops decline, while the cost of agricultural inputs—fertilizers and pesticides—increased. Urban workers were faced with declining wages and rising consumption costs. SAPs resulted in political protests and strikes in many countries.

The impact of SAPS on health-care services, particularly in Africa, were dramatic, as seen in the experience of Zambia, a country that had expanded its economy on the basis of income derived from its copper industry. When copper prices fell in the late 1970s, Zambia faced declining revenues and took on increased levels of debt to maintain economic growth and underwrite a range of government-support programs. By 1981, the country's external debt had more than doubled, to over US$2 billion. In the face of rising debt payments and a stagnating economy, the Zambian government entered into debt negotiations with the IMF and the World Bank.

The bank imposed a series of SAPs designed to liberalize the country's economy: removing import subsidies; raising interest rates; devaluing the currency; cutting governmental expenditures, including subsidies on food and fertilizer; and freezing wages. As a result, government-run health services throughout the country deteriorated during the 1990s. Bank-mandated conditionalities required Zambia to establish a wage cap, in order to reduce overall expenditures. Low wages and disruptions in pay caused many health workers to opt out of the health system. Skilled workers sought better-paying jobs in health systems in South Africa or abroad. In 1993, 3444 out of a total of 9000 trained nurses applied to work abroad.[22] Declining numbers of health workers caused the government to shut down health facilities. In the rural areas in particular, basic supplies, including antimalarial drugs and medical services, became widely unavailable. The accessibility of health care deteriorated, particularly in remote locales. In Zambia's Western Province in the late 1990s, the nearest health clinic to the Luampungu villages was 45 kilometers away. WHO estimated that in 2005, only half of all rural households in Zambia were located within 5 kilometers of a health facility.[23]

The World Bank's response to declining health services in Zambia and other countries was to implement user fees and essentially make health a commodity. The bank argued that user fees would generate revenues for resource-strapped health services and encourage patients to seek treatment in low-cost primary-health services, reducing the use of high-cost hospital services and the overconsumption of governmental health services. Yet, at their heart, user fees were part and parcel of a market-based approach to health care, intended to shift the burden of health care from governments to individual users. This policy was adopted, under pressure from the World Bank, by a number of African countries in 1987, at a meeting in Bamako, Mali.

If policy makers wanted to reduce overconsumption, they certainly achieved their goal. User fees, of top of declining income, led patients to stop seeking health care. In Zambia, the World Bank reported that following the imposition of user fees in 1993, visits to outpatient clinics in Lusaka, the capital city, fell by 60 percent, and to maternal delivery services by 20 percent.[24] Mothers stopped taking their children for vaccinations. Attempts to provide subsidized care for the most vulnerable populations through a community health-waiver scheme did little to offset the negative impact of fees on clinic attendance in rural areas.[25]

User fees did not solve the health-care crisis. Moreover, as an approach to health-care financing, user fees also failed to cover the costs of public-health programs aimed at protecting populations. Health education, epidemiological record keeping, sanitation, and even routine house spraying to prevent malaria, for which patients did not pay fees, declined or ceased to occur. The overall impact of the recession and SAPs on health in Zambia was revealed in the decline in immunizations and the rise in TB and malaria cases. Overall, the annual incidence of reported malaria cases nationwide rose steadily, from 121 per 1000 individuals in 1976 to 376 per 1000 in 2000.[26] Worse yet, the incidence of malaria in children under age five was over 900 per 1000. The incidence of TB rose from 300 to 600 per 100,000 between 1990 and 2005.

Across Africa, health services were devastated by SAPs. Health infrastructure deteriorated, and the availability of lifesaving drugs became sporadic. Health assistants, their salaries cut or, in many cases, not paid on a regular basis, set up private practices to make ends meet, reducing their availability within the government-run clinics. Increasingly, the only health services that functioned in many parts of Africa were those run by NGOs and, especially, by faith-based organizations. Yet these services were too few and too scattered to meet the health needs of rural populations. SAPs had succeeded in achieving their goal of reducing state involvement in health-care delivery, but at a heavy cost to the health of African populations. Tragically, the undermining of African health services from the mid-1980s through the early 1990s occurred as the epidemic of HIV/AIDS was picking up steam. By the time the international-health community began responding in a major way to this epidemic (part VII), the capacity of African health services to absorb and support prevention and treatment programs for AIDS was very limited.

Within this environment, hopes of building primary health-care systems, let alone addressing the underlying social determinants of health, as envisioned in the Alma-Ata conference, did not stand a chance. Neoliberal policies continued to shape global-health strategies into the twenty-first century (chapter 14). If the changing political and economic environment of the 1980s and 1990s made it unlikely that countries would be able to develop effective primary health-care approaches, they added support to the efforts of Grant, McNamara, and Warren to develop more-selective health strategies that would require less investment in people-centered health care while still saving the lives of millions of children.

Implementing Selective Primary Health Care

Grant's vision for a Child-Survival Revolution was translated into programs for the global immunization of children at a series of meetings in the early 1980s, beginning at the Rockefeller Foundation's conference center in Bellagio in 1984. A group of predominantly white European and American public-health leaders, nearly all of whom were men, dominated the initial Bellagio meeting. The location of the meeting and composition of the participants clearly highlighted the distance that still existed between the centers of international-health policy formation and the locations of those whose health would be affected by these same policies. Participants at the Bellagio meeting included many of those who had played a role in the successful smallpox-eradication campaign and who were wedded to a vision of international health built on the provision of effective medical technologies to prevent and eventually eliminate diseases. Many of these leaders would become major players in global health in the 1990s and 2000s (chapter 8). The Bellagio conference accepted Grant's goal of immunizing 80 percent of the world's children against six major diseases by 1990. Achieving this goal would require a massive commitment of financial resources; the development of effective cold-chain (temperature-controlled supply-chain) systems to protect the viability of vaccines as they traveled to remote locations across the globe; research to perfect new and existing vaccines; the political will of governmental leaders across the globe to support the campaigns; and occasional negotiations to temporarily end violent conflicts so that vaccinators could reach vulnerable populations. Ultimately, it would require the diplomatic skills of William Foege, who headed the child-survival program, to gain the cooperation of the various organizations, particularly WHO and UNICEF, that possessed conflicting visions of health and styles of administration.[27]

In the end, UNICEF and its allies achieved their goal. This success, combined with the expanded use of other elements of GOBI (growth monitoring, oral rehydration, breastfeeding, and immunization) dramatically reduced child mortality across the developing world by 1990. Only in a dozen countries in sub-Saharan Africa, where a large portion of child deaths were due to malaria, for which there was no vaccine, did under-age-five mortality remain above 175 per 1000. This was a remarkable achievement. Yet it was obtained at the expense of efforts either to strengthen the health services of developing countries or to address a wide range of health issues that could not be eliminated through immunizations. These included acute respiratory infections, malaria, tuberculosis, venereal disease, maternal mortality, nutritional-deficiency diseases, cardiovascular disease, and various forms of cancer. Nor did it begin to address the wider determinants of health. WHO estimated that 40 percent of the world's populations lacked adequate sanitation, and 25 percent lived without a clean water supply in 1990.[28] As a result, epidemics of cholera and plague remained recurrent threats in many developing countries. Epidemics that did occur rapidly overwhelmed seriously underresourced health services in many countries, allowing the diseases to spread over ever-larger areas. Finally, immunizations could not eliminate poverty, or the lack of educational opportunities, or poor housing, or unemployment. They could not ensure that families would have a minimal level of resources needed to maintain health and stave off disease. They could not ensure that those who became sick were able to access adequate health care. It would be unfair to blame those who led the Child-Survival Revolution for not addressing these wider needs. It was not their goal or responsibility. It nonetheless is true that without these broader improvements, populations across the globe remained unable to achieve or sustain freedom from disease and disability.

As the twentieth century came to an end, the continued vulnerability of the majority of the world's population to disease was made evident by the emergence and global spread of HIV/AIDS and a number of other newly surfacing diseases. As international-health leaders scrambled to contain new global-health threats, they crafted responses that continued and, in many ways, accelerated the movement away from comprehensive approaches to health care and toward an increasingly medicalized, selective approach to health. They created the set of ideas and practices which defined global health in the twenty-first century.

PART VII

Back to the Future

In October 2007, African ministers of health, representatives of major nongovernmental organizations, and the heads of private foundations, UNICEF, WHO, and the World Bank met in Seattle to explore new ways to combat malaria. The disease had made a major comeback after the termination of the WHO-led Malaria Eradication Programme in the 1960s and was killing an estimated 1 million children a year in Africa by the late 1980s. In response to this crisis, health ministers from around the world had met in Amsterdam in 1992 to mobilize efforts to once again drive back the disease. This led to the creation of the Roll Back Malaria Partnership in 1998 and to the widespread adoption of a set of interventions, including insecticide-treated bed nets, which were slowly reducing malaria morbidity and mortality across the globe. The Seattle meeting was convened to review what had been achieved and to find ways to accelerate this progress.

The meeting was important because it reflected changes that had occurred in the landscape of international health since the early 1990s. Like many previous international malaria conferences, this one was held far from the places where malaria was a problem. Yet the fact that the meeting was held in Seattle, at the headquarters of the Bill & Melinda Gates Foundation, not at WHO headquarters in Geneva, was significant. The Gates Foundation was established in 2000, with over US$200 million in assets. Finding ways to improve international health was one of its central missions, and the foundation quickly became a major funder of global-health interventions. The location of the Malaria Forum in Seattle symbolized not only the growing influence of the Gates Foundation in international health, but, more broadly, a shift in the nexus of international-health leadership, away from WHO and toward a new group of powerful institutional actors, including the Gates Foundation, the World Bank, and the newly formed Global Fund to Fight AIDS, Tuberculosis and Malaria.

The composition of the meeting was also symptomatic of changes that had occurred in international health. While three African ministers of health were present, they were there because their countries were sites for Gates-funded malaria-control initiatives, not as national representatives. The Seattle meeting was not intended to be a democratic forum, like the World Health Assembly. Increasingly, organizations that did not represent a consensus of member states were making decisions regarding which strategies would be used for improving global health. As a consequence, the locus of international-health decision making in the early twenty-first century was even more removed from the places and people receiving the vast majority of global-health assistance.

The Malaria Forum also included the heads of new public/private partnerships dedicated to developing vaccines and antimalarial drugs, as well as representatives from the pharmaceutical and chemical industries, including Novartis, Sumitomo Chemical, Sanofi-Aventis, and BASF Corporation.[1] It thus highlighted the emerging role of the private sector in international health, an increased commitment to research and development, and a growing emphasis on the use of biotechnologies to solve the world's health problems. The presence of the World Bank, the Global Fund to Fight AIDS, Tuberculosis and Malaria, the Exxon Corporation, and the Gates Foundation itself, reflected an expansion and diversification of international-health funding. The presence of NGOs instead of ministers of health marked the extent to which new investments in international health were being dispersed in ways that often circumvented national governments. Bilateral aid agreements continued, but more and more funds were being funneled through NGOs.

The composition of the Malaria Forum was important in another way. It included women, who now held leadership roles in several international-health organizations. Melinda Gates delivered the keynote address. In the audience was Margaret Chan, director-general of WHO; Ann Veneman, executive director of UNICEF; Regina Rabinovich, director of Infectious Diseases Development at the Gates Foundation; Joy Phumaphi, vice president of the Human Development Network at the World Bank; and Awa Marie Coll-Seck, executive director of the Roll Back Malaria Partnership.

Women had always been involved in efforts to improve the health of the world's populations (introduction). Yet for much of the twentieth century, with few exceptions, such as Martha May Eliot (chapter 3) or reproductive-health activist Margaret Sanger, women did not hold positions of authority in international-health organizations. International health was a male-dominated enterprise until the 1970s, when women began to play a greater role in family-planning programs. Beginning in the

1980s, this pattern began to change, as more and more women chose to enter the field of public health. Schools of public health went from being dominated by male students in the 1970s to enrolling equal numbers of men and women by the early 1990s. Over the next decade, women took up an increasing number of leadership roles in international-health and development organizations. The impact of this shift is difficult to measure, especially since much of what constituted global-health activities in the early twenty-first century strongly resembled approaches to health that had been developed before women became centrally involved in global-health leadership.[2] But the role of women in leadership positions had clearly grown.

The Malaria Forum in 2007 was also significant because, in their introductory speeches, both Bill and Melinda Gates called for the global eradication of malaria. This was a bold proposal that rejected the commonly held view of malariologists, which was that malaria eradication was impossible. Melinda and Bill Gates argued that new technologies, organizational structures, political support, and financial resources made it possible to again make eradication the goal. These resources had created a new optimism about what could be achieved in the field of international health. While their proposal was bold, it also marked a return to the strategy of disease eradication, an approach that had characterized the work of the Rockefeller Foundation in the early twentieth century and international-health organizations in the 1950s, 1960s, and 1970s.

Commitments to building up basic health services and personnel, while acknowledged at the Malaria Forum, received relatively little attention. Also missing was any mention of the importance of multisectoral approaches to reducing malaria: combining malaria-control interventions with efforts to improve education, especially among women; increase economic opportunities; and reduce poverty. In short, the social and economic determinants of malaria were not addressed at the forum. This was especially notable because the Roll Back Malaria Partnership's global plan had called for the integration of malaria-control programs with broader development efforts.[3] While little progress had been made in achieving this goal, it at least had been promoted.

The absence in Seattle of any mention of the need to address broader development problems reflected the Gates Foundation's very explicit demarcation of its role in global health. It saw itself as supporting the development of new biomedical technologies that would have a major impact on the health of the world's populations. It did not wish to be drawn into discussions of larger development issues. The absence of any such discussion was also characteristic of global health more broadly in the first decade of the twenty-first century. The need to address the structural determinants of health, so prominent at the Alma-Ata conference, remained part of the *discourse* of global health. But it received very little serious attention or funding. WHO's Commission on the Social Determinants of Health, which was established in 2005, worked hard to fill the gap and resurrect interest and commitments in addressing health inequalities related to social and economic forces. Yet there was little evidence that this approach had gained purchase among global-health organizations.

Viewed against the backdrop of the longer history of international health, the 2007 Malaria Forum represented both change and continuity. The site of the forum, the growing importance of nonrepresentative bodies in global-health decision making, the mobilization of new financial resources and approaches to fighting disease, the diversity of organizations and groups—including women—represented at the meeting, the prominence of nongovernmental organizations and the private sector, all reflected changes that had occurred in the landscape of international health since Alma-Ata. On the other hand, the lack of any reference to the structural determinants of malaria or the need to build basic health services, and the return to a strategy based on the use of biomedical technologies to eliminate diseases one at a time, were part of the old international-health concept. This mix of old and new characterized the field that became known as global health in the 1990s and early twenty-first century, as it moved back to the future.

Chapter 14 describes the conditions that gave rise to a new set of ideas, organizations, and practices that constituted what has become global health. It describes how, beginning in the early 1990s, growing concerns about a set of new global-health threats, the most important of which was HIV/AIDS, stimulated the emergence of new organizations and sources of funding, including the World Bank.

Chapter 15 describes how these trends were accelerated by the development of new drugs that could control HIV. It discusses the struggles that occurred between AIDS activists from around the world and public-health leaders, and how this eventually led to new commitments on the part of donor nations to raising hundreds of million of dollars for global health. It also led to the creation of new organizations, including the President's Emergency Plan for AIDS Relief and the Global Fund to Fight AIDS, Tuberculosis and Malaria. These new organizations greatly increased the funding of global-health activities. At the same time they imposed conditions on how global health should be practiced. These conditions changed the landscape of global-health interventions.

Finally, chapter 16 describes the growing role of evidence-based medicine and randomized control trials in global health and how these, together with new demands for accountability, contributed to a growing reliance on the use of biomedical technologies. Moving into the twenty-first century, in many ways global health returned to it roots, focusing its energies on developing and applying new technologies to combat global health problems.

CHAPTER FOURTEEN

AIDS and the Birth of Global Health

GLOBAL HEALTH EMERGED as a set of practices, organizations, and ideas in the early 1990s, as the world's health community faced new disease threats. The most important of these was HIV/AIDS. From the early 1980s to the 1990s, AIDS went from being a rare wasting disease, seen among groups of gay men in Europe and the United States, to a disease that affected growing numbers of men and women in sub-Saharan Africa, to a threat to all humanity. Linked to the growing AIDS epidemic, though initially independent of it, was an emerging epidemic of multidrug-resistant TB, produced by the failure of health systems around the globe to adequately treat TB patients. By the early 1990s, these new plagues were joined by epidemics of emerging diseases that spread rapidly across the globe. Dengue hemorrhagic fever, severe acute respiratory syndrome (SARS), and various deadly forms of influenza traveled along transportation routes that were the product of globalization: the increasing integration of the world's population into a single, global, economic system.

These events connected the health (and sickness) of people in Baltimore and Chicago with that of people in Nairobi and Delhi and led to a new awareness of the interconnectedness of global health. Plagues were no longer limited to the populations of underdeveloped countries. As a *Time* magazine article on the SARS epidemic noted in 2003: "It is becoming clear that what is taking place in Asia threatens the entire world. Epidemiologists have long worried about a highly contagious, fatal disease that could spread quickly around the globe, and SARS might end up confirming their worst fears. Microbes can go wherever jet airliners do these days, so it is a very real possibility that the disease has not yet shown its full fury."[1]

These fears were heightened after the 2001 terrorist attack on the World Trade Center in New York City, and the subsequent release of anthrax in the nation's capital. Fears that terrorists might employ biological weapons to attack the United States contributed to concerns about traveling microbes and the need for improved global-health surveillance and intervention. It also sparked a massive increase in the funding of emergency-preparedness measures, aimed in part at preventing infectious-disease disasters. For example, the Centers for Disease Control and Prevention's annual budget increased from US$2.1 billion in 1998 to US$7.7 billion in 2002. These fears also led to fresh outpourings of support for health programs in developing countries, aimed at heading off future plagues. Total funding for global-health activities rose from US$5.8 billion in 1990, to US$27.7 billion in 2011, to over US$30 billion in 2013.

Neither the fears nor the efforts to attack diseases overseas were new. The US invasion of Cuba in the nineteenth century (chapter 1) was stimulated in large measure by the need to prevent yellow fever from spreading to cities in the southern United States. The *Time* magazine article's reference to jet airlines carrying infectious agents across the globe reminds us of the airline map and the 1932 Cape Town conference's concerns about the global extension of yellow fever. Yet the speed with which microbes could travel across the globe had increased dramatically, as seen in the spread of SARS. The first case was reported in China on February 15, 2003. Less than a month later, on March 5, a patient had died from the disease in Toronto, Canada.

These fears were increased by the near-instantaneous speed with which information about epidemics traveled across the globe. The experiences of individuals in the middle of a major outbreak of dengue fever in Rio de Janeiro in 2010 were broadcast globally, via thousands of Twitter messages. Sitting at home in Baltimore, I could share the alarms and anxieties of people living thousands of miles away, as well as their occasional sense of humor: "If it makes you hot and takes you to bed it is dengue, not love." I could also view real-time reports of dengue cases occurring in Singapore on my iPhone, via a government-supported app called Dengue X.

Dread of the global spread of epidemic diseases was only one of the forces defining global health in the early 1990s. A second was a growing belief among international donor agencies and governments that diseases like AIDS and malaria were a major drain on the economic development of resource-poor countries. The idea that disease undermines development was not new. It had been a central argument used by public-health authorities to justify investments in health going back to the beginning of the twentieth

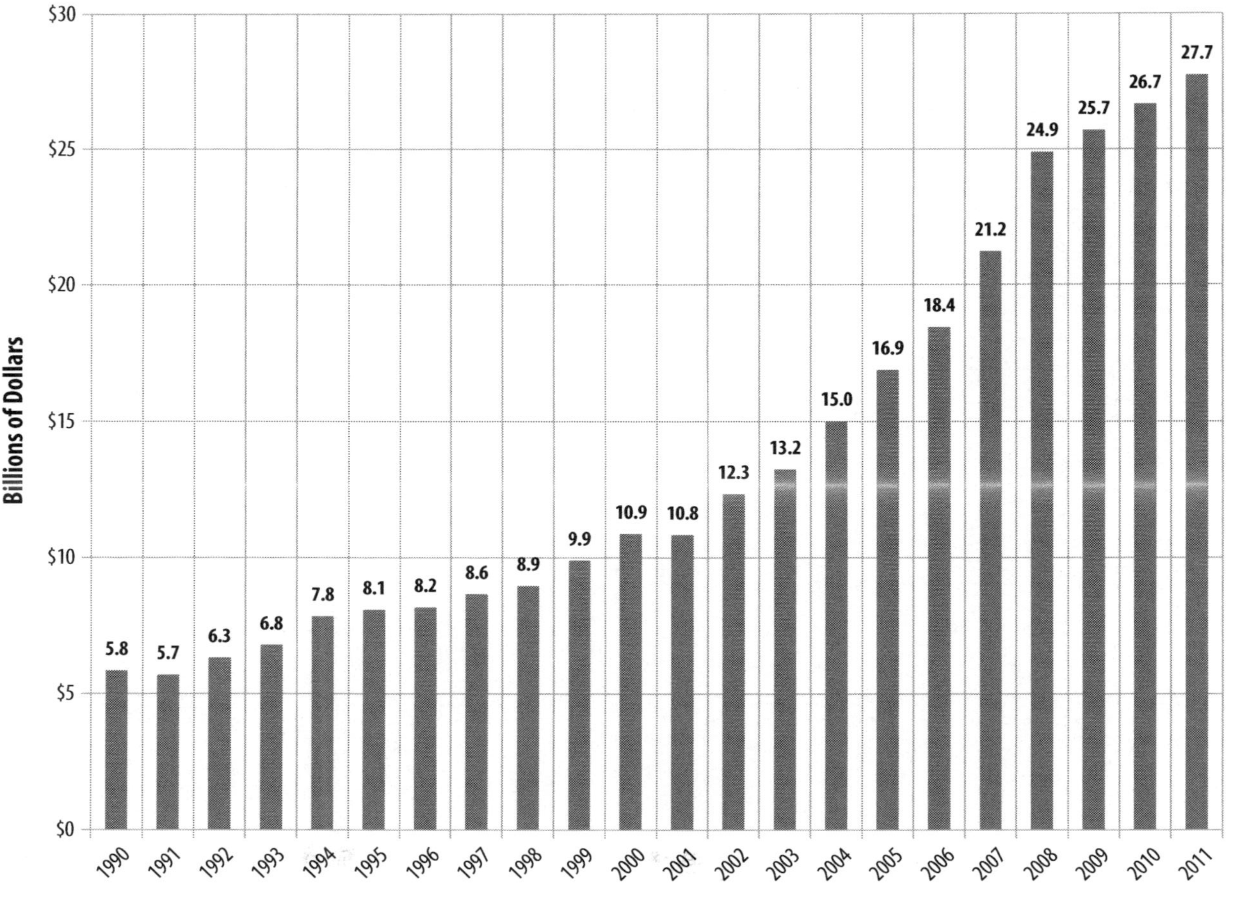

The growth of global-health funding, in billions. Institute for Health Metrics and Evaluation, *Financing Global Health 2012: End of the Golden Age?* (Seattle: University of Washington, 2012), 13.

century and earlier. Yet these arguments gained wider acceptance in the 1990s. Central to this new popularity was a recognition of health as an economic variable, both by economists at the World Bank and by the work of WHO's Commission on Macroeconomics and Health, established in 2000.

The landscape of global health was also defined by the conditionalities set by the organizations providing the bulk of global-health funding during the 1990s and early 2000s. These stipulations reflected the concerns of donor countries and organizations in the global north about the ineffectiveness and inefficiency of existing patterns of development aid for health. Critics of this aid, particularly in the United States, argued that too much aid money was being wasted on programs that were unable to demonstrate impact. Other saw resources being lost within bloated and corrupt governmental bureaucracies, which were unable to effectively deploy the resources they received or, worse, channeled them into private bank accounts. Critiques of how aid monies were being spent also flowed from neoliberal economists and politicians who sought to reduce the role of governments in the provision of health services for their citizens (chapter 13). These critics called for more privatized health systems, in which patients bore a larger share of their health costs, and for development assistance to be funneled through nongovernmental organizations. Neoliberal ideas, which had gained ascendancy during the Reagan and Thatcher years, had a much wider influence during the 1990s. The breakup of the Soviet Union and the decline of socialism as a counterweight to capitalist/free-market ideologies in the late 1980s were viewed as validating neoliberal policies, at the same time that they eliminated alternative sources of bilateral aid from socialist countries.[2] These changes increased the hegemony of neoliberalism in the world of global development.[3]

In line with neoliberal efforts to reduce the role of governments in health and avoid funding inefficient governments, the 1990s saw an explosion of both locally based and international NGOs. According to UNESCO, the number of NGOs linked to UN agencies rose from just over 9000 in 1980, to 17,419 by 1990, and then to nearly 28,000 by 2006.[4] But this was only the tip of the iceberg. The numbers did not include thousands of local NGOs, run by individuals in small storefronts or with only a briefcase, that operated locally in various countries across the globe. There were an estimated 100,000 NGOs in South Africa; 300,000 in Brazil; and over a million in India by 2005.[5]

NGOs came in all flavors. Some, such as Partners In Health and Doctors Without Borders, worked independently to bring improvements in health to peoples living in resource-poor settings, though they operated in very different ways (conclusion). Others were dedicated to raising funds to support ongoing health interventions or research. Beginning in the mid-1980s with the Ethiopian famine, the number of organizations dedicated to raising charitable contributions to help particularly hard-hit populations in the global south grew dramatically. Still others, including thousand of local-stakeholder organizations, worked at the local level, providing partners to implement policies designed by bilateral and multilateral organizations.

Many NGOs were a source of creativity, developing innovative grassroots programs that supplemented, and sometimes replaced, health services that were not furnished by the public sector in resource-poor countries. In many rural areas in Africa, the interventions provided by NGO-run health programs were often the only health services available. Yet NGO programs seldom replaced the need for a well-developed, government-run health system. NGO-operated health programs often targeted specific health problems and did not provide the basic infrastructure and medical personnel needed to sustain health care. NGO-based care also failed to provide systems for surveillance and case reporting that would prevent disease outbreaks from getting out of hand. Finally, NGO programs were most often supported by outside sources with funding that could be reduced or disappear completely. In short, reliance on NGO health programs could be a double-edged sword.

Donors concerns about the ineffectiveness of existing UN agencies, particularly WHO, in managing global-health funding also defined global health in the 1990s. For many involved in trying to initiate and manage global-health programs, the bureaucratic inefficiencies and internal politics of WHO had made it part of the problem, not the solution. They viewed WHO as a place where good ideas went to die. If the world was going to dedicate vast amounts of new financial resources to improving global health, different mechanisms had to be found for distributing these resources. This led to the creation of new organizations, such as the Global Fund to Fight AIDS, Tuberculosis and Malaria; new public/private partnerships, like the Global Alliance for Vaccines and Immunizations; and private foundations, including the Bill & Melinda Gates Foundation. These organizations demanded accountability. Funded programs needed to both keep careful track of the funds they received and demonstrate impact.

For its part, the World Health Organization, buffeted by the winds of neoliberalism, adapted to the times. It encouraged programs aimed at promoting economic growth, moved from primary health care to supporting disease-specific interventions, employed cost-effective calculations as a basis for policy choices, and supported market-driven solutions. Many of WHO's programs were designated as partnerships that included private foundations and multinational companies. WHO attempted to maintain some of its commitments to universalism and to serving the needs of the poor. It also made chronic noncommunicable diseases, including cardiovascular disease and mental illness—problems receiving little attention from other global-health organizations—a central concern. In addition, it launched campaigns against the bottlefeeding of infants and tobacco use. Yet WHO's limited financial resources prevented it from developing an effective counterbalance to the emerging neoliberal approach to global health that was promoted by the World Bank, the Global Fund, and major bilateral aid initiatives.[6]

Finally, the need for accountability led to an increased reliance on evidence-based interventions. Randomized controlled trials, long the gold standard for measuring efficacy within the pharmaceutical industry, became a central tool for evaluating the effectiveness of global-health interventions. Yet some interventions were easier to test than others. It was relatively easy, though clearly not without challenges, to test a new vaccine, vitamin supplement, insecticide-treated bed net, or TB drug. It was extremely difficult, however, to set up a trial to measure the benefits of a community-based primary health care system, clean water, or sanitation. The emphasis on quantifiable results therefore privileged approaches centered on the administration or distribution of technologies and contributed to an increasing commodification and medicalization of global health.

The remainder of this chapter traces the early history of these trends, as well as how they defined global health and transformed the ways in which interventions into the lives of other peoples were constructed and deployed at the turn of the twenty-first century.

The AIDS Epidemic and the Growth of Global-Health Funding

There can be little doubt that the emergence of AIDS as a threat to global health, as well as to economic growth and political stability of many parts of the world, was a driving force behind the expansion of global-health funding.[7] It took time, however, for this response to build. It also required an unprecedented mobilization of local activist groups from around the world to make AIDS funding a high priority for bilateral and multinational organizations. The global response to HIV/AIDS, unlike responses to earlier disease threats, was greatly influenced by the interests and actions of people living with the disease.

The dimensions of the AIDS problem outside Europe and America were not immediately apparent to global-health leaders. The absence of epidemiological capacity and basic health services in many developing countries, particularly in Africa, allowed the epidemic to percolate largely undetected for years. It was not surprising that the tests which identified the first Central African cases of AIDS were performed in Belgium, not in Africa. In addition, the denial by governments that AIDS was a problem limited local efforts to address the growing number of cases. This was particularly true in Africa. Early scientific and media claims that the disease had originated in Africa led a number of African leaders to push back and accuse the West of making Africa a scapegoat for the disease. Some argued that AIDS was a problem of homosexuals in Europe and America. Homosexuality, they wrongly believed, did not occur in Africa, so there was no AIDS there. They were tragically misguided.[8]

By the mid-1980s, however, several African countries, including Uganda and Sierra Leone, had begun to seek WHO assistance in controlling the disease. The organization's initial response was to dismiss these requests. Director-General Mahler viewed attempts to mobilize a focused attack on HIV/AIDS as a diversion from WHO's efforts to build primary health-care systems. He also argued that AIDS was of minor importance compared with other diseases, such as malaria.[9] Yet recognition that there was a simmering heterosexual epidemic of AIDS in Africa and other parts of the developing world was growing, and in 1986, WHO established a Special Programme on AIDS, which later became the Global Programme on AIDS (GPA).

The GPA's first director was Jonathan Mann. Mann had worked for Project SIDA in Kinshasa, where many of the early studies of HIV prevalence had been carried out. Mann, a health activist and tireless promoter of health and human rights, succeeded in attracting substantial outside financial support for the GPA, increasing its funding from US$30 million in 1987 to US$90 million in 1990. This made the GPA the largest single program in the history of WHO. Mann concluded agreements with 155 countries to establish AIDS programs for protecting blood supplies, training medical staff to care for AIDS patients and counsel those who tested positive for HIV, providing public education to modify sexual behaviors, and preventing discrimination. Yet progress in reducing the spread of HIV was limited, particularly in sub-Saharan Africa, by both internal struggles within WHO and the narrow focus of the GPA's initiatives.

By 1990, some within WHO felt that the AIDS program had become so large that it was becoming the tail that wagged the dog. The size of the GPA budget was symptomatic of a larger problem facing WHO. The organization was being starved of cash. Financial support for WHO was being directed increasingly to targeted programs, instead of to the organization's general fund.[10] By the early 1990s, special funds exceeded the general budget by US$21 million.[11] Categorical funding limited the ability of WHO's director-general and its Executive Board to develop programs and maintain the organization's leadership position in global health.

Not surprisingly, one of those who resented the GPA's dominance was WHO's new director-general, Hiroshi Nakajima, who had succeeded Mahler in 1988. Nakajima felt that too much WHO money was being spent on AIDS and not enough on other problems, such as malaria, tuberculosis, and diarrheal diseases. Yet because the GPA program was funded by external sources, Nakajima had limited control over its operation. He attempted instead to limit Mann's autonomy within WHO and to funnel GPA funds to other programs. He also tried to decentralize the GPA by shifting personnel and resources to regional WHO offices. Mann opposed this move, because he was convinced it would limit the GPA's ability to act quickly and decisively in responding to the evolving AIDS epidemic.[12] Difficulties working within WHO's structure led to Mann's resignation as head of the GPA in March 1990.

Early understandings of AIDs in Africa also hampered progress. In the beginning, the growing African heterosexual epidemic of AIDS was viewed by AIDS researchers from the global north as a behavioral problem. Studies of HIV among sex workers and truck drivers indicated that AIDS in Africa was a problem of Africans engaging in sex with multiple partners. The best way to combat the epidemic, therefore, was through sex education and the distribution of condoms. This early focus of AIDS research on African sexual behavior and other cultural practices precluded a broader inquiry in the social and economic conditions that were shaping the AIDS epidemic. Anthropologists, who were recruited to develop behavioral strategies for limiting the spread of HIV, were often complicit in promoting this approach and reinforcing African sexual stereotypes.[13] Behavioral-education programs had a limited impact in Africa, however, because they failed to address the structural conditions that produced vulnerability, particularly among women and young children.[14] Campaigns, such as "zero grazing" in Uganda, raised awareness of the dangers of multipartner sex but failed to address the underlying determinants of sexual behavior. In this respect, the GPA's response to AIDS resembled early efforts to limit fertility through the distribution of birth control technologies, without exploring the conditions that encouraged women to have multiple children. Like family planning, early AIDS campaigns in sub-Saharan Africa had marginal success in altering sexual behavior.

Beginning in the late 1980s, however, a series of studies focused on the role of poverty, economic exploitation, gender power, sexual oppression, and racism in creating social vulnerability to AIDS among groups and individuals.[15] Anthropologist Brooke Schoepf was one of the first to explore issues of sexual oppression and vulnerability and link these to the growing economic crisis in Africa during the 1980s. Schoepf highlighted how declining economic opportunities in rural areas forced women to seek employment in cities like Kinshasa. There they frequently found it necessary to engage in sexual activity in order to gain an economic foothold, whether it was to get a trading license, housing, or education. This placed them at risk of infection with HIV.

Schoepf also examined the difficulties women faced in negotiating safe sex with men, including their husbands. Men often insisted on not using condoms, because condoms interfered with their sexual pleasure, or because they wanted children, or because they believed that semen was essential to both the sexual fertility of their wives and the growth of their unborn child. Schoepf was quick to note that these cultural ideas were not the source of the problem. Rather, it was the dependency that women faced in relation to men. Married women feared that their husbands would divorce them if they did not have sex, even when one or the other partner knew that they were HIV positive. Within the deepening economic crisis, divorce represented a serious threat to a woman's survival. According to Schoepf, empowering women through education and the provision of economic opportunities would permit them to better control their sexual lives and resist HIV infection.[16]

By the early 1990s, AIDS had become a disease of poverty, with nearly 97 percent of the cases occurring in low-income countries, and the largest percentage of these cases being found in sub-Saharan Africa. Poverty and AIDS were linked in complicated ways that limited the effectiveness of educational and marketing strategies. Many in the AIDS community came to see the disease as a multisectoral problem, which could not be addressed by public-health interventions alone. The GPA, however, was still focusing on behavioral change and condom use.

In 1989, donor organizations called for a review of the Global Programme on AIDS. The results were published in 1992. That report praised the GPA for what it had achieved in raising global awareness of the AIDS problem and stimulating the development of prevention programs across the globe. But the report also criticized the GPA for its inefficiency in coordinating activities among different UN agencies, noting that these agencies had not systematically shared information or coordinated the development of their AIDS policies and programs. Duplications of effort and territorial rivalries also threatened to weaken the global response to AIDS. The report stated that some UN agencies had developed unilateral programs and activities that replicated those of the GPA and other agencies. The need to develop a coordinated multisectoral approach to AIDS among the various specialized UN agencies led to the decision to create a new Joint United Nations Programme on HIV/AIDS, which became known as UNAIDS. Peter Piot, a Belgian AIDS researcher who had worked with Mann in Project SIDA, was appointed as the first director. UNAIDS was officially launched in December 1995.[17]

Unlike the GPA, UNAIDS was not intended to provide funds directly to national AIDS programs, but to coordinate funding coming from other agencies. Meanwhile, outside contributions to the GPA quickly dried up. This created a gap in AIDS funding. Country programs found themselves strapped for cash. It was during this period that the World Bank became actively involved in supporting AIDS-related programs, beginning what would become a major buildup of its support for global health.

Enter the World Bank

The World Bank, like WHO, came into the battle against AIDS slowly. Funding health programs had not been a priority for the bank until the 1970s, when its president, Robert McNamara, visited West Africa and became convinced that river blindness was a roadblock to agricultural development in the region.[18] The bank subsequently agreed to support the river blindness–campaigns. Despite this support, the bank's economists did not wholeheartedly accept the argument that improvements in health could have direct economic benefits. One of their major objections to this way of thinking was that health interventions also drove increases in population, which, in the long run, could undermine development efforts (chapter 9). In 1986, the World Bank published a report by University of Michigan economists Robin Barlow and Lisa M. Grobar that examined the impact of malaria-control programs on economic growth in Sri Lanka. The report questioned whether malaria control could be justified on economic grounds, noting: "There is . . . the strong possibility, in view of the powerful demographic effects of malaria control, that the output increase will be swamped by population increase, and that an economic crisis will be produced."[19] The negative effects of uncontrolled population growth had been a central concern of the World Bank since the 1960s, and the bank had invested heavily in family-planning programs.

Nonetheless, in 1993, the bank announced a new commitment to funding health programs as part of its broader mission to promote global economic growth and reduce poverty. The bank's new approach was communicated in its 1993 World Development Report, titled *Investing in Health*, which asserted: "Good health, as people know from their own experience, is a crucial part of well-being, but spending on health can also be justified on purely economic grounds."[20] The report cited the Sri Lankan experience as evidence that investments in health paid economic dividends. So why did the bank's economists change their minds in 1993?

The World Bank's embrace of health as an economic variable was greatly influenced by the work of Harvard economist Christopher Murray. Murray and his colleagues constructed a new metric for measuring the "burden of disease" in terms of disability-adjusted life-years (DALYs) that were lost on account of a disease or another source of physical or mental disability. DALYs were not intended to explicitly measure the economic costs of ill health. The assignment of age-weights, however—which attributed different social values to the health disabilities of individuals at different stages of life, with the highest values going to those in their most productive years—implicitly linked DALYs to an econometric view of health that the World Bank's economists could understand.

DALYs had another virtue: they promised to provide a metric for making sound economic choices about how to allocate scarce development dollars. Health problems that caused the greatest disability, or disability life-years lost, or health interventions that led to the greatest disability life-years saved, should be given high priority. In actual practice, DALYs were seldom employed in this manner. Political factors often shaped health priorities. Moreover, an econometric view of health mandated that the benefits from reducing disability be measured in relation to the costs of doing so. Cost-effectiveness was the mantra that drove the World Bank's thinking. Nonetheless, DALYs provided bank economists with a way to measure and justify the bank's increased involvement in funding health programs. Not surprisingly, calculations of disability life-years lost or gained could be found throughout *Investing in Health*.[21]

Investing in Health was a landmark document in the history of global-health funding. Thus it is worth examining a bit closer. It was particularly noteworthy because of the way it attempted to strike a balance between supporting broad-based health strategies and advocating cost-efficient approaches to health-care delivery. In this way, it represented the two poles between which international-health efforts had swung since the beginning of the twentieth century. On the one hand, the report acknowledged the social determinants of health and called for broader investments in programs aimed at reducing poverty and increasing education as important strategies for improving global health. It also supported greater investments in public-health services, such as immunization programs. Third, it called for a reallocation of health-care resources from tertiary to primary care.

On the other hand, it criticized existing patterns of health-care funding and delivery in low- and middle-income countries. It made a case for the privatization of health care by increasing individual responsibility for health-care funding through such mechanisms as user fees, and by promoting a greater reliance on NGOs and the private sector. It also called for competition in the provision of health services. Governments should not and could not effectively provide for the health needs of their citizens. The World Bank's support for privatization and user fees was not new; it had been part of the contingencies the bank had imposed on debtor countries in return for providing new loans following the 1980s debt crisis (chapter 13).

The report also argued that investments in health care should focus on the provision of what it called "critical health services." By this it meant a package of programs aimed at particular health problems, as opposed to the more inclusive health services imagined at the Alma-Ata conference. For low- and middle-income countries, this package would include the following services:

1. short-course chemotherapy for tuberculosis;
2. management of the sick child;
3. prenatal and delivery care;
4. family planning;
5. treatment of STDs; and
6. limited care (treatment of pain, infection, and minor trauma).[22]

The critical-health-services approach was a variant of selective primary health care, a strategy that the bank had supported since the Bellagio meeting in 1984. Notably absent from the package at that time was any direct mention of AIDS. This would soon change.

Overall, despite its nod to the social determinants of health, *Investing in Health* was a manifesto for selective, cost-efficient approaches to health delivered through nongovernmental organizations and actors. The World Bank was announcing that it was going to put its financial weight behind global health, but in a way that reflected the bank's econometric/neoliberal approach to health financing. Having laid out a case for investing in health, the bank proceeded to ramp up its provision of loans, grants, and credits for health, nutrition, and population (HNP) programs. From only one HNP loan in 1970, the bank's investments grew to 154 active and 94 completed projects in 1997. In terms of a percentage of all World Bank investments, HNP programs grew from 1 percent of the bank's investments in 1987 to 24 percent (or US$5 billion) by 1997.[23] By the beginning of the twenty-first century, the World Bank had become the largest funder of global-health programs. Accordingly, the bank increased the number of staff devoted to health programming. Yet bank officials recognized that they did not possess the expertise needed to advise countries on their health policies or to develop new global-health initiatives. They therefore partnered with other organizations, including WHO, UNICEF, and, later, UNAIDS.

There were inherent tensions in the World Bank's relationship to these organizations, since the bank's approach to health did not always coincide with the goals and visions of the organizations with which it agreed to partner. This was particularly true of its dealings with WHO. Bank officials were critical of WHO's inability to effectively manage health programs. They regarded WHO as being hamstrung by political competition among member states and corrupted by the political nature of appointments to leadership positions. Because the organization was made up of representatives of individual nations, its programs had to be funneled through governmental ministries, which the bank viewed as inefficient. The bank's strained relations with WHO were particularly evident during the early 1990s, when efforts were made to create a new Children's Vaccine Initiative (CVI). The goals of CVI changed over time, but its intent was to develop and deploy effective vaccines to prevent all common childhood diseases around the globe, with the holy grail being a single, multivalent vaccine. The CVI, supported by a number of international organizations and private-industry partnerships, was plagued by institutional rivalries and by tensions between the United States and European countries over control of the initiative. Central to these tensions was the World Bank's distrust of WHO's leadership of CVI, which led the bank to attempt to wrest control of the initiative from Geneva.[24]

In general, the World Bank insisted that the programs it supported conform to its vision of health investments. Thus the bank joined other multinational and bilateral organizations in supporting the Roll Back Malaria Partnership (RBM), which was established in 1998 with the stated goal of decreasing the global occurrence of malaria by 50 percent by 2005. RBM's

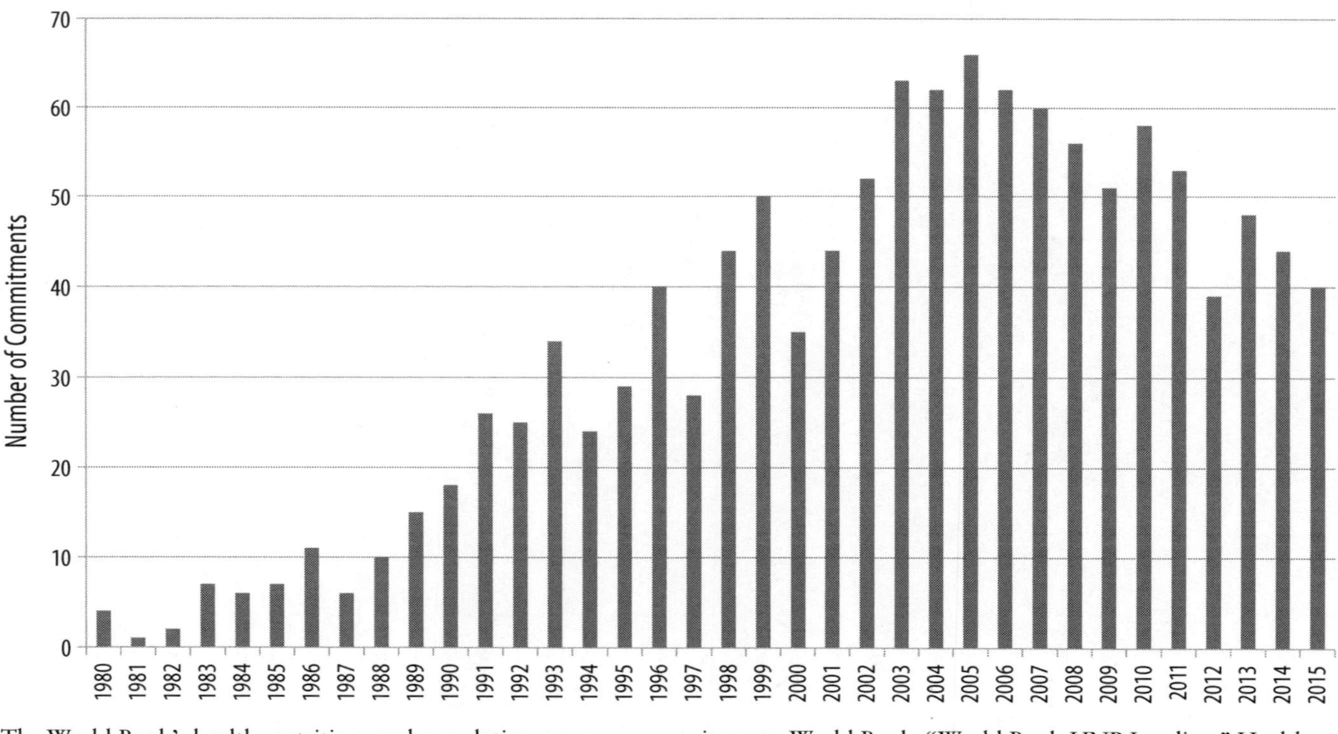

The World Bank's health, nutrition, and population program commitments. World Bank, "World Bank HNP Lending," Health, Nutrition, and Population Data and Statistics, http://datatopics.worldbank.org/hnp/worldbanklending/.

global strategy for reducing the burden of malaria focused on the use of insecticide-treated bed nets (ITNs), the timely and effective treatment of children with fevers, the prevention of malaria during pregnancy, and a rapid response to malaria epidemics. It provided loans and credits to countries that developed plans for implementing RBM-approved strategies. It also worked directly with finance ministers to convince them of the value of supporting increased spending on malaria control. Yet the bank insisted, during the early years of the RBM, that ITNs, the cornerstone of the RBM strategy, be sold rather than distributed for free to at-risk populations. For the bank, ITNs were the ideal neoliberal health intervention. They represented a commodity that could be marketed to populations at risk, shifting responsibility for the costs of health care away from the state and onto individuals. The bank also pushed for new forms of accountability from borrowing countries.[25] Finally, the Bank's lending arm placed limits on governmental spending in the countries they funded. These limits were part of the bank's structural adjustment programs. For some African countries, this meant that governments were not able to hire more health workers, even when donors were willing to finance human resources.[26] These limits also led the finance minister of Uganda to reject a US$36 million Global Fund grant for AIDS in 2003, fearing that the grant would violate the governmental spending limits it had agreed on with the World Bank and IMF as part of its structural adjustment agreement.[27]

The World Bank's conditionalities were characteristic of global-health funding from the 1990s. Increasingly, the new funding sources, which would contribute hundreds of millions of dollars to global-health programs, imposed requirements that encouraged selective approaches to public health; privileged the role of NGOs over national governments; increased the role of the private sector; and demanded new forms of accountability, or evidence-based public health.

Much of the bank's new commitment to health funding was directed toward supporting AIDS prevention and treatment programs, a major source of disability life-years lost. Between 1986 and 1997, the bank committed US$500 million in loans and credits to support freestanding projects and national AIDS programs throughout the world. In 2000, the bank launched the multicounty AIDS program in Africa (Africa MAP), to which it provided US$1 billion.[28] Funds were issued through loans and credits to support country-based AIDS strategies. In keeping with the bank's commitment to working through nongovernmental channels, however, it stipulated that World Bank monies be funneled through NGOs and community-based organizations. By 2013, the bank claimed to have funded 65,000 civil-society initiatives in sub-Saharan Africa.[29] Also in line with the bank's lending policies, Africa MAP placed a strong emphasis on monitoring and evaluation and a commitment to streamlining the bank's funding and implementation procedures.

By 2013, the World Bank had provided US$4.6 billion for HIV/AIDS-related activities. Where possible, the bank participated in the pooling of funds, as part of its partnership with UNAIDS. The bank was soon joined by new organizations that were dedicated to funding AIDS programs and inspired by the promise and challenges of new therapeutic responses to the AIDS epidemic and other diseases. These organizations—the Bill & Melinda Gates Foundation, the President's Emergency Plan for AIDS Relief, and the Global Fund to Fight AIDS, Tuberculosis and Malaria—played a major role in defining global health at the beginning of the twenty-first century.

CHAPTER FIFTEEN

The Global Fund, PEPFAR, and the Transformation of Global Health

WHILE THE AIDS EPIDEMIC was a stimulus to the growth of global-health funding in the 1990s, it was the development of new drugs, which could reverse the course of AIDS in individual patients, that boosted global-health funding to unimagined heights. Antiretroviral drugs (ARVs) greatly increased the need for larger investments in global health and led to the creation of new funding mechanisms. The most important were the Global Fund to Fight AIDS, Tuberculosis and Malaria (Global Fund), as well as the President's Emergency Plan for AIDS Relief (PEPFAR). These organizations provided billions of dollars to fight major global diseases. Yet the monies they supplied came with conditions that further defined the landscape of global health.

In 1995, the US Food and Drug Administration approved saquinavir, the first of a new class of ARVs called protease inhibitors, for the treatment of AIDS. This was followed in 1996 by the approval of neviraprine, the first in another class of ARVs known as non-nucleoside reverse transcriptase inhibitors. Later that year, at the International AIDS conference in Vancouver, researchers announced that a regimen containing a combination of three ARVs from at least two classes suppressed the AIDS virus and restored patients' immune systems.[1] This drug regimen, which became known as highly active antiretroviral therapy (HAART), transformed the world of AIDS treatment. It provided a potential magic bullet, another technological solution to a global-health crisis. The only problem was that it cost between US$10,000 and US$15,000 per patient per year, a price that made its widespread use in most of the world impossible. The price of the regimen quickly reinforced the geography of biomedical inequality. For those in richer nations who could afford the drugs or who had insurance that would pay for them, or for elites in poorer nations, AIDS quickly became a controllable chronic disease. In the United States, HAART replaced azidothymidine (AZT), which had only slowed the virus, as the first-line therapy for AIDS. As a result, AIDS-related mortality in the United States declined by 46 percent between 1996 and 1997, and another 18 percent between 1997 and 1998. Meanwhile, in Africa and the Caribbean, AIDS mortality continued to rise.[2]

This disparity triggered a struggle between international AIDS activists, demanding that HAART be made available globally, and public-health and development officials, who countered that HAART was not cost-effective and that AIDS programs in Africa should continue focusing on prevention strategies.[3] Objections to the provision of HAART in Africa also focused on Africa's lack of health infrastructure and the perceived inability of Africans to adjust to a complex drug regimen that required the carefully timed consumption of multiple drugs. In testimony before the US House of Representatives, Andrew Natios, head of USAID, stated: "If we had them [ARVs] today, we could not distribute them. We could not administer the program because we do not have the doctors, we do not have the roads, we do not have the cold chain. This sounds small and some people, if you have traveled to rural Africa you know this, this is not a criticism, just a different world. People do not know what watches and clocks are. They do not use Western means for telling time."[4]

The idea that the drugs could not be used because Africans could not tell time represented one of many cultural stereotypes that characterized Western perceptions of AIDS in Africa and shaped international AIDS policies during the 1990s. Those who maintained that Africans would be unable to take their drugs on time raised fears that introducing ARVs in Africa would lead to "therapeutic anarchy" and the development of drug resistance, which would undermine the drugs' effectiveness everywhere. Paul Farmer and his colleagues from Partners In Health, working in Haiti, where ARVs were used effectively to treat patients in a resource-poor environment, demonstrated the falsity of such "immodest claims." Médicins Sans Frontières achieved similar success with ARVs in the South African township of Khayelitsha, located outside Cape Town.

Yet the financial burden of ARVs remained a barrier to their use among the world's poor. The solution to this problem came out of the global south, where countries sought to find ways to drive down the price of ARVs. In the mid-1990s, a combination of public laboratories and private drug companies located in India and Brazil began producing generic formulations of ARVs, which led to a 70 percent drop in their cost. Other countries followed suit. In 1997, South Africa's parliament approved the Medicines and Related Substances Control Amendment Act, which permitted the production and parallel importation of generic versions of patented medicines without the patent holders' permission in the case of a national medical emergency.

This act quickly ran afoul of intellectual-property rights, enforced by the World Trade Organization (WTO), which had been founded in 1995 to "open trade for the benefit of all" by regulating trade and breaking down barriers and tariffs. Critics of the WTO argued that the organization served to promote the economic interests of wealthier industrial countries at the expense of commodity producers. It was frequently used to defend the intellectual-property rights of entertainment, pharmaceutical, and software corporations located in the global north. But intellectual-property rights prevented corporations in India, Brazil, South Africa, Thailand, and China from producing and marketing less-expensive drugs for resource-poor countries where they were desperately needed.[5]

The passage of the Medicines and Related Substances Control Amendment Act challenged intellectual-property rights and set off a firestorm of protests from the heads of northern pharmaceutical companies, who joined together to sue the South African government for rights infringements. The US government initially backed the suit, threatening economic sanctions against the South African government. At the 2000 International AIDS Conference in Durban, AIDS activists, including South Africa's Treatment Action Committee, loudly protested the failure of the United States and European countries, pharmaceutical companies, and the WTO to take actions that would make lifesaving AIDS drugs affordable and available globally. Soon afterward, under pressure from AIDS activists and the US Congress's Black Caucus, the Clinton administration reversed course and announced that the administration supported the Medicines and Related Substances Control Amendment Act. The lawsuits against the act were dropped in April 2001, and the WTO affirmed that the international agreement on trade-related aspects of intellectual-property rights (TRIPS) did not prevent the emergency measures empowered by the act. The prices of ARVs quickly plummeted, from US$10,000 to US$15,000 in the mid 1990s, to US$300 in 2002, and US$87 in 2007.[6]

Meanwhile, in South Africa, the government's Ministry of Health, influenced in part by President Thabo Mbeki's belief that AIDS was a product of poverty and apartheid rather than HIV, continued to oppose government-sponsored programs for the provision of AIDS drugs. Those seeking the ARVs would have to find them in the private sector or be enrolled in a clinical trial. Mbeki's evocation of history and apartheid was not without merit. It underlined the important political and economic conditions that produced both vulnerability and the cofactors that shaped the AIDS epidemics. Yet his refusal to accept that HIV was a direct cause of AIDS led many AIDS activists to reject Mbeki's intervention. Anyone invoking the broader structural conditions of AIDS were accused of "denialism." That reaction was understandable, given the life-and-death battle being fought over access to ARVs. But it nonetheless deflected attention from the nonmedical factors shaping the AIDS epidemic, while empowering a biomedical solution.

The decline of drug prices raised the possibility that treatment could be provided for the millions of people living with AIDS in Africa and other impoverished regions of the globe. Yet even at the reduced price of ARVs at the turn of the twenty-first century, the cost of mass-treatment programs remained formidable. The alliance of AIDS activists, academics, and politicians from the global north and south, who fought against the pharmaceutical industry's efforts to prevent the production and marketing of generic ARVs, also pressured donor governments and international organizations to provide the funds needed to make ARVs universally available. It was in response to this challenge that representatives of the G8 countries (France, the United States, the United Kingdom, Russia, Germany, Japan, Italy, and Canada), meeting in Denver in 1997, committed their countries to providing funds for an accelerated attack on AIDS. This promise was repeated and expanded to include TB and malaria when the G8 met in Okinawa in 2000. In a January 2001 *Lancet* article, economist Jeffrey Sachs, together with Amir Attran, calculated the amount needed to effectively bring HIV/AIDS under control worldwide would be US$7.5 billion annually, whereas over the previous decade, worldwide oversees development funding had never exceeded US$144 million annually.[7]

In early May 2001, UN Director-General Kofi Annan, in a speech to a special session on AIDS at the United Nations, called for the creation of a fund that would add from US$7 billion to US$10 billion to the current level of spending on AIDS, TB, and malaria. Representatives of the G8 group of countries endorsed the creation of such a fund at their annual meeting in Genoa the following July. This led to the creation of the Global Fund to Fight AIDS, Tuberculosis and Malaria in 2002.

The Role of the Global Fund in Defining the Terrain of Global Health

From the beginning, the Global Fund was intended to be a different kind of funding mechanism.[8] It would avoid the perceived limitations of existing institutions and avenues of funding by demanding accountability on the part of the governments and organizations that received funding, something the World Bank had begun to push for in the 1990s. The Gates Foundation, through its billion-dollar investment in the GAVI program, had also developed performance-based funding models that required grant recipients to demonstrate results and doled out monies in small tranches, with each subsequent payment dependent on demonstrated achievement. To enforce this model, the foundation implemented a system of audits, including surprise visits to development sites. Representatives at the G8 meeting in Okinawa committed themselves to supporting countries whose governments "have demonstrated a commitment to improve the well-being of their people through accountable and transparent management of resources devoted to development."

The Global Fund needed to ensure accountability and base its lending on concise performance measures. Yet no one involved in the creation of the Global Fund seemed to believe that funneling new monies through existing UN agencies would achieve the level of efficiency, transparency, and accountability that was needed to manage the new funds.[9] As a result of these concerns, the Global Fund was established as an independent funding institution, separate from existing UN agencies.

The Global Fund's mission was to raise financial resources for procuring drugs and other commodities needed by country-based control programs for AIDS, TB, and malaria. It was not intended to become another player in the design and implementation of control programs. The Global Fund placed a high value on country ownership of the programs it funded, and country ownership meant more than the participation of local governments. The fund insisted on the development of countrywide coalitions of stakeholders that included governments, NGOs, and representatives of civil society. These various stakeholders were to be part of a country coordinating mechanism (CCM), which was charged with designing and coordinating disease-control programs. The Global Fund's promotion of CCMs represented a commitment to grassroots participation. Yet it also reflected the fund's distrust of national governments.

The Global Fund launched its grants program in 2002 with haste and determination, on a time scale typical of the response of relief agencies to natural disasters. Indeed, the fund's organizers viewed themselves as responding to a worldwide health crisis. The details of how the Global Fund should be managed, including issues such as voting shares, review processes, the composition of CCMs, and relationships with partners, were only generally defined by its foundational documents. The desire to open up for business overrode the need for advanced organizational planning. Thus the

Global Fund began soliciting grant proposals before the full complement of structures and personnel were in place at its Secretariat in Geneva.[10]

The speed with which the Global Fund moved to disburse funds resulted in countries being given barely a month to respond to the first call for proposals. Nonetheless, the Global Fund's portfolio of supported projects grew in dramatic fashion. By the end of 2002, it had approved 56 proposals, worth US$567 million, in 37 countries, and it had begun to sign grant agreements and disburse funds. One year later, the Global Fund had approved a total of 224 projects in 121 countries, for a total value of US$2.1 billion. At the close of 2004, these totals amounted to 295 projects in 127 countries, at a value of US$3.1 billion. Disbursements followed a slower trajectory, but their pace accelerated rapidly from 2004 on. By June 2013, cumulative disbursements soared to almost US$135 billion. All of the dispersed funds came from voluntary contributions to the Global Fund from national governments.

The Global Fund's insistence on country ownership, while laudatory in principle, proved to be problematic in practice for several reasons. First, while the fund insisted that CCMs be representative and inclusive of all interested parties, just how these organizations were formed and who was included in them was not clearly defined. Second, in many cases these new structures were hastily constructed to meet the short lead time allowed for first-round proposals and did not work effectively. In addition, the relation of the newly formed CCMs to already-established organizations, such as the national AIDS councils and NGOs that had previously been recipients of external funds, was often unclear and a potential source of tension. In a number of countries, the CCMs were regarded by local stakeholders as being imposed by the Global Fund.

To further complicate the situation, the Global Fund initially awarded grants directly to NGOs rather than working through the CCMs, thereby sending mixed signals and raising questions about the role of the CCMs. One of the first grants made by the Global Fund was to a highly innovative patient-support program from the KwaZulu-Natal Province of South Africa. The South African minister of health claimed that the money should have gone to the South African National AIDS Council, which had been designated as the CCM shortly before the grant to KwaZulu-Natal was announced and had coordinated AIDS programs in-country. The minister went further, claiming that the Global Fund, in making the grant, was trying to bypass the democratically elected government of South Africa. The minister's reaction may have been triggered by ongoing controversies surrounding her management of the AIDS crisis in South Africa. But it was not completely without basis. The Global Fund did not wish to fund governments. Disagreements between the Global Fund and the South African government over who could legitimately receive external disease-control funds delayed the disbursement of funds until 2004. In subsequent rounds, the Global Fund attempted to work with countries to clarify the role and composition of the CCMs.

While the Global Fund encouraged country ownership of programs, resisting earlier models of development assistance in which external approaches were imposed by donor agencies, the fund's proposal-review process and project-monitoring mechanisms tended to undercut country ownership in a number of ways. First, in order to develop proposals that would meet the technical standards required by the Global Fund's review panels, many countries, particularly those in Africa, sought technical assistance from external agencies, primarily WHO and UNICEF. As a result, locally defined projects were recast in line with external advice. Second, the technical-review panels set up by the Global Fund were staffed by experts who lacked direct knowledge of local conditions in the countries applying for funds. Project evaluations and suggestions for revisions were therefore made on the basis of whether the proposals met a set of external technical criteria, rather than on whether they made sense in terms of local conditions. Finally, in order to ensure fiscal accountability, the Global Fund required that each country have a local fund agent (LFA), which would be responsible for monitoring project finances. Various options for providing LFAs were discussed by the fund's Transition Group; in the end, however, the fund decided to employ several established international accounting firms to serve as LFAs for many recipient countries. Subsequent reviews of Global Fund's activities revealed country-level concerns about the failure of the fund to employ existing monitoring mechanisms and the lack of country-specific expertise among appointed LFAs. The LFAs further undermined the principle of country autonomy and reinforced the externality of global-health decision making.

The fund also encountered other problems. The majority of grants approved by the Global Fund went to support the purchase of commodities and to specific projects aimed at combating AIDS, tuberculosis, and malaria. Yet from its inception, the Global Fund was aware that the success of the programs they financed depended on the existence of effective local systems for procuring and distributing purchased commodities. Inadequacies in health staffing, health infrastructure, and administrative capacity hampered fund-supported programs. The question of how—and even whether—to strengthen health systems plagued the Global Fund throughout each funding cycle. There was an inherent conflict between the fund's core organizing principle, which insisted that projects explicitly target the three diseases and show measurable results within the timeframe of project funding cycles, and the reality that this goal might require more long-term health-systems strengthening.

The Global Fund was aware of this conflict and attempted to resolve it by making some funds available for health-systems strengthening (HSS). Yet HSS proposals had a low success rate through the first seven rounds of funding.[11] The fund's Technical Review Panel, in examining its troubled relationship with HSS-oriented proposals, acknowledged that "the Global Fund system is not currently set up to generate strong HSS proposals nor to evaluate these effectively." Overall, from 2002 to 2013, only 3 percent of the fund's grants went toward health-systems strengthening. Most of this funding, moreover, was directed at improving the delivery of specific disease interventions and was routed through NGOs, rather than governmental health services.

The Global Fund succeeded beyond anyone's expectation in raising monies to support programs aimed at reducing the global burden of AIDS, tuberculosis, and malaria. Yet its financial contributions, like those of the World Bank, had collateral effects. First, by directing global financial resources toward the purchase of biomedical objects—bed nets and drugs—while largely ignoring broader support for the development of health services, infrastructure building, sanitation, clean water, and health training, the Global Fund contributed to the medicalization and commodification of global health.

The Global Fund's policies also reinforced the declining role of the state in the provision and direction of health programs. Working through CMMs, the Global Fund directed monies to decentralized programs, which, in many cases, were designed and run by a combination of local stakeholders and NGOs. Furthermore, through its insistence on various forms of external evaluation and monitoring, the fund supported performance-based lending mechanisms but, at the same time, further undermined the ability of health ministries to initiate and manage health initiatives. The Global Fund's emphasis on accountability also placed new demands on the limited administrative capacity of countries receiving international aid. Increased accounting requirements strained local manpower resources, forcing health workers to spend large amounts of time reporting on projects instead of planning and implementing them.

Finally, by targeting AIDS, tuberculosis, and malaria, the Global Fund consumed a large portion of the funds available for global-health activities.

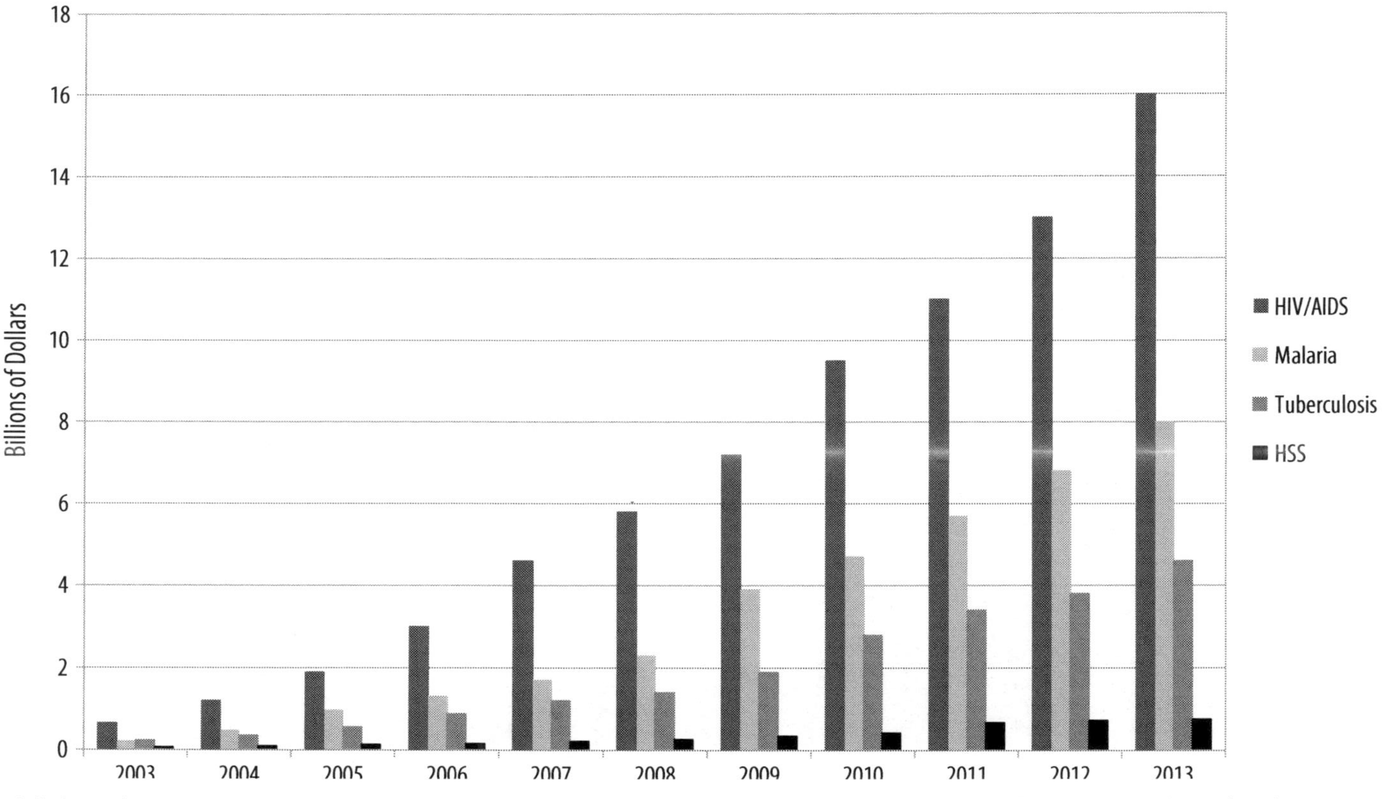

Global Fund commitments, 2002–2013. Global Fund for AIDS, Tuberculosis and Malaria, "Cumulative Assigned Funding by Disease, 2013," Funding and Spending, www.theglobalfund.org/en/about/funding/spending/ (no longer available).

While prioritizing these three diseases could be justified on many grounds, Global Fund made it difficult for programs aimed at other health problems to obtain resources. It also hindered efforts to build basic health services. Programs to prevent respiratory and intestinal infections, which continued to cause high rates of mortality among children in low-income countries, and efforts to prevent maternal mortality struggled to maintain, let alone increase, their funding levels. Campaigns aimed at a long list of protozoan, bacterial, viral, and helminthic diseases, which WHO labeled "neglected tropical diseases," were also underfinanced. These diseases, including schistosomiasis, onchocerciasis, and dengue fever, effected an estimated 1.4 billion people world wide. The focus on the "big three" diseases also deflected attention from a growing epidemic of noncommunicable diseases, including various forms of cancers and mental illnesses in the developing world.

WHO reported that there were 14 million new cases of cancer and 8.2 million cancer-related deaths in 2012; 60 percent of these new annual cases occurred in Africa, Asia, and Central and South America.[12] Yet cancer services in many low-income countries were severely handicapped by insufficient infrastructure, lack of access to treatment, and the absence of professionally trained cancer specialists.[13] Mental illnesses were the leading causes of disability-adjusted life-years worldwide in 2011, amounting to 37 percent of healthy years lost from noncommunicable disease. Depression alone accounted for one-third of this disability. Noncommunicable diseases took a heavy financial toll, as well. A report by the World Economic Forum estimated the global cost of mental illness at nearly $2.5 trillion (two-thirds in indirect costs) in 2010, with a projected increase to over $6 trillion by 2030.[14] The total costs of all noncommunicable diseases, including cardiovascular disease, chronic respiratory disease, cancer, diabetes, and mental health, were projected to be $47 trillion in 2010.[15] Yet these health problems, for which there are no easy technological fixes, received only a fraction of the monies spent on AIDS, tuberculosis, and malaria.

Recognition of the need to fund health initiatives not supported by the Global Fund led the World Bank to propose the creation of a Global Financing Facility (GFF), designed to bridge the funding gap for addressing preventable death and disease among children and women. The bank set a target of $33 billion for the GFF, which was launched in July 2015. The goal of the GFF was to prevent an estimated 3.8 million maternal deaths, 101 million child deaths, and 21 million stillbirths over 15 years.[16]

My point is not to discount the importance of AIDS, tuberculosis, and malaria, or to question global-health priorities. Rather, I want to indicate that the rising tide of global-health funding from the 1990s was uneven, and that the Global Fund, by attracting large amounts of money in the form of contributions from donor nations, added to the unevenness of the impact that rising global-health funding had. It doing so, it shaped the landscape of global health.

PEPFAR, Bilateral Aid, and the Politics of Global Health

In 2003, President George W. Bush proposed that $15 billion be provided over five years to increase funding for AIDS-prevention programs in Africa, as well as to care for children orphaned by AIDS and support the treatment of people living with AIDS. The commitment to the President's Emergency Plan for AIDS Relief included $5 billion for existing bilateral programs throughout the world, $1 billion for the Global Fund ($200 million per year), and $9 billion for new programs in 14 countries in Africa and the Caribbean.[17] The decision to create a separate bilateral program for AIDS, rather than providing additional financing for the newly established Global Fund, reflected the Bush administration's ongoing concerns about the ineffectiveness of multilateral organizations and its desire to tailor support to meet US strategic interests.[18]

Channeling AIDS funding through PEPFAR also protected US economic interests. PEPFAR legislation prevented the financing of generic drugs, despite their use having been authorized by the Clinton administration and certified by WHO prequalification programs. This policy clearly benefited US pharmaceutical companies. But by increasing the cost of drugs, it limited the numbers of people who could be treated. It was no coincidence that Bush appointed Randall Tobias, the former chairman of the giant pharmaceutical corporation Eli Lilly and Company, as the US global AIDS coordinator, despite the fact that Tobias had no experience administering AIDS programs. Pushback from activist groups and the European Union eventually led the United States to reverse its stand on the use of PEPFAR funds to purchase generic drugs.[19]

Like financial assistance from the World Bank and the Global Fund, PEPFAR funding came with conditions that shaped the emerging landscape of global-health assistance. In funding PEPFAR, the Republican-led US Congress required that at least one-third of the prevention funds be directed toward abstinence programs and that church-based programs not be required to distribute condoms. Abstinence was part of the ABC strategy—abstinence, be faithful, and use a condom—which was credited with having driven down the prevalence of AIDS in Uganda during the 1990s. Yet many in the global AIDS-prevention community believed that the emphasis on abstinence was unrealistic and essentially driven by conservative Christian groups in the United States. Of the 61 million people served by PEPFAR-supported outreach programs between 2004 and 2007, over 40 million were in programs only promoting abstinence and/or being faithful. A study published in 2015 indicated that these programs had had virtually no impact on slowing the AIDS epidemic in Africa.[20]

Like the World Bank and the Global Fund, PEPFAR preferred to work through NGOs rather than through the public-health sector in the African countries in which it operated. This decision had multiple consequences. The reliance on NGOs to develop and implement programs clearly facilitated the rapid buildup of projects, which was in accord with PEPFAR's mandate to provide emergency assistance. Supporters of the policy also argued that it ensured greater accountability. Working through local political bureaucracies, they argued, would have taken longer and made the oversight of funds more challenging. True or not, the rollout moved with amazing speed.

Within the first two years, PEPFAR provided ARV therapy for more than 800,000 adults and children, HIV testing for nearly 19 million people, services to women to prevent the mother-to-child transmission of HIV, and care and support services for approximately 4.5 million adults, orphans, and other vulnerable children.[21] According to PEPFAR, its programs also supported numerous services designed to reduce the risks faced by women and girls. These efforts were focused on increasing gender equity, addressing male norms, reducing violence and sexual coercion, increasing income generation for both women and girls, and ensuring legal protection and property rights. By the end of 2007, PEPFAR had spent more than $6 billion on HIV care, prevention, and treatment in the 12 focus countries in sub-Saharan Africa. And, by one estimate, the deaths of 1.2 million people (10.5%) had been averted because of PEPFAR's activities.[22]

Yet PEPFAR's reliance on NGOs also caused problems. To begin with, in many places it led to the development of parallel health structures, which were only loosely tied to the public sector. The PEPFAR approach was also highly verticalized, focusing narrowly on the scale-up of HIV services and avoiding the larger needs of the health system. This resulted in the creation of parallel medical-supply, data-collection, and management systems dedicated only to HIV. In Mozambique, this vertical approach resulted in the construction of 23 new ARV-focused "day hospitals" beside crumbling primary-health-care infrastructure in existing public, urban health-system compounds.[23] In principle, the ARV scale-up was integrated into the public-sector national health system. But in practice, each NGO partner tasked with the scale-up in their region had a different approach to collaborating with the Ministry of Health, staffing clinics, collecting data, selecting a geographical focus, and providing services. The result was a fragmented approach that was typical of PEPFAR activities across Africa.

This pattern of program fragmentation also occurred in Uganda. Susan Whyte and her colleagues described the multiple channels through which ARVs entered Uganda. These included numerous small- and medium-sized drug-treatment trials funded by donors, including PEPFAR, that provided treatment for free, but only to those who fit particular inclusion categories. These programs also left uncertainties as to whether treatment would be continued after the trails were concluded. Gazetted treatment centers, which were primarily fee-for-service facilities, formed a second avenue of care. Among these was the PEPFAR-funded Joint Clinical Research Center, which was the oldest and largest treatment center in sub-Saharan Africa. Private practitioners provided a third route. The actual number of these private outlets was unknown, but an estimated 40 such units existed in Kampala. A fourth channel involved leakage from the other systems, as workers with access to drugs found ways to supply them on the side the family members and others. There was very little, if any, coordination among these various pathways.[24]

The disarticulation of PEPFAR-funded AIDS programs in various African countries was seen by some as a positive development, because in the early stages of the ARV rollout, there was much to be learned about the best way to provide services. It could also be argued that such variations reflected adjustments to local conditions. As we have seen with other disease-control programs, one size seldom fit all. Yet the lack of integration of HIV services with national health services discouraged the development and implementation of national AIDS policies. They produced instead an archipelago of autonomous care facilities dependent on external funding and existing independently beside a collapsing public-health sector. To make matters worse, better compensation and improved working conditions within the private, NGO-run AIDS clinics drew health workers away from governmental clinics and hospitals, contributing to these facilities' further decline.

The cost of creating independent treatment programs that ignored the deteriorating condition of governmental health services became apparent when PEPFAR began to require AIDS services to be integrated into national health systems, as part of a movement toward sustainability after 2008.

Just as had occurred with similar efforts to integrate malaria-control programs into basic health service in the 1970s, program organizers quickly realized that the deplorable state of most national health services made integration a challenge.[25] This realization resulted in a new emphasis on health-systems strengthening, which became a buzzword in global-health funding in the second decade of the early twenty-first century. Yet much of what fell under HSS amounted to planning, coordinating, and assessing activities, not building programs. The following were PEPFAR's five-year HSS goals in 2014:[26]

- Years 1–2
  - Work with countries to identify how HIV/AIDS activities can contribute to broader HSS efforts
  - Support both headquarters and country teams to obtain the skills necessary to carry out programming with a HSS perspective, and provide the technical assistance necessary to increase these skills in partner countries
  - Increase multilateral engagement around accomplishing HSS HIV/AIDS activities
  - Continue efforts to identify internationally accepted indicators for HSS
  - Support policy change needed to attain HSS goals defined in Partnership Frameworks
- Years 3–5
  - Implement harmonized HSS indicators within PEPFAR programming
  - Expand and coordinate, as appropriate, multilateral efforts to support HSS
  - Strengthen country governance and financing of the health system to advance the goals of country ownership and sustainability

This classic example of "development-speak" made no mention of hard-dollar contributions to building up health infrastructures or providing medical training for health-care workers. There were no quantitative indicators related to providing what Paul Farmer calls "the four S's: Staff, Stuff, Space, and Systems."

Finally, critics of PEPFAR's programs asserted that the organization was more concerned with producing numbers that demonstrated its impact than in measuring actual outcomes. They argued that the number of people assisted by PEPFAR funds was not reflective of the program's actual impact. Of particular concern was the loss of patients to treatment. Retention was a major problem for ARV programs in sub-Saharan Africa. A 2009 review of African ARV programs reported that approximately 35 percent of the patients had disappeared from treatment after three years.[27] Opponents of ARV use in Africa pointed to this high dropout rate as evidence that Africans were unable to manage their ARV treatments. They again raised the possibility of therapeutic anarchy in sub-Saharan Africa, leading to the emergence of drug resistance.

But treatment dropout was a direct response to the continued high cost of drugs, not to patient noncompliance. For example, in 2003, despite a dramatic drop in the price of ARVs, the drugs still cost US$28 a month in Uganda, which was far more than most Ugandans could afford. Unless a person was enrolled in an externally funded drug trial, drugs had to be purchased. For families who could accumulate the funds necessary to begin treatment for one of their members, sustaining this treatment for years was a tremendous financial burden. Families needed to make extremely difficult choices about whether to devote their limited income to saving the life of a family member or to covering their many other expenses. Sometimes they had to choose *between* family members when, as was often the case, more than one of them was infected with HIV. Those infected with HIV and dependent on their family members for treatment support recognized the sacrifice being made on their behalf and often felt the guilt of inflicting financial hardships on their families. Susan Whyte and her colleagues described how the stress experienced by one Ugandan man with AIDS, whose wife was also HIV positive and in need of treatment, led him to commit suicide by jumping off a bridge into the Nile in order to save his family the necessity of choosing whose treatment to support, and whose life to save. The high cost of sustaining the treatment month after month and the financial burden this created often led patients to drop out.[28] Recording the number of people enrolled in PEPFAR-funded programs did not account for this grim reality. While some PEPFAR-funded programs have been concerned with patient retention, the drive for numbers has worked against the development of sustainable treatment programs. Similar problems plagued the distribution of insecticide-treated bed nets through the Roll Back Malaria programs. The need to meet coverage targets (60% of children under five sleeping under nets by 2005), limited efforts to develop effective communication strategies that would insure the nets were used properly.

PEPFAR, the Global Fund, and the World Bank have provided massive amounts of new funding to combat major threats to global health. Millions of people have benefited from this funding. Yet these organizations have also produced a pattern of global-health assistance that has focused on attacking particular problems through a disarticulated system of nongovernmental organizations, which has left basic health services relatively impoverished.[29] They also contributed to the increasing medicalization of global health, as we will see in the next chapter.

CHAPTER SIXTEEN

Medicalizing Global Health

As ANTIRETROVIRAL DRUGS became increasingly available for people living with AIDS in Africa and other parts of the global south, it raised the possibility that AIDS, like other infectious diseases, could be prevented or even eliminated through the application of biomedical technology. Developing medical solutions for AIDS prevention was but one of a number ways in which global-health interventions became increasingly medicalized, or "pharmaceuticalized," in the early twenty-first century.[1] Biomedical technologies, validated through randomized controlled trials (RCTs), became the gold standard for global-health interventions. This trend was reinforced by the increasing demand of donor organizations for evidence-based interventions and accountability. It was also supported by the massive investments of the Bill & Melinda Gates Foundation into global health.

Beginning in 2009, the provision of ARVs came to be seen as supplying not only an effective treatment for AIDS, but also a highly effective form of prevention. This possibility was initially raised by a study published in *Lancet* in 2009. Its authors claimed that ARV therapy could so reduce infectivity as to make possible, in time, the virtual elimination of HIV transmission. This report was followed two years later by a study that showed that in HIV-discordant couples, ARV therapy had reduced the rate of transmission to the uninfected partner by 96 percent. In 2012, a study published in the *New England Journal of Medicine* took the finding of earlier studies one step further and demonstrated that the provision of ARVs to HIV-negative individuals could provide protection against transmission. These and other findings led the US Centers for Disease Control and Prevention to support the pre-exposure prophylactic use (PrEP) of ARVs as a form of

HIV prevention. PrEP was subsequently adopted by PEPFAR as a part of its prevention efforts.[2]

PrEP was a promising breakthrough in the prevention of HIV. Yet it raised troubling questions. How could one justify the prophylactic use of AIDS drugs when so many people living with AIDS lacked access to ARV treatment? To what extent did the prophylactic use of ARVs undermine programs that had been developed in many countries to address the underlying causes of transmission by reducing the vulnerability of women and the social and economic conditions that fostered unsafe sexual practices? Was PrEP really a magic bullet for AIDS? Could it be employed without extensive efforts to educate those receiving the drugs and monitoring their adherence and disease progression? Retaining infected people in treatment programs was challenging, even when they knew that abandoning the treatment could be life threatening. Doing so for people who were not infected would be a greater challenge. Finally, would the prophylactic use of ARVs undermine their effectiveness in treating AIDS patients?

The medicalization of AIDS prevention was encouraged further by studies that suggested male circumcision could effectively reduce the risk of HIV infection in areas of high transmission and low rates of circumcision. The link between male circumcision and lower rates of HIV infection was first suggested in the 1980s, and subsequent observational studies reinforced this connection. Yet questions remained about whether the results of these studies were affected by confounding factors. To resolve this question, a series of RCTs were conducted in South Africa, Kenya, and Uganda in the early 2000s. Results from the South African trial were reported in 2005. The study showed a significant reduction of risk in circumcised males and concluded that circumcision offered protection against HIV infection "equivalent to what a vaccine of high efficacy would have achieved." The trials in Uganda and Kenya showed similar results, so they were subsequently stopped for ethical reasons, it being unfair to subject control groups, who did not receive the treatment, to much greater risks. In 2007, WHO and UNAIDS reviewed the trial results and concluded that male circumcision be considered an effective intervention to reduce HIV transmission. WHO's recommendation noted, however, that "male circumcision should never replace other known effective prevention methods and should always be considered as part of a comprehensive prevention package, which includes correct and consistent use of male or female condoms, reduction in the number of sexual partners, delaying the onset of sexual relations, and HIV testing and counseling."[3]

Similarly, in 2009, an Expert Group on Modeling the Impact and Cost of Male Circumcision for HIV Prevention projected that in high prevalence/low circumcision regions, male circumcision could, over a 10-year period, reduce HIV incidence from 30 to 50 percent, at a cost of between US$150 and US$900 per case averted. Yet the report noted that the benefits of male circumcision could be undermined by risk-compensation behaviors—in other words, by encouraging unprotected sexual activity on the part of those who had been circumcised. It therefore concluded, "There is a clear need for intensive social change communication campaigns, aimed at the whole population, to prevent increases in risk behaviors."[4]

The small-print message in these studies was that while male circumcision could be an effective means of reducing HIV transmission, it did not replace the need for strategies aimed at behavioral changes and education. Male circumcision was not a magic bullet. Those supporting male circumcision, however, often promoted it as if it were. Financial support for male-circumcision programs was rapidly mobilized. In fiscal year 2007, PEPFAR allocated approximately US$16 million to support male-circumcision activities. That figure increased to nearly US$26 million in fiscal year 2008. These funds supported activities in Botswana, Ethiopia, Kenya, Lesotho, Malawi, Mozambique, Namibia, Rwanda, South Africa, Swaziland, Tanzania, Uganda, and Zambia.[5] In 2013, PEPFAR projected that their partners would circumcise some 4.7 million men in eastern and southern Africa. Together with PrEP, male circumcisions programs contributed to the medicalization of AIDS prevention. They also were evidence of the growing role of RCTs in global-health decision making.

The Randomization of Global Health

By the beginning of the twenty-first century, the use of RCTs to demonstrate the effectiveness of health interventions had become a dominant feature of global health. The growing use of RCTs was fueled by a rapidly expanding academic-research enterprise funded by multilateral organizations, bilateral donors, and the manufacturers of pharmaceuticals and biomedical devices. In the United States, governmental funding awarded through USAID, the Centers for Disease Control and Prevention, the Department of Defense, and the National Institutes of Health, together with support from philanthropic organizations, including the Gates Foundation, funded a growing army of global-health researchers conducting RCTs of various biomedical technologies in the global south. NGOs dedicated to running RCTs collaborated with development organizations to test planned interventions.

One of the largest such organizations, Innovations for Poverty Action, boasted on their website that they conducted "randomized evaluations because they provide the highest quality and most reliable answers to what works and what does not."[6]

The use of research trials to test public-health interventions was not new. Trials of TB drugs, pesticides, birth control devices, and vaccines contributed to the development of international-health interventions in the 1950s, 1960s, and 1970s. In the early twenty-first century, however, the scale of research trials grew dramatically, driven by scientific advances and a growing demand by new institutional donors for evidence-based interventions validated through RCTs. It was no longer enough to show improvements in health among targeted populations. These populations needed to be matched with control groups who did not receive the intervention, in order to demonstrate that the intervention actually had an impact.

The expansion of global-health research was abetted by the increasing reliance of schools of public health on research dollars to support their missions. As early as 1976, a Milbank Fund report on *Higher Education for Public Health* expressed concern that schools of public health had become so dependent on federal research funds that "their policies and programs are determined by dollars available and they no longer control their own destiny."[7] Since the 1990s, research grants have supported larger and larger percentages of faculty salaries, and overall school budgets have become increasingly dependent on overheads from grants. R01s (NIH's standard research-project grants), the currency of career advancement in medicine, were increasingly playing the same role in public health. This shifted research priorities away from developing and implementing public-health interventions and toward testing or evaluating interventions—increasingly in the form of biomedical technologies—through RCTs. Public-health students in the top graduate schools in the United States were being trained to conduct research, rather than how to build and sustain programs.[8]

Global-health research was frequently undertaken in collaboration with local researchers. These collaborations, however, were often unequal partnerships. While local researchers in Brazil and India, which possessed relatively strong institutions and resources, played a significant role in shaping research agendas, their counterparts in resource-poor countries in Africa, where hundreds of millions of dollars were spent on trials related to AIDS, malaria, and other diseases, had less influence. African researchers made significant contributions to these projects and published papers derived from their work. In addition, these projects were approved by local institutional review boards (IRBs). Yet researchers and institutions located in the global north controlled the funding sources and, thus, the overall design, direction, and choice of interventions to be studied. Local IRBs, moreover, were slow to develop in many countries in which global-health research projects were conducted in the 1980s and 1990s. Controversies arising from the use of placebos in West African trials designed to test a modified and cheaper regimen of an AIDS drug, which had previously been shown to reduce vertical HIV transmission in Europe and the United States, sparked efforts to increase local oversight of RCTs in Africa. Yet the financial stakes involved in allowing foreign researchers to conduct trials sometimes undermined local institutional oversight.[9]

The early trials on AIDS vaccines in Africa reveals the ways in which biomedical research agendas were determined outside of Africa. As Johanna Crane has shown, early phase-1 AIDS vaccine trials, designed to test vaccine safety, were conducted in Uganda in cooperation with Ugandan researchers in 1999. The vaccine that was tested had been developed to prevent HIV subtype B. Subtype B is one of eleven major subtypes of HIV-1, the most common and pathogenic of the HIV viruses. Each subtype has a geographical range. Subtype B is the dominant form of HIV in Europe, the Americas, Japan, Thailand, and Australia. Subtype D is the dominant HIV virus in East Africa. In other words, the early phase-1 trials conducted in Uganda tested a vaccine designed to attack a virus not found in Uganda. Subtype B was used because developers were looking for a vaccine that would protect populations in the countries where the vaccines were manufactured and in places where there was a large potential market that could afford the vaccine. Ugandans were, in effect, used as guinea pigs to test the safety of a vaccine that would not benefit them. This raised ethical questions and protests that led, over time, to an expansion of vaccine research to include non–subtype B viruses. In addition, Crane noted that the early ARV drugs rolled out in Africa were also based on research using the subtype B virus. Fortunately, these drugs appear to have been effective in combating non–subtype B infections.[10]

The fact that research protocols and the drugs and vaccines tested in Africa were designed to meet the needs of non-African populations did not mean that the trials were of almost no benefit to Africans. As one Uganda researcher involved in the phase-1 vaccine testing noted, the trials provided opportunities to build research infrastructure and train local scientists. Yet this support was dependent on continued participation in externally funded research programs.[11] Marissa Mika's work on the history of Uganda's

National Cancer Institute starkly reveals the extent to which the fortunes of the institute and the patients it served fluctuated with the ebb and flow of outside funding.[12]

Patients also benefited from research trials, gaining access not only to potentially lifesaving drugs and vaccines, but, in many cases, to other forms of biomedical care that were not available through existing health services. In addition, they had access to other programs that provided food supplements and care for children. As anthropologist Nguyen Vinh-Kim has argued, AIDS trials in West Africa created a kind of therapeutic citizenship, in which those who were fortunate enough to be included in a clinical trial or treatment program gained access to a range of benefits and forms of security that were associated with social programs in many countries in the global north. These trials and treatment programs became surrogates for governments that were too weak to provide such services.

Vinh-Kim also noted that therapeutic citizenship only extended to those who met a metric based on CD4 counts (an immune-system blood test) and the fact that they lived in a catchment zone defined by a clinical trial or by NGO treatment-program organizers. Those who did not fall within a catchment area or did not meet inclusion criteria were left out. Access to ARV trials and programs were limited, and creating therapeutic citizens involved triage. The numbers of people on ARV treatment grew dramatically after 2000. Yet the epidemic continued to expand faster than resources could be mobilized to meet to rising demands for treatment.[13]

Finally, the benefits of inclusion in clinical trials existed only as long as the research trials continue. Adrianna Petryna has shown that subjects involved in drug trials in Eastern Europe and Latin America often lost access to the drugs being tested after the trials ended.[14] It is too soon to know whether similar fates will await those enrolled in AIDS trials and treatment programs in Africa and elsewhere. In the case of AIDS, the loss of therapeutic citizenship is almost certainly a death sentence.

The growth of evidence-based global health and RCTs had collateral effects. First, it privileged certain kinds of interventions over others. RCTs were designed to test the efficacy of biomedical interventions and privileged interventions that could be easily counted, just as Robert McNamara had proposed to UNICEF's Jim Grant in the early 1980s. They contributed to what anthropologist Vincanne Adams has called an "audit culture" in global health. Supporters of RCTs insisted that all forms of interventions could be studied using experimental research methods. But as Adams notes: "The reality is that the bar of RCT is simply too high for many lived realities and determinants of health in most of the world. And then there are all the things that can't be counted, or that, even if they could be counted, in being counted would be turned into misrepresentations of their true relationship to health." By privileging those things that can be counted, such as the numbers of pills taken, vaccines administered, and nutritional supplements provided, evidence-based public health contributed to the search for biomedical solutions to health problems and to the further medicalization of global health.[15]

It should be noted that RCTs have also been employed to measure the health impact of various health-education interventions, insurance schemes, microcredit programs, and community-based women's groups on health outcomes. For example, RCT's were used in a series of studies to evaluate the role of participatory women's groups in reducing maternal, perinatal, neonatal, and infant mortality rates in Nepal, India, Bangladesh, and Malawi.[16] This extension of RCTs to evaluate nonbiomedical interventions reveals how powerful a tool RCTs have become for global health. Yet there remain limits to what can be measured. The more diffuse or broad-based the intervention, the more difficult it is to define variables and account for the role of confounding factors. In addition, the impact of some interventions, such as community-based health-care programs, extend beyond quantitative reductions in mortality rates. They help empower communities and improve overall health in ways that are difficult to capture in terms of reductions in mortality.

Enter Bill Gates

The push for the discovery of new biomedical technologies to solve the health problems of populations living in resource-poor countries received major encouragement from the Bill & Melinda Gates Foundation. From its establishment in 2000, the foundation supported a wide range of initiatives in the area of global health. These addressed problems related to vaccine and drug development, maternal and child health, family planning, and disease control and eradication. Above all else, the foundation had a laser focus on encouraging the development of new biomedical technologies to solve global-health problems, an approach that would be expected from the former founder and CEO of Microsoft. Through its various activities, the foundation has had a major impact on global health, injecting hundreds of millions of dollars into health interventions and raising awareness of the importance of improving the health of peoples around the world. Yet, like the World Bank, the Global Fund, and PEPFAR, the Gates Foundation's largess has shaped the landscape of global health in the early twenty-first century by promoting evidence-based technological solutions to health problems, while paying little attention to the development of basic health services or addressing the social determinants of health.

To advance the development of new biomedical technologies, the foundation encouraged the creation of new public/private initiatives, which drew on the resources of private industry to advance research and development. The most successful of these was the Global Alliance for Vaccines and Immunizations (GAVI). GAVI was the Gates Foundation's first venture into global health. It was launched in response to the widespread critique by global-health professionals that developing new vaccines for the global south was stalled, due to a lack of funding and the unwillingness of pharmaceutical companies to invest in developing vaccines that had little commercial potential, because they were intended for populations who could not afford to pay for them.[17]

To break this roadblock, the Gates Foundation committed US$750 million to fund vaccine development. In the context of global-health funding at the time, this was an astronomical amount of resources. GAVI brought together three groups that had been largely disconnected. This included basic-science researchers working in university labs and governmental facilities, such as the National Institutes of Health; corporate product developers who took the fruits of scientific discovery and converted them into licensed, mass-produced commodities; and in-country distributors who gathered those vaccines and delivered them to be injected into the arms of children. The foundation's money provided grants to basic scientists to seek out new vaccine models; subsidized the process through which pharmaceutical companies moved promising vaccines to market; worked on developing better distribution systems; and brought those who were involved in immunization programs to the table to discuss which diseases needed to be prioritized. This was no small achievement. It meant bringing together large pharmaceutical companies and WHO, two entities that viewed each other with distrust. As William Muraskin noted: "Many in WHO saw industry as little more than merchants of death, who put profits above saving lives. . . . In turn, many in the pharmaceutical firms saw WHO as 'a bunch of raving socialists' who didn't understand that 'legitimate profit' was the only engine of progress."[18]

The Gates Foundation claimed that GAVI saved millions of lives by providing new vaccines for major childhood diseases. Yet it is difficult to assess this claim. To begin with, GAVI, during its early years, was much more effective in developing new vaccines than in building distribution systems.

The foundation has never been comfortable with health-systems strengthening. Consequently, getting the vaccines into the arms of children proved to be a challenge. In addition, data on child mortality in most resource-poor countries is very difficult to collect and interpret. Children die from multiple causes, and it is hard to identify which health interventions spared a life. In addition, those saved from preventable diseases by vaccines may subsequently die from one of the many other diseases for which there were no medical magic bullets.

A Gates-funded vaccine for type A meningitis, which was once responsible for more than 80 percent of Africa's meningitis cases, further revealed the limits of the foundation's vaccine-centered approach to disease prevention. The vaccine was introduced in 2010 by WHO, the Bill & Melinda Gates Foundation, and PATH, a Seattle-based health-technology group. Over 220 million doses against type A meningitis, made by the Serum Institute of India for US$0.50 each, were injected, virtually eliminating the threat of this disease in the region. The vaccine clearly had an immediate impact on the health of people living in the region of West Africa in which meningitis epidemics have regularly occurred. Yet this same region is now facing the threat of a new strain of meningitis, raising the specter of deaths occurring among those who were saved by the Gates-funded vaccine.[19] It is likely that epidemics caused by various strains of meningitis will continue to occur in this region until the living conditions that encourage the seasonal spread of the disease are eliminated.

None of this is to discount the success of the Gates Foundation in kick-starting the production of new drugs and vaccines. Rather, it is to point out that vaccines, as powerful as they may be in preventing disease, are not a panacea. A child can be immunized against every disease for which there is a vaccine and still die a premature death, due either to a condition for which there are no magic bullets, or simply to the absence of basic health services.

The larger impact of the Gates Foundation on global health lies in the extent to which it has come to dominate global-health funding and, thus, priority setting. Researchers and development NGOs in the global north chase after Gates funding, defining their research and program priorities to fit the technological parameters of the foundation's core mission. This has meant that discovering and implementing new biomedical technologies has become a high priority, further medicalizing global health. Faculty and students in schools of public health from Seattle, to Baltimore, to Cambridge, to London are discussing "innovation" and "platforms" and "grand challenges" for new technologies, rapid diagnostic tests, delivery systems, and vaccines for Africa, in addition to plans for "mHealth" (mobile devices for health) and electronic medical records.[20] There is less talk about training health workers, or building and provisioning badly needed clinics and hospitals, let alone addressing the social determinants of health.[21]

Chasing Polio

The Bill & Melinda Gates Foundation has also invested heavily in interventions designed to eradicate diseases, including malaria and polio. Polio eradication has been a top priority of the foundation. The Global Polio Eradication Initiative (GPEI), begun in 1988, was built around the delivery of a vaccine. It represented another effort to improve the health of the world's populations through the application of a simple biomedical technology. As such, it was an attractive target for the foundation.[22] Optimism that derived from the success of the smallpox-eradication campaign inspired the GPEI to set the year 2000 for the global eradication of polio. When that goal was not achieved, the target date was changed to 2005. When polio was still not eradicated in 2005, no new targets were set, but WHO insisted that eradication was near at hand. The decision to push on was encouraged by the Gates Foundation's decision to donate US$100 million in a matching gift to the campaign in 2007. This figure was increased to US$355 million in 2009.[23] Six years later, the GPEI was still, in the words of anthropologist Svea Closser, "chasing polio."[24]

The history of the GPEI raises a number of questions about the wisdom of pursuing disease-eradication strategies and about the limitations of medicalized solutions to global-health problems. It is also further evidence of how much biology matters when it comes to eradicating disease. Advocates of polio eradication argued that the disease was a good target, because it had no animal reservoir and because there was an effective vaccine for protecting children. In fact, there were two vaccines: the Salk inactivated polio vaccine (IPV), which was administered through injections, and the Sabin oral polio vaccine (OPV). These vaccines were already part of the child-survival initiative and WHO's Expanded Programme on Immunizations. They were thus widely available. The vaccines had produced a dramatic decrease in polio cases in Latin America, particularly Brazil, by the early 1980s. This success led Brazilian Ciro de Quadros to push for the hemisphere-wide eradication of the disease. Quadros convinced UNICEF's Jim Grant that polio eradication would advance the broader use of childhood immunizations and brought UNICEF on board in 1984. That same

In 1988, the forty-first World Health Assembly, then consisting of delegates from 166 member states, adopted a resolution for the worldwide eradication of polio. Note that the overwhelming majority of those present were men. WHO.

year, Albert Sabin persuaded Rotary International, which had supported childhood immunizations since the 1970s, to commit US$120 million to polio elimination in the Americas. That organization subsequently raised over US$200 million.

The early results of the polio-eradication campaign were promising. The Americas were declared polio free in 1997. Progress was also being made in Oceania and Africa during the 1990s. In South Asia, however, where 75 percent of all polio cases occurred, progress was slow. There were neither surveillance systems nor routine immunization programs in place when the campaign began. By 2000, 22 countries were still reporting cases of polio. Over the next three years, however, with renewed resources, the number of endemic countries was reduced to six, in which roughly 700 cases occurred. Eradication seemed within reach, but this turned out to be an illusion. As in all eradication programs, the last few cases were the hardest and most costly to try to eliminate.

Polio eradication had inherent problems. Polio was not like smallpox, and the polio vaccine was not like the smallpox vaccine. With smallpox, there was little question about who was infected. This made effective case finding, surveillance, and containment possible. Polio, on the other hand, circulated invisibly. It was caused by an enterovirus that replicated in the gut and was spread into the environment through human defecation. It was an oral fecal disease, which spread easily, especially in countries with poor sanitation. It only caused paralysis in one in every 100–200 people infected, however. Thus most cases were either asymptomatic or had symptoms that were difficult to diagnose. Undetected cases could shed the virus for years. This meant that very high percentages of people (80%–86%), had to be vaccinated to break transmission. Thus the containment strategy employed so successfully in the smallpox campaign was not possible with polio. The asymptomatic nature of most cases also meant that outbreaks could occur and go unnoticed for extended periods of time.[25]

The polio vaccine was also a problem. The campaign chose to use the Sabin oral vaccine, because it was easier to administer and cheaper than the injectable Salk vaccine that was widely used in the United States and European countries. The OPV had significant limitations. First, unlike the smallpox vaccine, which provided immunity for 95–98 percent of those vaccinated with a single shot, the polio vaccine had to be given at least three times, spread a month apart. In areas in which the population had weakened immune systems, due to malnutrition and the presence of co-infections, as many as 10 doses might be required. This meant that vaccinators had to track and vaccinate children for an extended period of time. In the context of places where vaccination was resisted by local populations for political or cultural reasons, or where warfare or other political disturbances made access to at-risk populations challenging, gaining and sustaining access and cooperation became more difficult when multiple doses had to be administered. The smallpox campaign was successful in the face of opposition and conflicts precisely because vaccinators required limited access to the populations that needed to be vaccinated (chapter 8). People only needed to be convinced or coerced into being vaccinated one time. The OPV had an additional problem. In rare cases, the attenuated virus in the OPV regained its ability to circulate and triggered an outbreak. In 2013, there were 44 vaccine-derived cases of polio in Pakistan.

Polio thus failed to meet Perez Yekutiel's first two criteria for disease eradication (chapter 8), published in the *World Health Forum* in 1981, seven years before the polio-eradication campaign was launched. This campaign did not have epidemiological characteristics that facilitated effective case detection and surveillance in the advanced stages of the eradication program, and there was no completely effective control measure for break-

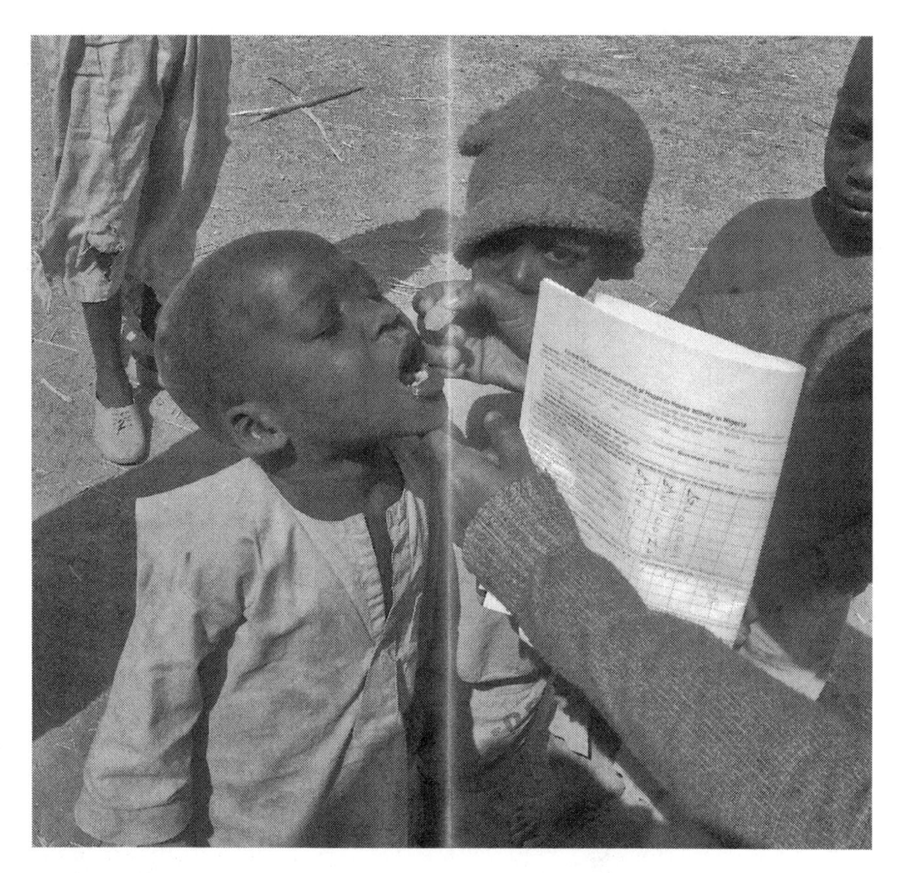

Polio eradication campaign in Kano State, northern Nigeria, January 2013. Centers for Disease Control and Prevention/Binta Bako Sule, Nigeria.

ing transmission. OPV could work in places in which immune systems permitted immunity to be conveyed in three doses over three months, and where there was not widespread opposition or political disruptions. It did not work well where 8–10 doses had to be administered, or where there were significant sources of opposition and disruption, as there was in places like northern Nigeria, Pakistan, and Afghanistan. It did not help that in Pakistan, the US Central Intelligence Agency used vaccinators to help track down Osama bin Laden.

Despite these limitations, the number of endemic countries was reduced to three by 2013: Nigeria, Pakistan, and Afghanistan. Yet the reintroduction of the virus from these three foci to other countries that had been declared polio free continued to occur, raising the possibility of renewed transmission over a larger region. In 2014, WHO declared that the renewed spread of polio was a global-health emergency. An increasing number of global-health experts began to question whether polio could ever be eradicated. Donald Henderson, who led the smallpox-eradication campaign, was an early and vocal opponent of polio eradication. The inability to achieve eradication also raised questions about the merits of spending vast amounts of money chasing polio. In 1988, it was estimated that it would cost US$150 million to eradicate the disease by 2000. By 2011, US$9 billion had been spent, and each year this figure increases by US$1 billion.[26]

Critics of the program asked whether the money would have been better spent strengthening health systems, and whether it would be more cost-effective to simply include the polio vaccine as a part of routine immunization programs.[27] Eradication supporters replied that polio-eradication programs strengthened health systems by providing surveillance capacity, bolstering immunization programs, and supplying laboratory facilities. This argument had been made repeatedly since the 1950s with regard to the advantages of eradication programs. Nigeria's experience with Ebola in 2014 would seem to support these claims. Polio-eradication teams on the ground in that country provided the surveillance network, which helped contain transmission and prevent a larger outbreak from occurring. On the other hand, as Carl Taylor argued in 1997, increasing some of the health infrastructure to support immunization was insufficient to meet the health needs of local communities: "Training people and providing resources to investigate cases of acute flaccid paralysis are far short of training people and empowering them to access local health needs and work with communities to prioritize and deliver interventions to different age groups with a variety of strategies."[28]

Critics have also questioned whether polio was really a priority of many of the countries that were drawn into the campaign. There were 350,000 cases of polio worldwide in 1988, when the campaign began. That same year, there were an estimated 250 million cases of malaria and 1 million deaths from this disease. Tuberculosis claimed hundreds of thousands of lives across the globe. In addition, millions of cases of respiratory and diarrheal diseases afflicted children living in the global south and could be life threatening. If given a choice, some argued, national governments would not have opted to devote their scarce health resources to combating polio. Supporters countered that the funding, coming from private foundations interested in polio-eradication, represented new money that would not have been available for routine health programs.[29] Critics of the GPEI are likely to increase as the goal of eradication continues to remain on the horizon.

They have raised important questions that reflect the continued struggle between advocates of opposing strategies for improving global health.

Vitamin A: Saving Lives, Millions at a Time

Perhaps the most stunningly successful recent example of the application of a biomedical technology to solve a complex public-health problem—in other words, the medicalization of global health—has been the widespread use of vitamin A supplementation to reduce overall child mortality. Vitamin A is a naturally occurring substance in many foods. Yet in many parts of the world, children receive insufficient amounts of this vitamin, due to inadequate diets. The health costs of vitamin A deficiencies had long been recognized by public-health officials. British nutritionist Wallace Aykroyd began his research studying vitamin A deficiencies and night blindness in Newfoundland in the early 1930s (chapter 3). During the 1960s, vitamin A supplementation programs were established to combat night blindness and keratomalacia in a number of countries in the global south. Yet there was

Vitamin A capsule distribution in Indonesia. Joanne Katz.

evidence that vitamin A deficiency could have a more systemic impact on the health of children.

Interest in the systemic consequences of vitamin A deficiency increased following the observation by Johns Hopkins ophthalmologist Alfred Sommer and his colleagues that Indonesian children with "mild" xerophthalmia (night blindness, Bitot's spots) died at far higher rates than their nonxerophthalmic peers. This led Sommer to conduct a study of the relationship between vitamin A supplementation and childhood mortality in Aceh, Indonesia. Sommer subsequently reported that children from ages one to five receiving a dose of 200,000 IU of vitamin A every six months had a 34 percent reduction in mortality. The intervention was particular effective in reducing mortality from measles and diarrheal diseases. The study was criticized for its methodology, yet it stimulated a series of followup trials by researchers at Johns Hopkins and in several other countries. These studies confirmed the positive impact of vitamin A on reducing childhood mortality. This led UNICEF and WHO to recommend that children with measles be given vitamin A in regions where low levels of the vitamin were present. This was later broadened to include all children under age five in countries where there were vitamin A deficiencies.[30] UNICEF subsequently claimed that the distribution of vitamin A capsules (VACs) was essential to achieving the UN Millennium Development Goal of reducing the deaths of children under age five by two-thirds. Today, nearly 80 percent of children under 60 months of age receive periodic VACs in the developing world. It has indeed become a modern magic bullet for saving the lives of children.

Yet the story of Vitamin A is more complicated than this narrative of success would suggest. A closer examination of the development and use of vitamin A supplementation tells us a great deal about the process by which certain technologies become global-health interventions, the role of RCTs in global health, and the benefits and pitfalls of our growing reliance on technological or biomedical solutions to address complex health problems.

From the beginning, not everyone was convinced of the value of vitamin A supplementation in reducing child mortality. It was not clear why vitamin A supplementation decreased under-age-five child mortality. The best guess was that it reduced mortality from measles and diarrhea. Importantly, the studies that showed the greatest effect of VACs were conducted in countries with high rates of vitamin A deficiency and high measles rates. Some asked, what percentage of the impact of VACs on mortality resulted from reductions in mortality from measles, a disease that could also be prevented through vaccination programs?[31] The second question was whether the distribution of VACs was the optimal method for addressing vitamin A deficiencies. In 1987, a report published by the UN Standing Committee on Nutrition noted that there were three approaches to addressing vitamin A deficiencies: vitamin A capsules, food fortification, and horticultural and public-health measures. The report noted that VACs and fortification could only improve vitamin A status, while intervention horticulture and public-health measures could produce wider benefits in health and nutrition. Capsule distribution was recognized as a short-term solution.[32]

In 1992, an International Conference on Nutrition was convened by the FAO and WHO in Rome. The meeting's final report urged countries to

> ensure that sustainable food-based strategies are given first priority, particularly for populations deficient in vitamin A and iron, favoring locally available foods and taking into account local food habits. Supplementation of intakes with vitamin A, iodine, and iron may be required on a short-term basis to reinforce dietary approaches in severely deficient populations, utilizing primary health care services when possible. . . . Supplementation should be progressively phased out as soon as micronutrient-rich food based strategies enable adequate consumption of micronutrients.[33]

Again, vitamin A supplementation was seen as a stopgap measure. In 1993, the International Vitamin A Consultative Group, which was founded in 1975 and funded by USAID, proposed that donors supporting short-term, universal capsule-distribution programs be asked to also fund complementary food-based programs linked to a simple diet-monitoring system. The monitoring would determine when capsule distribution was no longer necessary.[34]

Despite these recommendations for developing multiple approaches to improving vitamin A intake, VAC programs became the norm for improving vitamin A deficiencies, as well as a substitute for other nutritional strategies in many countries. A crucial factor in this development was a meta-analysis of 10 previous trials of vitamin A supplementation, funded by the Canadian International Development Administration and published in 1993. A team of researchers, led by the distinguished University of Toronto professor of nutritional sciences, George H. Beaton, conducted the study. Their analysis concluded that vitamin A supplementation produced an overall 23 percent reduction (RR of 0.77) in under-age-five child mortality. It supported the claims of advocates of vitamin A supplementation and encouraged increased donor support for VAC programs. The World Bank, UNICEF, the Canadian International Development Agency, and USAID all accepted the report's findings and agreed to support country-based programs aimed at providing universal vitamin A supplementation for children from 6 to 60 months old in countries with high child-mortality rates.

Yet the Beaton et al. report included a number of caveats that raised questions about the nature of the benefits that could be achieved by VAC programs. First, it noted that while the overall effect was a 23 percent reduction in mortality (RR of 0.77), the mortality effect was pronounced for diarrheal disease (RR of 0.71) and demonstrable for deaths attributed to measles (RR of 0.47), but it might be absent or very small for deaths attributed to respiratory disease or to malaria (RR of 0.94). In other words, the benefits of VACs could be dependent on the levels of diarrheal disease and measles in the target populations.[35] Second, the authors were unable to determine whether populations lacking evidence of clinical manifestations of vitamin A deficiency but presenting biochemical evidence of major vitamin A depletion would be responsive to improvements in vitamin A status. They concluded, "We *think* the answer is 'yes' but we lack hard evidence."[36]

The report thus suggested that VACs were not a magic bullet for reducing child mortality, but they were effective in reducing mortality in a particular subset of children: those living in areas where diarrheal disease and measles were a major cause of mortality, and who were exhibiting clinical manifestations of vitamin A deficiency. VACs might be expected to have much less of an impact on mortality among children who did not fall into these two categories. The report also suggested that the value of VACs might decline over time as the underlying causes of mortality changed: "It seems logical to assume that as public health measures significantly reduce young child deaths attributable to diarrhea and measles, the relative impact of vitamin A on mortality will be diminished."[37] Similarly, it might be suggested that if the vitamin A status unrelated to supplementation improved sufficiently to eliminate clinical manifestations of vitamin A deficiency, the impact of VACs on mortality might also be reduced or eliminated.

These possibilities raised important questions about the lifespan of biomedical interventions. At what point does one reassess the value of an intervention that has been shown to be effective? There was a tendency within global health to continue funding apparently successful programs. Why question success? But it might be necessary to think about whether interventions needed to be reassessed over time. This was especially true when considering the costs of an intervention. In 2007, an estimated 500 million VACs were distributed annually. Policy recommendations assumed that the cost of each capsule was around US$0.10. Yet one study indicated that the true costs of distributing the capsules—including labor, marketing, training, and administration, in addition to the cost of the capsule—while varying from country to country, was, on average, around US$1. This meant that US$500 million was being spent annually for vitamin A supplementation, a not inconsequential amount.[38]

These caveats received little attention, and advocates of VAC programs touted the report's overall finding of a 23 percent reduction in child mortality. The criteria for adopting VAC programs focused not on levels of vitamin A deficiency, or the presence of measles and diarrheal diseases, but on overall under-age-five mortality. Countries with an overall child mortality rate of 100 per 1000 individuals were targeted for inclusion; this threshold was subsequently lowered to 70 per 1000.[39] In other words, VACs were promoted as a universal technology for reducing child mortality.[40]

The questions raised by the Beaton et al. study remain unanswered and have led some to question the value of continuing high intermittent doses of vitamin A supplements. In a 2014 article published in the *International Journal of Epidemiology*, Tulane University nutritionist John Mason and his colleagues noted that since the original studies done in the 1980s and early 1990s, no subsequent study had shown comparable reductions in child mortality.[41] To the contrary, all of the studies conducted since the 1993 Beaton et al. meta-analysis reported an overall 23 percent reduction in child mortality have shown significantly lower protection levels. This included the massive deworming and enhanced vitamin A (DEVTA) study conducted in India between 1999 and 2004, involving 1 million children, which reported an RR of 0.94, or no impact. Mason et al. concluded from this that "it seems very likely that the overall effect of VACs on young child mortality has decreased over time, and by the 2000s become negligible." Echoing the caveats in the Beaton et al. report, Mason and his colleagues suggested that this decline might reflect the widespread availability of measles vaccines, largely eliminating the major pathway by which vitamin A appeared to reduce child mortality, as well as by reductions in the severity of diarrheal disease through oral rehydration and improvements in sanitation.

Mason et al. also argued that low serum retinol levels, reflecting a subclinical vitamin A deficiency, may have adverse health effects but are not adequately prevented through intermittent high doses of vitamin A in VAC programs. Regular intakes of vitamin A, through the consumption either of natural foods that have high levels of vitamin A or of vitamin A–enriched foods, are more effective in reducing or preventing low serum retinol levels. The authors proposed that the current policy of focusing all resources on VAC programs should be replaced with one that supports a more diverse set of interventions. This recommendation, they noted, was not new, but rather was a return to the position taken by international nutrition committees in the 1980s and early 1990s. Proponents of vitamin A supplementation acknowledged that VAC programs should be monitored to determine whether vitamin A levels had increased significantly, due to increases in the consumption of foods rich in vitamin A or dietary supplements. If serum retinol levels had increased, VAC programs should be reduced or eliminated. Nonetheless, supporters warned against the premature elimination of such programs.[42]

Finally, the Mason et al. article raised another important issue. It noted that despite the widespread acceptance of the efficacy of VAC programs, there had been no large studies to measure program impacts. This, the authors noted, was a persistent problem in nutrition. It was also a problem for many programs based on RCTs. Efficacy and impact were not the same thing. This was well understood, yet few public-health programs actually conducted impact studies to measure the effect of an intervention once it was rolled out into a community on a large scale. Such studies were difficult to conduct and were often discouraged for ethical reasons. How could one conduct a trial of an intervention of proven efficacy that prevented a portion of the population from gaining access to the intervention? African trials of the HIV drug zidovudine, which had been shown to greatly reduce the vertical transmission of HIV in Europe and America, were widely criticized because zidovudine was withheld from women in the control arm of the study.[43] If we knew vitamin A supplements could save lives, how could we justify withholding them from a group of vulnerable children? There were also institutional disincentives for conducting such studies. They were difficult to fund, since they would produce no new scientific data. They also ran the risk of raising questions about an intervention that has gained wide acceptance and substantial funding—and also might be earning significant profits for pharmaceutical and medical-device companies. Professional reputations might also be at stake.

Even where there was evidence that an intervention might not be producing results comparable to those achieved in controlled-efficacy trials, there was resistance to conducting impact studies. For example, in the case of insecticide-treated bed nets (ITNs), there was abundant evidence during the early years of ITN roll out that actual-use patterns reduced their expected impact in real-life settings. Yet no impact studies were conducted until 2011.[44]

This brings us back to the DEVTA study. This study not only directly questioned the value of VACs, but its sheer size, including more children than all the previous studies combined, meant that its inclusion in subsequent meta-analyses would lower the risk ratios for VACs. A great deal was at stake for VAC supporters. Johns Hopkins researchers were quick to criticize the DEVTA study for being methodologically flawed. In an article published in *Lancet*, Alfred Sommer and Keith West, together with Emory University nutritionist Reynaldo Martorell, claimed that the study "was neither a rigorously conducted nor acceptably executed efficacy trial: children were not enumerated, consented, formally enrolled, or carefully followed up for vital events. . . . Coverage was ascertained from logbooks of overworked government community workers (*anganwadi*), and verified by a small number of supervisors who periodically visited randomly selected *anganwadi* workers to question and examine children who these workers gathered for them. Both *anganwadi* worker self-reports, and the validation procedures, are fraught with potential bias that would inflate the actual coverage". They noted further that "each of the 18 study monitors was responsible for overseeing the work of 463 *anganwadi* workers and the status of 55,000 children. Their alleged coverage reached or exceeded that of intensive efficacy trials, yet the researchers spent substantially less than US$1 million. That comes to US$0.20 in field-research costs per child per year ($1 million per 1 million children over five years)—roughly a thousandth what a rigorously done field-efficacy trial costs. Although an expensive trial does not guarantee quality, a trial that does not spend adequately raises serious questions about its validity."[45]

For Sommer et al., the DEVTA study clearly did not meet the requirements of a well-run efficacy trial. It lacked the resources to adequately monitor exposed populations, and thus to ensure coverage and compliance. Such trials, as one of Sommer's colleagues explained to me, are very expensive to mount. There was a high likelihood that these weaknesses could have caused errors that reduced evidence of efficacy. These objections delayed publication of the DEVTA study's findings, which only appeared in *Lancet* in 2013, nearly 10 years after the completion of the study.[46]

The critiques of Sommer and his colleagues challenged the validity of the DEVTA trial's findings. Yet they also raised questions about the production of global-health knowledge. The requirement that studies be generously funded in order to be considered valid highlighted the inequalities that existed in the field of global-health research, for it privileged studies conducted by researchers in the global north over those carried out by researchers in the global south. In effect, the growing role of RCTs in global-health decision making reinforced the tendency for global-health interventions to be developed outside of those countries where such interventions were implemented. But did the objections raised by Sommer et al. mean that the DEVTA study's findings had little value in assessing vitamin A supplementation strategies? DEVTA organizers claimed that these did not, and that the study's methodology was sound.[47] But even if the DEVTA study had methodological problems—failure to ensure compliance among exposed groups, difficulties in running the trials with limited financial resources, and dependence on overworked community health workers—the study may still have value, not as a test of the efficacy of vitamin A, but as a test of the impact of supplementation programs.

The very conditions Sommer et al. and others identified as methodological flaws were conditions that were part of the real world in which vitamin A supplementation often occurred. Overburdened health workers do not have the resources to test the capsules for vitamin A content or possible adulteration. Nor are they able ensure that the VACs are always delivered on time and taken properly. In other words, the DEVTA study was a test of the effectiveness of vitamin A supplementation when it was applied to a very large, at-risk population, using existing health resources. Its findings may support those who have opposed using VACs, not because it showed that VACs were ineffective in reducing mortality, but because it demonstrated that in real-life settings, they did not have the impact that was expected.

The most vocal opposition to the use of VACs came from a group of influential nutritionists in India, led by Coluthur Gopalan. India had one of the earliest national vitamin A supplementation program, aimed at eliminating night blindness in children. Yet Gopalan and his colleagues rigorously rejected the extension of these programs to prevent under-age-five mortality. They questioned the science behind the use of VACs, pointing to studies that had shown low or no impact and arguing that VACs were not universally beneficial but had different effects on different populations. VACs needed to be tested in local settings, and the nutritionists pointed to such studies in India, including the DEVTA study, which suggested that VACs were of little value in India. They also argued that the main benefactors of VAC programs were the pharmaceutical companies that manufactured the supplements.

As Sarah Wallace has shown, Gopalan had bigger fish to fry.[48] The battle over the universal use of VACs was also about a growing tendency in global health to rely on short-term, Band-Aid approaches to health while ignoring the need to build basic health services and to develop agricultural systems that could provide adequate supplies of vitamin-enriched foods. In effect, Gopalan and his colleagues were opposed to the increasing medicalization of global health. His criticisms echoed earlier criticisms of BCG trials in India in the 1950s (chapter 3). Given the deplorable state of much of India's health system, the country's massive lack of sanitation, and its poorly developed food systems, one can understand Gopalan's concerns.

Simple technologies, however effective in the short term, do not make a long-term solution. They need to be coupled with serious efforts to address the larger structural problems.[49] VAC supporters did not deny this. They acknowledged that improvements in agriculture, combined with the fortification of foods, would eventually eliminate the need for VAC programs. There were efforts under way in countries such as the Philippines, Nepal, and Guatemala to develop alternative sources of vitamin A, and VAC researchers were working on protocols for determining when VACs could be withdrawn.[50] The question was whether the widespread use and reliance on VACs discouraged these various efforts to achieve long-term improvements in food supplies.[51]

Vitamin A supplementation, insecticide-treated bed nets, antiretroviral drugs, male circumcision, childhood immunizations, and polio vaccinations have all been shown to be effective interventions that improve the health of populations across the globe. They have the potential to save lives—millions at a time. But they should not be seen as solutions in their own right, or as substitutes for building effective health systems and addressing the underlying structural causes of ill health in society. Yet increasingly, over the course of the twenty-first century, these technologies have become just that. Under pressure from donor organizations for measurable-impact programs, from neoliberal economic strategies that encourage the commodification of health, and from the changing landscape of public-health training, with its growing reliance on funding tied to scientific discovery, global health has become centered on developing, deploying, and measuring the impact of technologies. During the course of the twentieth century, the pendulum, which has swung between narrow technological approaches to eliminating diseases and broader-based efforts to build health systems and address the underlying structural causes of health, has moved decidedly toward the former. The consequences of this shift were visible in 2014 in the villages and urban slums of Liberia, Sierra Leone, and Guinea.

Responding to Ebola

I OPENED THIS BOOK with a discussion of the Ebola outbreak that began in Guinea in 2013 and what it revealed about the state of global-health assistance. I will close by returning to West Africa to examine the international response to the epidemic—a response that reflected many of the challenges and conflicting visions that have marked the history of global health.

The global response to the West African Ebola outbreak, which had killed over 8000 people by January 2015, unfolded slowly. A team of physicians working for Doctors Without Borders/Médicins Sans Frontières (MSF) confirmed the presence of Ebola in Guinea in March 2014. By April, the disease was spreading beyond the country's borders, yet the World Health Organization was reluctant to label it an epidemic. WHO spokesperson Gregory Hartl played down the importance of the outbreak at an April news conference in Geneva, saying, "Ebola already causes enough concern, and we need to be very careful about how we characterize something which is up until now an outbreak with sporadic cases."[1] WHO Director-General Margaret Chan was reluctant to raise the alarm about what was occurring in West Africa, in part because she had been criticized for her earlier warning about the threat of the H1N1 virus in 2009, and in part because a number of earlier outbreaks of Ebola in East and Central Africa had been successfully contained.

MSF, which had sent teams to the region to open clinics and emergency hospitals soon after Ebola was confirmed, strongly disagreed, warning, "We are facing an epidemic of a magnitude never before seen in terms of the distribution of cases in the country."[2] MSF called for a much more robust response from the world community. WHO, however, did not declare that the outbreak was a global-health emergency until August 9, 2014, and not until October did significant additional aid begin reaching the three most affected countries. There is no doubt that the delayed global response contributed to the spread and severity of the epidemic.

With the discovery of the first case of Ebola in the United States in September 2014, the Ebola outbreak suddenly became a global health emergency. This one case stimulated increased US commitments to fighting the disease in West Africa, including the promise of several thousand US military troops to help with logistics. The US Congress was asked for US$4.6 billion dollars, most of it to go to strengthening disease preparedness in the United States, global surveillance systems, and vaccine and drug development. Private foundations committed over US$100 million to support control efforts in West Africa. When aid began pouring in, it traveled along well-trodden colonial pathways, with the United States sending money and 2400 military troops to Liberia, the British sending resources primarily to its former colony in Sierra Leone, and the French doing the same for its former colony in Guinea. Channeling aid in this way ignored the regional nature of the epidemic and the need to coordinate relief activities across borders.

In the meantime, dozens of international NGOs arrived, working as best they could to help stabilize the situation. There was very little coordination among these various groups, however, and no one was in charge. WHO, which should have played this role, lacked the resources to do so. After the 2008 financial crisis, its overall budget had been cut to roughly US$2 billion, or less than that of some major medical centers in the United States. Most outside contributions to WHO were directed to categorical programs, such as polio eradication. WHO's emergency-response office had had its budget cut by 51 percent. Again underplaying the need for regional coordination, Director-General Chan preferred to have the countries themselves coordinate activities within their own borders. The three most affected countries, however, had a limited capacity to mobilize or coordinate the relief response.

Lacking coordination, NGOs employed different protocols, duplicated efforts, and, in some cases, overbuilt facilities, setting up beds that remained empty. Early foreign assistance led to the creation of Ebola case-management centers but did not provide trained staff to run them.[3] In Liberia, where most of the US funding was directed, almost all of the support went to building facilities in the capital, Monrovia, while rural areas, where many cases were occurring, received little attention. International actors, moreover, seemed unable to adapt quickly to a rapidly changing situation. Resources were being allocated to activities that were no longer appropriate. In Monrovia, more case-management facilities were being built, despite the existence of adequate isolation capacities and a drop in cases in the capitol. Only 28 patients were treated in the 11 treatment centers built by the US military.[4] To make matters worse, CDC officials, who were supposed to oversee control efforts, circulated in and out of Monrovia on two-week cycles, requiring local health authorities to devote precious time to bringing new CDC workers up to speed.[5] The NGO response was, in many ways, characteristic of the wider pattern of disarticulated health programs that have marked the global-health arena in the twenty-first century.

Finally, as in so many past international-health efforts, the peoples of the region were viewed by many aid providers as part of the problem, rather than the solution. Local knowledge, capacities for self-help, community building, and caring were undervalued, ignored, or only slowly recognized. In Liberia, as the country reeled from the rising toll of cases that overwhelmed its health infrastructure, and international assistance was just beginning to trickle in, local people stepped up and intervened where they could to fight the epidemic. As a *New York Times* reporter observed, volunteer Ebola watchdog groups sprang up in many neighborhoods in Monrovia.[6] These groups were typically overseen by local elders and led by educated youths, who drew on a long history of community organizing created to survive war, poverty, and governmental neglect.

With little or no outside help in the early months, the watchdog groups educated their communities about Ebola, a disease new to this part of Africa, and collected money to set up hand-washing stations at key spots. They kept records of the sick and the dead. Many also placed households under quarantine and restricted visits by outsiders. When the sick were turned away at the gates of treatment centers because of a lack of beds, people inside homes began protecting themselves better, covering their arms in plastic shopping bags as they cared for ailing relatives. The gear became known as Liberian PPEs, or personal protective equipment, a reference to the more impermeable suits worn by health workers. These watchdog groups helped slow the progress of the epidemic, yet their participation received little or no international media coverage. Typical of this silence was a *60 Minutes* story on the Liberian epidemic by CBS reporter Laura Logan.[7] The story focused exclusively on the role of foreign aid workers. Not a single Liberian was interviewed.

The immediate goal of international relief efforts was to provide training, supplies, and infrastructure to slow the spread of Ebola. There was wide agreement, however, that the long-term solution to the problem—and to preventing similar health crises from happening in the future—lay in rebuilding the fractured health systems of the three affected countries. To what extent did relief efforts contribute to this goal? And to what extent did they simply provide a temporary bandage to prevent the disease from spreading globally?

It is still too early to answer these questions. I believe we can gain a partial answer, however, by examining the activities of two of the most visible humanitarian health organizations that played a major role in responding to this and other global-health crises: Médicins Sans Frontières (MSF) and Partners In Health (PIH). Both organizations were active in combating Ebola by providing health care, building clinics, and training medical staff. Yet here, and in other places, they operated in very different ways. These differences reflected the fault lines that have divided the world of global health since the beginning of the twentieth century.

Médicins Sans Frontières / Doctors Without Borders

MSF was created by a group of French physicians who had participated in health relief during the Biafran civil war in Nigeria in 1971. It was, and is, committed to providing emergency medical assistance wherever it is needed and to giving witness to the suffering of people and the conditions that produce it. It has intervened to assist injured and sick people in the aftermath of natural disasters, such as the earthquake in Managua in 1972, the tsunami in Indonesia in 2010, and the earthquake in Haiti that same year. The organization has also provided medical assistance to refugees in war-torn regions of the globe, including in Kosovo in the 1990s and Southern Sudan from 1983 to 2005, and combatted major disease outbreaks, such as the cholera epidemic in Haiti in 2010 and previous Ebola outbreaks in Central and West Africa. Its leadership and members have been advocates for providing affordable drugs to the poor and were centrally involved in the battle over access to generic AIDS drugs in South Africa and elsewhere. MSF often works in places where there are no other sources of health care, or where health-care services are inadequate.

Over the years, MSF has grown in size, undergone schisms, and developed considerable operational capacity, with specialized sections handling various forms of logistics, transport, and management related to their medical-relief activities, epidemiological services, and operational programs to deal with issues as varied as street children, mental health, and the trauma of sexual violence. There are now multiple national sections, with headquarters in cities across the global north, including Brussels, Amsterdam, Paris, Barcelona, and Geneva. While MSF members have become masters at mobilizing rapid emergency relief, claiming to be able to be on the ground in a crisis situation within 24 hours, they have also become involved in longer missions, lasting, in some cases, for decades.

Through it all, MSF has maintained certain guiding principles. It is decidedly apolitical and does not take sides in political disputes. It will engage in advocacy, but not partisanship. Thus MSF supported South Africa's Medicines and Related Substances Control Amendment Act and the defense against the suit by pharmaceutical companies trying to prevent the use of generic drugs. But MSF members were not signers of the petitions or participants in the lawsuits. MSF defends its right to provide care where it is needed. While it tries to work with local governmental authorities where possible, this not a prerequisite for its involvement, and it often works independently. MSF's goal is to not only provide medical assistance, but also to draw attention to the conditions that produced a medical need. By doing so, it seeks to encourage others to provide support and brings international pressure to bear to change those conditions.

Because MSF often work in places where governments are weak, it is forced to take on roles that are normally played by individual nations, and to engage in forms of what anthropologist Peter Redfield refers to as "mobile sovereignty," administered through Toyota Land Cruisers, satellite phones, and laptop computers.[8] These roles are attenuated, however. MSF cannot be a substitute for the state. It does not extend its mission to include broader efforts at improving public health through sanitation or the provision of clean water, or offering wider development assistance that might impact positively on the health of the people MSF is trying to help. MSF is a provider of emergency medical care.

MSF missions are organized to facilitate rapid responses, rather than to build sustainable health systems. It hires local support staff but depends on medically trained volunteers from the global north, who can be mobilized quickly, to direct most of its work. Similarly, MSF response teams employ simple technologies and medicines designed to be used in a wide range of resource-poor settings, rather than interventions that are tailored to local conditions. These practices allow MSF teams to hit the ground running.

While MSF missions have lasted for years, the organization has limited resources and is committed to being ready to respond to the next crisis. This means that no mission is permanent. MSF defends its independence, its ability to move, and its global commitment. It does not build for the long term. As an epidemiologist working for MSF in Uganda told Redfield: "When MSF does a program, the idea is to do it well and then go. We don't believe much in sustainability."[9] When projects are done, MSF tries to turn them over to national agencies or other organizations more suited to operating long-term programs. The reality is that in many places this is not possible, and the programs MSF has built eventually collapse. For example, MSF was operating for over two decades in Liberia during and after the civil war, running a series of clinics, and it withdrew from the country in 2012, just before the Ebola outbreak occurred. The clinics MSF had turned over to the state lacked the equipment and personnel needed to contain the epidemic.

MSF is very good at what it does. Its response to the Ebola outbreak was rapid. Having been involved in earlier outbreaks in Central Africa, MSF members knew what needed to be done. In addition, MSF already had teams on the ground in Sierra Leone. Unlike other organizations, MSF provided assistance to all three affected countries, establishing seven Ebola case-management centers, as well as providing 600 beds in isolation and two transit centers. MSF's interventions involved 300 international aid workers, along with over 3000 locally hired staff. From the time MSF teams began operations, in March 2014, through January 2015, they admitted 7000 patients, among whom 4400 had Ebola. Nearly 1900 of these survived.[10]

It is difficult to know how long MSF will stay in the region. It is likely that given the collapse of local health services under the burden of the epidemic, the organization will remain in place for years to come. But if its history is any guide, MSF will eventually pull up stakes and move on to the next crisis. MSF is continually engaged in a process of triage on a global scale, deciding who is most in need of their help. When MSF workers depart, they will leave behind a significant number of local staff trained in infectious disease–control practices. They will also leave behind some health facilities. But it will be up to others to sustain these health resources by continuing to train health workers and staffing and supplying health facilities. It is unclear who will do this. Without substantial amounts of aid to strengthen health systems, the work that MSF has done is likely to gradually disappear.

Partners In Health

Partners In Health is not MSF, as it volunteers and directors frequently emphasize. It sees itself as fulfilling a very different mission. It is not built to rapidly provide emergency medical services across the globe, wherever they are needed. PIH reacts more slowly, carefully choosing its targets. But once PIH begins a mission, it is in it for the long term. Its goal is to build good health services in places where no or few services exist. Its members believe that health care is a human right and reject the idea that the poor should accept minimal packages of health care, limited to immunizations, vitamin A supplements, and oral rehydration. Finally, PIH believes that health care requires broader commitments to meeting the social and economic needs of the people it serves.

PIH was begun in 1987 by physicians Paul Farmer and Jim Yong Kim and health activist Ophelia Dahl. It grew out of Farmer's experiences as a medically trained anthropologist working in Haiti on the early emergence of AIDS. In what remains a classic study in medical anthropology, Farmer described what had happened to a community of farmers who had been displaced by the building of a dam, which had deprived them of their land and livelihoods. The book, *AIDS and Accusation*, traced the early spread of HIV into the community, placing this situation within the longer political and economic history of Haiti and the Caribbean and using his analysis to demolish claims about Haiti as a vector for the hemispheric spread of the disease.[11] Farmer's vision of the health needs of communities like the one he studied and the structural conditions that created them shaped the vision of PIH.

Farmer and his colleagues returned to Cange, the village in which he had studied, and built a community-based health project known as *Zanmi Lasante*, which is Creole for "Partners in Health." The project became a model for PIH's subsequent work. It was built in conjunction with other NGOs and the Haitian Ministry of Health. It hired and trained large numbers of *accompagnateurs* (community health workers) to work within the community: preventing illness; monitoring medical and socioeconomic needs; and delivering quality health care to people living with chronic diseases, such as HIV and tuberculosis. The use of community health workers not only helped provide effective health care but also generated much-needed employment in the region. PIH is concerned with the overall well-being of the patients and communities in which it works, taking a broad view of health that is more attuned to the rural-hygiene efforts of the 1930s and the original primary health-care declaration at Alma-Ata. But PIH and Farmer do not attack the political and economic structures that produce poverty in places like Cange, the structures that Farmer so clearly identified in *AIDS and Accusation*. Excoriation of the World Bank (now headed by Farmer's friend and PIH cofounder, Jim Yong Kim), which funded the dam that flooded the valley from which the farmers of Cange came, has been replaced with a softer evocation of the need to address the direct social determinants of ill health: poverty, lack of sanitation, and jobs. The central mission of

PIH is, above all, providing access to care. Like the rural-hygiene experiments in the 1930s, PIH's commitment to fundamental social and economic betterment has been limited.

Over the years, *Zanmi Lasante* has grown from a small clinic to a first-class medical center that employs over 4000 people, almost all of them Haitians, including doctors, nurses, and community health workers. It now has a 104-bed, full-service hospital, with two operating rooms, adult and pediatric inpatient wards, an infectious-disease center, an outpatient clinic, a women's health clinic, ophthalmology and general-medicine clinics, a laboratory, a pharmaceutical warehouse, a Red Cross blood bank, radiographic services, and a dozen schools. Working with the Ministry of Health, it has expanded its operations, building or refurbishing and supplying closed or run-down clinics across the central plateau of Haiti and training growing numbers of health workers. This system provided health care and other social services to 2.6 million people in the region in 2015. PIH also built a hospital to serve the needs of the poor in Port-au-Prince after the 2010 earthquake that destroyed most of the medical facilities in the capitol. The hospital served the needs of 185,000 people and also provided medical training for Haitian nurses, medical students, and resident physicians. PIH has now been working in Haiti for thirty years. This is not triage. It is commitment.

PIH has replicated *Zanmi Lasante* in other countries, most notably in Rwanda, where it now operates a network of community-based health services across a large portion of the country, working in partnership with the Rwandan Ministry of Health and other NGOs. Along the way, PIH had taken on some of the "immodest claims" that have characterized the attitudes of other global-health organizations regarding what is possible in resource-poor environments. For example, PIH demonstrated that complex ARV therapies could be administered effectively among populations who were viewed by many as incapable of managing their own health care (chapter 15). Similarly, working in Peru and Russia, PIH demonstrated that multidrug-resistant TB could be effectively treated in poor communities.

The PIH model has been very successful. Yet it unclear whether additional replications are possible. Farmer and PIH have raised hundreds of millions of dollars to support their present activities. Without similar financing, the organization would not be able to reproduce the PIH model elsewhere. More importantly, despite PIH's success in expanding its operations in Haiti and Rwanda, replicating its model over wide areas of the globe, in places where populations lack basic health services, would require far more resources than have been generated by the current growth in global-health spending. Farmer and his colleagues fundamentally reject such calculations, arguing that they are based on funding levels that, while much larger than in the past, could be greatly expanded if wealthy countries devoted even a small portion of what they allocate for military assistance to health instead. The United States, for example, spent nearly US$6 billion on foreign military assistance in 2014.

PIH's response to the Ebola outbreak in West Africa was guided by its principles and experience in Haiti, Rwanda, and other countries in which it had committed its resources. It did not rush in to provide emergency assistance.[12] It was not until mid-September 2014 that Farmer led a team of PIH volunteers to assess the situation and provide immediate assistance in the form of training local health staff in dealing with Ebola cases in rural areas of Liberia and Sierra Leone. PIH viewed quelling Ebola as a first step in a much longer process aimed at rebuilding the rural health systems of both countries. In Liberia, it partnered with a local NGO, Last Mile Health, which was begun before the epidemic by a Farmer protégé, Raj Panjabi, who had been raised in Sierra Leone but, during its civil war, moved to the United States and studied medicine at Harvard. Last Mile Health is committed to providing health care in remote parts of the country where health services rarely reach, places where people need to travel miles to access care. In Sierra Leone, PIH collaborated with another NGO, Wellbody Alliance.

The PIH/Last Mile Health/Wellbody Alliance coalition plans to build a force of 500 health workers to staff 47 health centers in Liberia and Sierra Leone. It will also train an additional corps of 800 community health workers, who will serve in villages and focus on education, surveillance, and monitoring. In the future, the coalition will help the local Ministries of Health transition from the Ebola response to a more robust health system. As Panjabi stated, "Ebola started in the rainforest and it could have stopped there if we had a health care system in place."[13] It is too early to assess how successful PIH will be in these two countries. But given the organization's record of sustaining the commitments it begins, it is likely to be there for decades to come.

The contrasting approaches of MSF and PIH reflect, in many ways, the competing strategies that have characterized the efforts of wealthier nations and multinational organizations located in the global north to improve the health of people living in resource-poor countries in the global south. For much of the twentieth and early twenty-first centuries, global health has been about providing simple medical technologies (including vaccines, drugs, bed nets, vitamin A capsules, birth control devices, and oral-rehydration packets) to treat or prevent specific health problems (such as smallpox, malaria, hookworm, or malnutrition) and, in this way, save lives, millions at a time. These interventions have been attractive because they are relatively easy to apply and can be delivered rapidly, with little involvement of local populations in their design or application. They can have an immediate impact, which serves political as well as humanitarian interests. They are also thought to be cost-effective, measurable, and able to bring about improvements in health without addressing the more complex and messy social and economic conditions that generate health problems. These interventions do not require long-term investments of time, personnel, and financial resources in the places where they are applied. Instead, they can be put into place without needing to know much about the locales in which they are used. They have also been encouraged by a crisis mentality, which has characterized much of global health: "We need to act fast and cannot wait for more long-term, fundamental changes." MSF is, in many ways, representative of this approach to global health.

Occasionally, alternative visions of health, similar to that espoused by PIH, have been explored and applied on a limited basis. These approaches have focused on the building of health-care systems that address a wide range of health problems. They reflect a broader definition of health, which encompasses the general well-being of populations, not simply the absence of disease. They have also been attentive to the underlying social and economic determinants of health and are marked by efforts to integrate health care into wider development efforts. In addition, they have valued the participation of local populations in the design and application of health interventions. They are about long-term commitments. Above all, health is viewed as a basic right.

For a range of reasons, this vision has failed to gain wider backing. It is viewed as too expensive to build on a global scale. The results of this approach are more difficult to measure. This alternative approach to global health also requires addressing complex economic and social conditions that are deeply imbedded and are often supported by local and global political interests. It demands engagement with and the empowerment of people whose visions of health and of the world are very different from those hoping to bring about change. It requires collaboration across professional disciplines that view the world in dissimilar ways. Physicians and public-health authorities need to listen to what engineers, economists, and anthropologists have to say, and visa versa. Our schools of medicine and public health do little to support these strategies. Finally, at various times this alternative vision has been seen as politically suspect. In the 1950s, conservative political leaders identified it and the strategies it generated with socialism. More recently, neoliberal economists and politicians, who view health as an individual responsibility and reject rights-based arguments for the provision of health care, have labeled these strategies as neither efficient nor cost-effective.

Global health today contains elements of both visions. Yet it has overwhelmingly become subject to technical approaches that have their origins in colonial medicine at the beginning of the twentieth century. The landscape of global health is dominated by thousands of vertical, NGO-run programs dedicated to addressing specific health problems, yet failing to provide for the basic health needs of the world's poor.

This is not to dismiss the value of Médicins Sans Frontières or the approach it represents. The provision of lifesaving drugs, vaccines, and simple technologies like ITNs has saved millions of lives. Yet without basic health care, many existing and emerging health problems will go unattended. Functioning, community-based health systems create a platform for providing access to technologies. They also build links with their communities, facilitating communications and the sharing of health information that, for instance, might have limited the spread of Ebola and might curb the extent of future epidemics.

Many will argue that building such systems, let alone addressing the structural causes of ill health, is too expensive, and that we need to use the limited resources we have in the most cost-efficient manner to save lives. Even if the exhortations of Farmer and others—that we need to rethink the limits of the available financial pie—are rejected, the Ebola epidemic forces us to question the logic of the "too expensive" argument and the fallacy of cost-effective rationales.

Before we are done, the Ebola epidemic may well cost more money than would have been spent strengthening the health systems of the affected countries. At the end of December 2014, the World Bank projected that the price tag for containing the epidemic could be US$3–$4 billion.[14] In addition, the epidemic may cost the economies of the affected countries additional tens of billions of dollars and require larger commitments of foreign aid from donor nations. Investing in building health systems would have avoided much of this expense and have had an impact on a much wider range of health problems.

PIH's health-care model may be difficult to replicate on a large scale. It is also not a complete answer. It is, in many ways, a medical approach to health that, despite its development rhetoric, has not seriously attempted to make health care part of a broader, multisectoral approach to improving the overall well-being of the populations its serves. In some ways the Jamkhed project in India (chapter 12) is a closer match to a broad-based community approach to health and development, though it remains an NGO-run program, with limited connections to the Indian health system. PIH, nonetheless, begins to offer an alternative strategy to the current collection of vertical programs that save lives but fail to provide for the basic health needs of the majority of the world's populations.

There are signs of growing interest in developing community-based health-care systems and community health worker programs. The Gates Foundation is currently funding initiatives designed to promote the delivery of health services at the community level. To date, however, the foundation's focus has been on providing technologies that will improve the performance of health-care systems, rather than expanding health-care coverage. In one state in India, for example, it has invested in improving performance management within the Bihar government's primary health-care system. The foundation has done this by providing mobile training tools for community health workers; digital systems for recording patient data and monitoring family-planning and immunization programs; and digital payment systems. In short, the Gates Foundation is searching for ways to deploy technology to improve performance.[15] Economist Jeffrey Sachs, who chaired WHO's Macroeconomics of Health Commission, has called for the recruitment and training of 1 million community health workers for sub-Saharan Africa.[16] The World Bank, under Jim Yong Kim's leadership, has made a commitment to the concept of universal health coverage. In September 2015, 267 economists from 44 countries, led by Lawrence H. Summers of Harvard University, signed the Economists Declaration on Universal Health Coverage, which calls on global policymakers to prioritize a pro-poor pathway to universal health coverage as an essential pillar of sustainable development.[17] These initiatives are still a long way from building community-based primary health-care systems, but they indicate an awareness that there is a need to focus on the basic health needs of the world's poor.

Progress toward universal health coverage has also been made in several African countries. Since 2003 in Ethiopia, a health-extension program has trained and deployed over 38,000 health workers to deliver primary health-care services in rural communities. The Ethiopian health minister between 2005 and 2012, Tedros Adhanom Ghebreyesus, who initiated this development, was elected director-general of WHO in 2017 and has called for a new commitment to building primary health services. In Ghana, a tax-funded national health insurance system, known as the National Health

Insurance Scheme, covers 95 percent of the diseases that affect Ghanaians, enabling financial protection and expanding coverage. Rwanda's health-care-strengthening partnership with Partners In Health has contributed to a two-thirds reduction in child mortality since 2003 and increased that country's average life expectancy by 10 years in the past decade. Progress is possible. Building on these initiatives, however, will require major new commitments on the part of international donors and national governments.[18]

Ironically, the likelihood of primary health-care strengthening happening in Liberia, Guinea, or Sierra Leone is threatened by news that a vaccine for Ebola has been developed. If history tells us anything, it is that finding a vaccine will save lives, but it will also delay efforts to build health systems and alleviate the structural causes of diseases. As WHO Director-General Margaret Chan warned in response to news of the potentially effective Ebola vaccine, "there is no replacement for very strong and good, resilient health systems with the capability for surveillance."[19] We need to ask whether Ebola will be one more health crisis for which the world of global health finds a magic bullet that saves lives but fails to address the conditions that generated the epidemic in the first place, leading eventually to the next global health crisis.

ACKNOWLEDGMENTS

This volume draws much of its material from a large and growing body of literature on the history of international and global health. I am indebted to the many historians, sociologists, political scientists, and anthropologists who have contributed to this literature.

The idea for the book and its central arguments evolved over a number of years from discussions I have had with students in the seminar I have taught on the history of international health and development at Emory University's Rollins School of Public Health and, since 2003, at the Johns Hopkins Bloomberg School of Public Health. Many of the students who took the course at Emory and Johns Hopkins had worked in the field of global health before entering graduate school. Their work experiences enriched the seminar and contributed to my own education.

My understanding of global-health history has also been expanded by my interactions with colleagues from both schools, as well as from the Centers for Disease Control and Prevention and the Carter Center in Atlanta. Many of these men and women have devoted their careers to improving the health of other peoples. The insights gained from these interactions—in seminars, at conferences, and over beers—have been invaluable. Among the many people who have shared their knowledge and experience were Kent Campbell, Stanley Foster, Frank Richards, Rick Steketee, Bernard Nahlen, Jim Setzer, Claudia Fishman, Jim Curran, Peter Brown, Marcia Inhorn, Deb Mcfarland, Ruth Berkelman, Howard Frumkin, John McGowan, Jennifer Hirsch, Basu Protik, Peter Winch, Al Sommer, and Lori Leonard. We have not always agreed about the pathways that global-health efforts should take, but I hope they will feel that I have fairly represented these pathways in this volume.

I often thought about turning my international health and development course into a book. But there were always other projects that consumed my research and writing time, so I kept putting off writing a book on global health. The decision to finally push forward with this project was encouraged by my editor at Johns Hopkins University Press, Jackie Wehmueller.

Jackie was also instrumental in convincing me to write my history of malaria, *The Making of a Tropical Disease.* Jackie is a wonderful editor and friend, whose critical eye and unflinching honesty in pointing out strengths and weaknesses in my work have guided me in writing this volume. In the end, it may not be the book she envisioned, but it is nonetheless stronger for her guidance.

Much of this book was written during a sabbatical leave in 2013. Sabbaticals are rare events within the Johns Hopkins School of Medicine, and I want to thank the school and my dean, Paul Rothman, for granting my sabbatical request.

A number of people have read and commented on parts or all of this manuscript. These included Jeremy Greene, Julie Livingston, Joanne Katz, Henry Perry, Sara Berry, Katherine Arner, Julia Cummiskey, Kirsten Moore-Sheeley, and Lotte Meinert. Part 4 was presented at the University of Pennsylvania History and Sociology of Science Seminar, part 3 at the Critical Global Health Seminar at the Johns Hopkins Bloomberg School of Public Health, and part 7 at the Anthropology Seminar at the University of Stellenbosch. I also assigned early drafts of the book's chapters to students in my History of Global Health and Development seminar in spring 2015 and received helpful feedback from a number of students, especially Naoko Kazuki and Heidi Morefield. I have done my best to incorporate the comments of all these readers into the final manuscript.

I want to express special gratitude to Socrates Litsios, who worked for years at WHO and was a witness to many of the events described in the later chapters of this book. Socco is also an historian who produced two volumes of the official history of WHO, along with many other articles and books on the history of international health. Over the years he has generously shared data on a wide range of topics, along with drafts of published and unpublished papers. He has been a friend and colleague and I am greatly in his debt.

Finally, I want to thank Carolyn, my wife and partner for life, whose continued love and support has made it possible for me to write this and earlier books.

NOTES

Introduction: Ebola

1. This narrative is based on a series of news accounts describing the beginnings of the Ebola outbreak in southeastern Guinea. We still do not know exactly what happened during these early days of the epidemic. This, as best I can tell, is close to the events that occurred. See Holly Yan and Esprit Smith, "Ebola: Who Is Patient Zero? Disease Traced Back to 2-Year-Old in Guinea," CNN, updated January 21, 2015, www.cnn.com/2014/10/28/health/ebola-patient-zero/index.html; Childby Bahar Gholipour, "Ebola 'Patient Zero': How Outbreak Started from Single Child," *Live Science*, October 30, 2014, www.livescience.com/48527-ebola-toddler-patient-zero.html; Jeffrey E. Stern, "Hell in the Hot Zone," *Vanity Fair*, October 2014, www.vanityfair.com/politics/2014/10/ebola-virus-epidemic-containment/.

2. Centers for Disease Control and Prevention, "Initial Announcement," March 25, 2014, www.cdc.gov/vhf/ebola/outbreaks/2014-west-africa/previous-updates.html.

3. Michael McGovern, "Bushmeat and the Politics of Disgust," *Cultural Anthropology Online*, Ebola in Perspective series, October 7, 2014, http://culanth.org/fieldsights/588-bushmeat-and-the-politics-of-disgust/.

4. Sharon Alane Abramowitz, "How the Liberia Health Sector Became a Vector for Ebola," *Cultural Anthropology Online*, Ebola in Perspective series, October 7, 2014, http://culanth.org/fieldsights/598-how-the-liberian-health-sector-became-a-vector-for-ebola/. Abramowitz noted that in Guinea, the government invested only 1.8% of the GDP in health care. Liberia's public-health sector, even before the Ebola crisis, was virtually nonexistent. What health services were available were provided largely by NGOs, which ran three-quarters of the government-owned health facilities. The costs of this reliance were laid bare when, in 2006–2007, Médecins Sans Frontières (MSF) withdrew its branches from Liberia in order to move its resources to other areas deemed to be in greater need of MSF's emergency services. This resulted in the closure of critical regional and urban hospitals, and the abrupt closure of 30 World Vision clinics in the Monrovia.

5. Joseph D. Forrester, Satish K. Pillai, Karlyn D. Beer, Adam Bjork, John Neatherlin, Moses Massaquoi, Tolbert G. Nyenswah, Joel M. Montgomery, and Kevin De Cock, "Assessment of Ebola Virus Disease, Health Care Infrastructure,

and Preparedness—Four Counties, Southeastern Liberia, August 2014," *Morbidity and Mortality Weekly Report* 63, 40 (2014): 891–900.

6. Anthony Fauci, "Lessons from the Outbreak," *Atlantic*, December 17, 2014.

7. One study suggested that 54% of the rural private medical practitioners were informally trained and either had unrecognized diplomas or were not medically qualified. See Rashmi Kumar, Vijay Jaiswal, Sandeep Tripathi, Akshay Kumar, and M. Z. Idris, "Inequity in Health Care Delivery in India: The Problem of Rural Medical Practitioners," *Health Care Analysis: An International Journal of Health Care Philosophy and Policy* 15, 3 (2007): 223–233.

8. Moreover, in terms of their ability to properly diagnose and treat standardized patients, the difference between the trained and untrained medical providers was minimal. In both cases, correct diagnoses were rare and incorrect treatments were widely prescribed. See Jishnu Das, Alaka Holla, Veena Das, Manoj Mohanan, Diana Tabak, and Brian Chan, "In Urban and Rural India, a Standardized Patient Study Showed Low Levels of Provider Training and Huge Quality Gaps," *Health Affairs (Project Hope)* 31, 12 (2012): 2774–2784.

9. Parthajit Dasgupta, "The Shameful Frailty of the Rural Healthcare System in India," Future Challenges, February 2, 2013, https://futurechallenges.org/local/the-frailty-of-rural-healthcare-system-in-india/.

10. Paul Farmer, Arthur Kleinman, Jim Yong Kim, and Matthew Basilico, *Reimagining Global Health: An Introduction* (Berkeley: University of California Press, 2013).

11. Randall M. Packard, *The Making of a Tropical Disease: A Short History of Malaria* (Baltimore: Johns Hopkins University Press, 2007).

12. I first taught the course in a summer school program run by Tufts University on their lakeside campus in Talloires, France, in 1992. All of the courses in the program were supposed to have a connection to the program's location in France. So, on the first day of the course, I led my students onto a patio overlooking the lake and asked them what they thought a course that examined efforts to improve health and nutrition in resource-poor countries in Asia, Africa, and Latin America had to do with this location. The answer was that Talloires had been the site of important meetings related to the launching of the Child-Survival Revolution in the 1980s. More generally, places like Talloires—be it the Rockefeller Conference Center on Lake Como in Bellagio, Italy, or the World Health Organization's headquarters on Lake Geneva in nearby Switzerland—were where major decisions affecting the health of peoples living in the global south had been made for over a century. These locales were where interventions into the lives of other peoples were designed.

13. Examples of recent scholarship in this area are Warwick Anderson, *Colonial Pathologies: American Tropical Medicine, Race, and Hygiene in the Philippines* (Durham, NC: Duke University Press, 2006); Richard Keller, *Colonial Madness: Psychiatry in French North Africa* (Chicago: University of Chicago Press, 2007); Mariola Espinosa, *Epidemic Invasions: Yellow Fever and the Limits of*

Cuban Independence, 1878–1930 (Chicago: University of Chicago Press, 2009); John Farley, *To Cast Out Disease: A History of the International Health Division of the Rockefeller Foundation (1913–1951)* (Oxford: Oxford University Press, 2004); Iris Borowy, *Uneasy Encounters: The Politics of Medicine and Health in China, 1900–1937* (New York: Peter Lang, 2009); Iris Borowy, *Coming to Terms with World Health: The League of Nations Health Organization, 1921–1946* (New York: Peter Lang, 2009); John Farley, *Brock Chisholm, the World Health Organization, and the Cold War* (Vancouver: University of British Columbia Press, 2008); Socrates Litsios, *The Third Ten Years of the World Health Organization, 1968–1977* (Geneva: World Health Organization, 2008); Marcos Cueto, *The Value of Health: A History of the Pan American Health Organization* (Washington, DC: Pan American Health Organization, 2007); Steven Palmer, *Launching Global Health: The Caribbean Odyssey of the Rockefeller Foundation* (Ann Arbor: University of Michigan Press, 2010); William Muraskin, *The Politics of International Health* (Albany: State University of New York Press, 1998); Sanjoy Bhattacharya, *Expunging* Variola*: The Control and Eradication of Smallpox in India, 1947–1977*, New Perspectives in South Asian History 14 (New Delhi: Orient Longman, 2006); Packard, *Making of a Tropical Disease*; Marcos Cueto, *Cold War, Deadly Fevers: Malaria Eradication in Mexico, 1955–1975* (Baltimore: Johns Hopkins University Press, 2007); Ka-che Yip, *Disease, Colonialism, and the State* (Hong Kong: Hong Kong University Press, 2009); Sunil S. Amrith, *Decolonizing International Health: India and Southeast Asia, 1930–65* (Cambridge: Cambridge University Press, 2006); Matthew James Connelly, *Fatal Misconception: The Struggle to Control World Population* (Cambridge, MA: Belknap Press of Harvard University Press, 2008); Nitsan Chorev, *The World Health Organization between North and South* (Ithaca, NY: Cornell University Press, 2012).

14. Farmer et al., *Reimagining Global Health.*

Part I: Colonial Entanglements

1. "The Cape Town Conference: Report of the International Conference of Representatives of the Health Services of Certain African Territories and British India, Held in Cape Town, November 15th to 25th, 1932," *Quarterly Bulletin of the Health Organization of the League of Nations* 2 (1933): 3–115.

2. The extent and duration of the entanglement between international health and colonial medicine has been only partially appreciated by historians. While they have acknowledged the interplay of these subjects, they have failed to recognize the extent to which these two phenomena form a single history. This failure is largely the result of a division of labor within the field of history. Histories of colonial medicine had been largely the remit of historians working within a regional context; they have been trained as African or South Asian historians, or Imperial historians. Histories of international-health organizations, by contrast, have been the purview of historians of public health, international organizations, and medicine and science, who have little interest in

colonial history. There are exceptions to this pattern, however. For example, Helen Tilley's magisterial study of scientific research in Africa, *Africa as a Living Laboratory: Empire, Development, and the Problem of Scientific Knowledge, 1870–1950* (Chicago: University of Chicago Press, 2011), explores the intersection of international scientific research and African colonial medicine. Deborah Neill, in *Networks of Tropical Medicine: Internationalism, Colonialism, and the Rise of a Medical Specialty* (Stanford, CA; Stanford University Press, 2012), similarly shows how international research activities linked colonial medical enterprises with the emergence of international knowledge communities. Sunil Amrith's *Decolonizing International Health: India and Southeast Asia, 1930–65*, Cambridge Imperial and Post-Colonial Studies Series (New York: Palgrave Macmillan, 2006) links the history of international-health activities and organizations to colonial developments in India and Southeast Asia. Nancy Stepan's *Eradication: Ridding the World of Diseases Forever?* (London: Reaktion Books, 2011) clearly sets the origins of eradication strategies within colonial settings. See also Warwick Anderson, *Colonial Pathologies: American Tropical Medicine, Race, and Hygiene in the Philippines* (Durham, NC: Duke University Press, 2006); Warwick Anderson, *The Collectors of Lost Souls: Turning Kuru Scientists into Whitemen* (Baltimore: Johns Hopkins University Press, 2008); Steven Paul Palmer, *Launching Global Health: The Caribbean Odyssey of the Rockefeller Foundation* (Ann Arbor: University of Michigan Press, 2010). This chapter seeks to integrate these two literatures and, by so doing, reframe discussions of the history of interventions into the health of other peoples during the first half of the twentieth century.

3. I have chosen to start this story at the beginning of the twentieth century, with an examination of how the activities of early international-health organizations were entangled with and shaped by colonial medicine. In doing so, I have excluded an examination of earlier efforts at international cooperation around health issues, including the international sanitary conferences of the nineteenth century, because these earlier activities did not attempt to *directly* intervene in the health of peoples living in poorer regions of the world. Instead, they were about preventing the spread of epidemics across the globe by means of better communications and quarantine measures.

Chapter One: Colonial Training Grounds

1. A great deal has been written about the history of colonial medicine. Among the many recent books on this subject are Warwick Anderson, *Colonial Pathologies: American Tropical Medicine, Race, and Hygiene in the Philippines* (Durham, NC: Duke University Press, 2006); John Farley, *Bilharzia: A History of Imperial Tropical Medicine* (Cambridge: Cambridge University Press, 1991); Mark Harrison, *Public Health in British India: Anglo-Indian Preventive Medicine, 1859–1914* (New York: Cambridge University Press, 1994); Maryinez Lyons, *The Colonial Disease: A Social History of Sleeping Sickness in*

Northern Zaire, 1900–1940 (New York: Cambridge University Press, 1992); David Arnold, *Colonizing the Body: State Medicine and Epidemic Disease in Nineteenth-Century India* (Berkeley: University of California Press, 1993); Randall M. Packard, *White Plague, Black Labor: Tuberculosis and the Political Economy of Health and Disease in South Africa* (Berkeley: University of California Press, 1989); Myron Echenberg, *Black Death, White Medicine: Bubonic Plague and the Politics of Public Health in Colonial Senegal* (Portsmouth, NH: Heinemann, 1992); Jonathan Sadowsky, *Imperial Bedlam: Institutions of Madness in Colonial Southwest Nigeria* (Berkeley: University of California Press, 1999); Meagan Vaughn, *Curing Their Ills: Colonial Power and African Illness* (Stanford, CA: Stanford University Press, 1991).

2. Mariola Espinoza, *Epidemic Invasions, Yellow Fever, and the Limits of Cuban Independence, 1878–1930* (Chicago: University of Chicago Press, 2008), 41.

3. Ibid., 11–30.

4. His men first attacked the filth that was most conspicuous—on the streets. The bodies of dead animals were dragged away and buried. The sick, who had been expelled from the hospitals when the Spanish officers commandeered these buildings for use as barracks, were no longer seen inhabiting the parks and plazas. Other teams attacked more out-of-the-way places. Residents of private homes became familiar with Gorgas's workers and inspectors. His men scrutinized backyards and trash piles. Business establishments were carefully examined and required to stay clean. Gorgas's biographer concluded that "probably no city has ever been so thoroughly cleaned inside and out as Havana was." See John M. Gibson, *Physician to the World: The Life of General William C. Gorgas* (Tuscaloosa: University of Alabama Press, 1989), 58–59.

5. Espinoza, *Epidemic Invasions*, 15.

6. Ibid., 34–45.

7. Both tasks involved laborious work. Before houses could be fumigated, they had to be made airtight; cracks had to be sealed. Once this was done, a dutch oven was place at the center of the house and filled with sulphur and small amounts of alcohol. The sulphur was lit and smoked for three to four hours. This killed nearly all of the mosquitoes in the residence. It also stained fixtures and fabrics, making it unpopular, especially among well-to-do homeowners. To accommodate these wealthier residents, Gorgas substituted pyrethrum for the sulphur and alcohol. This eliminated the staining but increased the time and labor required to eliminate the mosquitoes, since pyrethrum only intoxicated mosquitoes. This meant that the mosquitoes had to be swept up and burned after fumigation was completed. House spraying depended on a centralized reporting system, which was to be notified of all cases of yellow fever. Reported cases were then verified by sanitation-officer visits.

8. Health officers were given the authority to impose fines, which were collected by the Cuban courts, and the proceeds were deposited in the Cuban treasury. The fine was US$5, but if the nuisance was remitted, the fine was

returned. Courts dealt with householders if the first notice did not lead to either payment or abatement. Out of 2500 fines levied in the last nine months of 1901, only 50 were imposed and the monies deposited in the treasury.

9. Gregorio M. Guitéras, "The Yellow Fever Epidemic of 1903 at Laredo, Texas," *Journal of the American Medical Association* 43, 2 (1904): 115–121.

10. William Crawford Gorgas, *Sanitation in Panama* (New York: D. Appleton, 1918), 182–205.

11. In a 1905 letter to the surgeon general of the US Army, Gorgas noted that "while I proposed to the Commission a scheme which, if carried out, would have given a well-rounded sanitary organization, such as we had in Havana, I myself do not lend such great importance to such matters as street cleaning, collection of garbage, disposal of night soil, &c, as long as we can do the mosquito work and have official reporting of contagious diseases." Similarly, in a letter to the noted sanitarian Charles V. Chapin, Gorgas argued that the while the Panama Canal Bill made reference to the provision of water and a sewage system in Panama City and Colón, "the two major diseases are yellow fever and malaria," and provisions for care of the sick and the elimination of mosquitoes "will have a great deal more to do with the health of Panama than the most important sewage and water systems. Yet the dirt theory is so firmly impressed on people generally that they will agree to spend several millions from this point of view, and not say a word about the really vital and important sanitation."

12. There were, of course, public-health campaigns that employed various levels of coercion in nineteenth-century Europe and America.

13. Steven Palmer, *Launching Global Health: The Caribbean Odyssey of the Rockefeller Foundation* (Ann Arbor: University of Michigan Press, 2010).

14. Commenting on the challenges of using smoldering sulphur to eliminate adult mosquitoes in infected houses in Havana, Gorgas stated that "with a large force of ignorant men engaged in this work, constant watchfulness has to be used to see the fires do not occur." See William Gorgas, *Sanitation in Panama*, 54.

15. Some of Gorgas's colleagues were more vocal in their racial views. J. C. Perry, who was assigned to Panama as part of the US Public Health Service in 1914, described the native sections of Colón: "The third section of the city . . . contains the stores and residences of the greater portion of the inhabitants. . . . These habitations are filthy in the extreme, and it is difficult to understand how people can live in such insanitary surroundings with any semblance of health. Probably the whites could not, but the negroes who occupy this section do not seem to suffer to an appreciable extent." See J. C. Perry, "Preliminary Report on Sanitary Condition of Colón and Panama, and the Isthmus between These Points," *Public Health Reports (1896–1970)* 19, 10 (1904): 351–357.

16. In the concluding chapter to his *Sanitation in Panama*, Gorgas argued that humans had escaped from the tropics to avoid disease, establishing great civilizations in the temperate zones. For centuries, the tropics remained places uninhabitable by the white races. Recent victories over tropical disease, how-

ever, had laid the tropics open to reconquest by the white race. As a result, some of the most luxuriant plants on the earth's surface would be available for use by "civilized man." To this end, Gorgas suggested that as great an accomplishment as the completion of the Panama Canal was, in the future the sanitary phase of the work would be considered more important than the actual construction of the canal. He concluded the book with the following assertion: "The discovery of the Americas was a great epoch in the history of the white man, and threw large areas of fertile and healthy country open to his settlement. The demonstration made in Panama that he can live a healthy life in the tropics will be an equally important milestone in the history of the race."

17. In her review of Gorgas's yellow fever work in Panama, Alexandra Stern notes that Gorgas was much more concerned about the health of white workers than for the health of the large number of West Indian blacks who formed the bulk of the canal's work force. She points to the disparities that existed in access to health care and preventive services in the Canal Zone between "gold" (primarily white) and "silver" (primarily nonwhite) employees. The latter, unsurprisingly, suffered higher mortality rates. The number-one killer among these workers was pneumonia. She observes that between 1904 and 1910, deaths from pneumonia contributed to at least 25% of the total deaths and were suffered by silver workers, primarily hailing from Jamaica, Barbados, and Colombia. See Alexandra Minna Stern, "Yellow Fever Crusade: US Colonialism, Tropical Medicine, and the International Politics of Mosquito Control, 1900–1920," in *Medicine at the Border: Disease, Globalization, and Security, 1850 to the Present*, ed. Alison Bashford (Houndmills, Basingstoke, Hampshire, UK: Palgrave Macmillan, 2006), 41–59. Also see Julie Greene, *The Canal Builders: Making America's Empire at the Panama Canal* (New York: Penguin, 2009).

18. Letter from Ronald Ross to Colonel Gorgas, June 10, 1904, William Crawford Gorgas Papers, University of Alabama Archives, Tuscaloosa.

19. Gorgas noted: "Our men worked continuously in the rain, and came home soaked through at night. The majority of them were so poor that they did not have a change of clothing. They therefore slept in their wet clothes." See William Gorgas, "Recommendation as to Sanitation Concerning Employees of the Mines on the Rand, Made to the Transvaal Chamber of Mines," *Journal of the American Medical Association* 62, 24 (1914): 1855–1865. Julie Greene, in her otherwise compelling portrait of labor conditions in the Canal Zone, incorrectly states that Gorgas supported research on pneumonia among West Indian workers but did little to address the problem. See Greene, *Canal Builders*, 136.

20. Gorgas also noted that the percentage of workers who had previously worked on the canal increased during this time, suggesting that experienced workers developed some immunity to the disease. Gorgas, "Recommendations as to Sanitation," 1857. It should also be noted that the overall living conditions among West Indian workers remained deplorable, if less crowded.

21. William Gorgas, "Tropical Sanitation and its Relationship to General Sanitation," *Journal of the American Medical Association* 65, 26 (1915): 2209.

22. Elizabeth Fee, *Disease and Discovery, A History of the Johns Hopkins School of Hygiene and Public Health, 1916–1939* (Baltimore: Johns Hopkins University Press, 1989), 18–19.

23. Gorgas, "Recommendations as to Sanitation," 1855–1865.

24. Randall M. Packard, "The Invention of the 'Tropical Worker': Medical Research and the Quest for Central African Labor on the South African Gold Mines, 1903–1936," *Journal of African History* 34, 2 (1993): 271–292.

25. Lois F. Parks and Gustave A. Nuermberger, "The Sanitation of Guayaquil," *Hispanic American Historical Review* 23, 2 (2012): 197–221.

26. Dorothy Porter and Roy Porter, "What Was Social Medicine? An Historiographical Essay," *Journal of Historical Sociology* 1, 1 (1988): 90–109.

27. Farley, *Bilharzia*.

28. Anderson, *Colonial Pathologies*.

29. Ibid., 119.

30. C. Reynaldo Ileto, "Cholera and the Origins of the American Sanitary Order on the Philippines," in *Imperial Medicine in Indigenous Societies*, ed. David Arnold (New York: Manchester University Press, 1988), 125–148.

31. Anderson, *Colonial Pathologies*, 107.

32. Raquel Reyes, "Environmentalist Thinking and the Question of Disease Causation in Late Spanish Philippines," *Journal of the History of Medicine & Allied Sciences* 69, 4 (2004): 554–579.

33. Victor C. Heiser, *An American Doctor's Odyssey* (New York: W.W. Norton, 1936), 35.

34. Mary C. Gillett, "U.S. Army Military Officers and Public Health in the Philippines in the Wake of the Spanish-American War, 1898–1905," *Bulletin of the History of Medicine* 64, 4 (1990): 580.

35. Anderson, *Colonial Pathologies*.

Chapter Two: From Colonial to International Health

1. For recent discussions of the origins and motivations behind the Rockefeller Foundation's IHB, see Anne-Emanuelle Birn, *Marriage of Convenience: Rockefeller International Health and Revolutionary Mexico* (Rochester, NY: University of Rochester Press, 2006); John Farley, *To Cast Out Disease: A History of the International Health Division of the Rockefeller Foundation (1913–1951)* (New York: Oxford University Press, 2004).

2. Steven Palmer, *Launching Global Health: The Caribbean Odyssey of the Rockefeller Foundation* (Ann Arbor: University of Michigan Press, 2010).

3. Ibid., 22–54.

4. Farley, *To Cast Out Disease*, 63.

5. Palmer, *Launching Global Health*, 4.

6. Ibid., 136–139.

7. Birn, *Marriage of Convenience*, 84–85.

8. The design and execution the foundation's hookworm program in Mexico was shaped by its desire to influence political developments in the state of Veracruz, which was the location of large agricultural and petroleum industries. When political unrest forced the IHB to abandon Veracruz, they moved to neighboring areas to be close to Veracruz, even though surveys revealed that hookworm was not a serious problem in these areas. They quickly wrapped up activities in these neighboring areas when Veracruz opened up again, abandoning prevention efforts. See Anne-Emanuelle Birn and Armando Solórzano, "Public Health Policy Paradoxes: Science and Politics," *Social Science & Medicine* 49, 9 (1999): 1197–1213.

9. Ibid.

10. Samuel Taylor Darling, "The Hookworm Index and Mass Treatment," *American Journal of Tropical Medicine and Hygiene* 2, September (1922): 397.

11. Samuel Taylor Darling, Marshall A. Barber, and H. P. Hacker, *Hookworm and Malaria Research in Malaya, Java, and the Fiji Islands* (New York: Rockefeller Foundation International Health Board, 1920), 3.

12. Ibid., 37.

13. Samuel Taylor Darling, Marshall A. Barber, and H. P. Hacker, "The Treatment of Hookworm Infection," *Journal of the American Medical Association* 70, 8 (1918): 499–507.

14. Lois F. Parks and Gustave A. Nuermberger, "The Sanitation of Guayaquil," *Hispanic American Historical Review* 23, 2 (2012): 197–221.

15. Marcos Cueto, "Sanitation from Above: Yellow Fever and Foreign Intervention in Peru, 1919–1922," *Hispanic American Historical Review* 72, 1 (1992): 1–22.

16. Ibid., 16.

17. Ibid., 21.

18. Elizabeth Fee, *Disease and Discovery, A History of the Johns Hopkins School of Hygiene and Public Health, 1916–1939* (Baltimore: Johns Hopkins University Press, 1989), 73.

19. Marcos Cueto, *The Value of Health: A History of the Pan American Health Organization* (Washington, DC: Pan American Health Organization, 2007), 66.

20. Socrates Litsios, pers. comm.

21. Randall M. Packard and P. Gadehla, "Land Filled with Mosquitoes: Fred L. Soper, the Rockefeller Foundation, and the *Anopheles gambiae* Invasion of Brazil, 1932–1939," *Parassitologia* 36, 1–2 (1994): 197–214. Soper largely ignored the existence of previous work on malaria in Brazil conducted by Brazilian researchers.

22. In 1924, Darling was elected president of the American Society for Tropical Medicine and Hygiene. In the following year, the League of Nations invited Darling to join the Malaria Commission on its tour of Egypt, Palestine, and Syria. The trip proved fatal, as he died in a car accident near Beirut.

23. Trinidad was another IHB training ground where the Rockefeller Foundation's officers were sent to learn hookworm-control methods before being posted to other countries. See Palmer, *Launching Global Health*, 210–211.

24. Similar attitudes would continue to encourage vector-control approaches to disease control, as John Farley noted in *Bilharzia: A History of Imperial Tropical Medicine* (Cambridge: Cambridge University Press, 1991).

25. Warwick Anderson, *Colonial Pathologies: American Tropical Medicine, Race, and Hygiene in the Philippines* (Durham, NC: Duke University Press, 2006).

26. Paul F. Russell, "Malaria in the Philippine Islands," *American Journal of Tropical Medicine* 13 (1933): 174, cited in Anderson, *Colonial Pathologies*, 223.

27. Russell traveled to West Africa in summer 1952 to survey malaria-control activities. His diaries indicate that he held a very low view of African leaders and their abilities to run their countries. For example, on July 12, in Ilaro, Nigeria, he recorded: "The local African is certainly not ready for self government but is getting it rapidly. This will result in another Liberia. But the problem of the USA, U.K., and other Western powers is to safeguard essential raw materials. Perhaps this can be done even if the political administration under Africans becomes the joke it has long been in Liberia. May even be easier to get the raw materials because bribery becomes the rule rather than the exception and a few powerful well-bribed politicians can usually 'deliver the goods.' " See the entry for July 12, 1952, Paul F. Russell Diaries, Rockefeller Archives Center, Sleepy Hollow, New York.

Part II: Social Medicine, the Depression, and Rural Hygiene

1. League of Nations, "The Cape Town Conference: Report of the International Conference of Representatives of the Health Services of Certain African Territories and British India, Held in Cape Town, November 15th to 25th, 1932," *Quarterly Bulletin of the Health Organization of the League of Nations* 2 (1933): 101–102.

2. League of Nations, "Report of the Pan African Health Conference held at Johannesburg, November 20th to 30th, 1935," *Quarterly Bulletin of the Health Organization of the League of Nations* 5, 1 (1936): 113.

3. Ibid., 141.

4. Ibid., 199.

5. See Dorothy Porter and Roy Porter, "What Is Social Medicine? A Historiographical Essay," *Journal of Historical Sociology* 1, 1 (1988): 90–109.

Chapter Three. The League of Nations Health Organization

1. Iris Borowy, "Crisis as Opportunity: International Health Work During the Economic Depression," *Dynamis* (Granada, Spain) 28, 1 (2008): 29–51.

2. At the same time, Cumming resisted efforts to have the bureau involved in promoting regional health programs. As a result, prior to the 1940s, the PASB limited its activities to collecting statistics, disseminating health information, and

setting health and sanitation standards, particularly with regard to the prevention of infectious diseases. See Anne-Emanuelle Birn, "'No More Surprising Than a Broken Pitcher'? Maternal and Child Health in the Early Years of the Pan American Sanitary Bureau," *Canadian Bulletin of the History of Medicine* 19, 1 (2002): 27.

3. The former German colonies of Southwest Africa, Tanganyika, Togo, and the Cameroons had become Mandate Territories after the war, ostensibly under the authority of the League, but each had been placed under the direct supervision of another colonial power. Only South Africa and Liberia were members of the League.

4. Iris Borowy, *Coming to Terms with World Health: The League of Nations Health Organization, 1921–1946* (New York: Peter Lang, 2009), 145–148.

5. More controversially, it suggested that Europeans were responsible for the introduction of TB into Central Africa. See Borowy, *Coming to Terms*, 109–110.

6. Ibid., 301–305.

7. Ibid., 135.

8. Marta Balinska, *For the Good of Humanity: Ludwik Rajchman, Medical Statesman* (Budapest: Central European University Press, 1995).

9. At the Pasteur Institute, Rajchman worked under Russian scientist Elie Metchnikoff, who shared the Nobel Prize in Physiology and Medicine with Paul Ehrlich in 1908.

10. At the institute, he worked closely with Sir Robert Smith, a public-health pioneer in Great Britain. He subsequently taught in London and was named head of the city's central laboratory on dysentery.

11. This was viewed as an interim move, while the League worked to establish a more permanent Health Section through the incorporation of the existing International Office of Public Hygiene into the League. When this proved impossible to negotiate, the League established a temporary Health Section, with Rajchman as its first head.

12. For more on the early history of malaria control and international disputes over strategies, see Randall M. Packard, *The Making of a Tropical Disease: A Short History of Malaria* (Baltimore: Johns Hopkins University Press, 2007).

13. League of Nations Health Organization, Malaria Commission, *Report on Its Tour of Investigation in Certain European Countries, 1924* (Geneva: Publication Department, League of Nations, 1925).

14. Sydney Price James, "Remarks on Malaria in Kenya," *Kenya and East African Medical Journal* 6, 4 (1929): 96.

15. Packard, *Making of a Tropical Disease*, 128–130.

16. Socrates Litsios, "Revisiting Bandoeng," unpublished paper, 2013, 10.

17. Paul Weindling, "Social Medicine at the League of Nations Health Organization and the International Labour Office Compared," in *International Health Organizations and Movements, 1918–1939*, ed. Paul Weindling (Cambridge: Cambridge University Press, 1995), 142.

18. Etienne Burnet, "General Principles Governing the Prevention of Tuberculosis," *Quarterly Bulletin of the Health Organization of the League of Nations* 1, 4 (1932): 489–663. The report was individually authored by Burnet, who was deputy director of the Pasteur Institute in Tunis, but was vetted by a Reporting Committee, set up by the Health Committee to study the problem of tuberculosis, and by the Health Committee as a whole.

19. Borowy, *Coming to Terms*, 329.

20. See Michael Watts, *Silent Violence: Food, Famine, and Peasantry in Northern Nigeria* (Berkeley: University of California Press, 1983).

21. Randall M. Packard, *White Plague, Black Labor: Tuberculosis and the Political Economy of Health and Disease in South Africa* (Berkeley: University of California Press, 1989).

22. "It is essential that there should be co-ordination of the work of all agencies concerned in rural sanitation. This co-ordination implies the co-operation of the technical personnel concerned (agricultural experts, architects, hygienists, engineers, medical men, doctors of veterinary medicine, etc.)." See League of Nations, Health Organization, *European Conference on Rural Hygiene*, League of Nations Official Journal (Geneva: League of Nations, September 1931), 1888.

23. Ibid., 1896.

24. Ibid., 1885.

25. Paul Weindling, "The Role of International Organizations in Setting Nutritional Standards in the 1920s and 1930s," *Clio Medica* 32, September (1995): 319–332.

26. Kenneth J. Carpenter, "The Work of Wallace Aykroyd: International Nutritionist and Author," *Journal of Nutrition* 137, 4 (2007): 873–878.

27. League of Nations, Health Section, "The Economic Depression and Public Health," *Quarterly Bulletin of the Health Organization of the League of Nations* 1, 3 (1932): 457–458. Using data collected principally in Germany, and comparing the purchasing power of employed workers in 1927–28 with that of unemployed workers in 1932–33, the memorandum found that the family of an unemployed man could just about remain adequately fed if his wife shopped very prudently and all other expenses were pared down, so that 60%–70% of the family's governmental allowance could be spent on food.

28. Weindling, "Role of International Organizations," 325.

29. Wallace R. Aykroyd and Etienne Burnet, "Nutrition and Public Health," *Quarterly Bulletin of the Health Organization of the League of Nations* 4, 2 (1935): 329.

30. Ibid., 448.

31. League of Nations, *Final Report of the Mixed Committee of the League of Nations on the Relation of Nutrition to Health, Agriculture, and Economic Policy* (Geneva: League of Nations, 1937), 21.

32. Ibid., 38.

33. Ibid., 36.

Chapter Four: Internationalizing Rural Hygiene and Nutrition

1. Marta Balinska, *For the Good of Humanity: Ludwik Rajchman, Medical Statesman* (Budapest: Central European University Press, 1995), 80.

2. League of Nations, Council Committee on Technical Co-Operation between the League of Nations and China, *Report of the Technical Agent of the Council on His Mission in China* (Geneva: League of Nations, 1934).

3. Balinska, *For the Good of Humanity*, 86. Balinska rightly points to the failure of WHO to continue this language following World War II and to its return to the language of "technical assistance."

4. Ibid., 91.

5. While Amrith has suggested that this was a development that marked the decolonization of international health in the 1930s, the phenomenon has a longer pedigree, exemplified by the careers of many health officers of the Rockefeller Foundation (part I). See Sunil S. Amrith, *Decolonizing International Health: India and Southeast Asia, 1930–65* (New York: Palgrave Macmillan, 2006).

6. Margherita Zanasi, "Exporting Development: The League of Nations and Republican China," *Comparative Studies in Society and History* 49, 1 (2006): 152.

7. Ibid., 157–163.

8. Elizabeth Fee and Ted Brown, "Andrija Štampar, Charismatic Leader of Social Medicine and International Health," *American Journal of Public Health* 96, 8 (2006): 1383.

9. Andrija Štampar, "Health and Social Conditions in China," in *Serving the Cause of Public Health: Selected Papers of Andrija Štampar*, ed. Mirko D. Grmek (Zagreb: Andrija Štampar School of Public Health, 1966), 129. Originally published in the *Quarterly Bulletin of the Health Organization of the League of Nations* 5 (1936): 1090–1126.

10. Štampar, "Health and Social Conditions," 149.

11. Frank Ninkovich, "The Rockefeller Foundation, China, and Cultural Change," *Journal of American History* 70, 4 (1984): 799–820.

12. Jan Hoffmeyr referred to Africa as a "magnificent natural laboratory" in his address to the South African Association's annual meeting in July 1929. See Helen Tilley, *Africa as a Living Laboratory: Empire, Development, and the Problem of Scientific Knowledge, 1870–1950* (Chicago: University of Chicago Press, 2007).

13. Ninkovich, "Rockefeller Foundation," 811.

14. Ka-Chi Yip, "Health and Nationalist Reconstruction: Rural Health in Nationalist China, 1928–1937," *Modern Asian Studies* 26, 2 (1992): 395–415.

15. Ibid., 403.

16. Socrates Litsios, "Selskar Gunn and China: The Rockefeller Foundation's 'Other' Approach to Public Health," *Bulletin of the History of Medicine* 79, 2 (2005): 295–318.

17. Both Litsios and Ninkovich, who have described the program, based their accounts on the foundation's grant authorizations; in other words, on what the foundation approved, not on what was actually done. See Ninkovich, "Rockefeller Foundation," 813.

18. Anne-Emanuelle Birn, *Marriage of Convenience: Rockefeller International Health and Revolutionary Mexico* (Rochester, NY: University of Rochester Press, 2006), 129–131.

19. Ibid., 162–166.

20. Ibid., 157–158.

21. Soma Hewa, *Colonialism, Tropical Disease, and Imperial Medicine* (Lanham, MD: University Press of America, 1995), 122–159.

22. Amrith, *Decolonizing International Health*, 29–32.

23. John L. Hydrick, *Intensive Rural Hygiene Work and Public Health Education of the Public Health Service of Netherlands India* (Java, "Netherlands India" [Indonesia]: Batavia-Centrum, 1937).

24. Eric A. Stein, "'Sanitary Makeshifts' and the Perpetuation of Health Stratification in Indonesia," in *Anthropology and Public Health: Bridging Differences in Culture and Society*, ed. Marcia Inhorn and Robert Hahn (New York: Oxford University Press, 2009), 1–29.

25. Amrith, *Decolonizing International Health*, 29–32.

26. Jan Breman, *Control of Land and Labour in Colonial Java* (Dordrecht, The Netherlands: Foris, 1983).

27. Stein, "'Sanitary Makeshifts,'" 550.

28. Ibid., 551.

29. Michael Worboys, "The Discovery of Colonial Malnutrition," in *Imperial Medicine and Indigenous Societies*, ed. David Arnold (Manchester, UK: Manchester University Press, 1988), 211–212.

30. Robert McCarrison, "Rice in Relation to Beri-Beri in India," *British Medical Journal* 1, 3297 (1924): 414–420.

31. Robert McCarrison, "A Paper on Food and Goitre," *British Medical Journal* 2, 3797 (1933): 671–675.

32. The problem of highly milled rice was taken up at the League of Nations Intergovernmental Conference on Far Eastern Countries on Rural Hygiene, held in Bandoeng, Java, in 1937, about which more will be said below. Delegates there agreed that "the habit of using highly milled rice in lieu of under-milled or home-pounded rice was spreading" and that "it should be condemned from the standpoint of nutrition." The delegates fell short of recommending a ban on highly milled rice, recommending instead that more research was needed on the nutritional, commercial, economic, and psychological aspects of the problem. Aykroyd, who attended the conference as a representative of the Indian government, took up the challenge, along with several of his Indian colleagues.

33. Wallace R. Aykroyd, B. G. Krishnan, R. Passmore, and A. R. Sundararajan, *The Rice Problem in India*, Indian Medical Research Memoirs 32 (Calcutta: Thacker, Spink, 1940), 81–82.

34. Amrith, *Decolonizing International Health*, 36.

35. Aykroyd et al., *Rice Problem in India*, 64.

36. Worboys, "Discovery of Colonial Malnutrition," 208–225.

37. Ibid., 221.

38. Socrates Litsios, "Revisiting Bandoeng," *Social Medicine* 8, 3 (2014): 113–128.

39. League of Nations, Health Organization, *Report of the Intergovernmental Conference of Far Eastern Countries on Rural Hygiene* (Geneva: League of Nations, 1937), 65–68.

40. Ibid., 50.

41. Ibid., 51.

42. Theodore M. Brown and Elizabeth Fee, "The Bandoeng Conference of 1937: A Milestone of Health and Development," *American Journal of Public Health* 98, 1 (2008): 42–43.

Part III: Changing Postwar Visions of Health and Development

1. World Health Organization, *Constitution of the World Health Organization*, www.who.int/governance/eb/who_constitution_en.pdf.

Chapter Five: Planning for a Postwar World

1. "Third World" was a term coined by French demographer Alfred Sauvey to describe countries that did not fit into either the capitalist/NATO block of nations or countries that were part of the Communist block. It was a cold war term. Over time, however, Third World became a euphemism for countries, most of which had been European colonies, that were relatively impoverished and underdeveloped in comparison with wealthier nations in Europe and North America. The former were to become the target of development assistance and international-health work.

2. Frederick Cooper and Randall M. Packard (eds.), *International Development and the Social Sciences: Essays on the History and Politics of Knowledge* (Berkeley: University of California Press, 1997), 7–8.

3. Randall M. Packard, "Malaria Dreams: Postwar Visions of Health and Development in the Third World," *Medical Anthropology* 17, 3 (1997): 279–296; Amy Staples, *The Birth of Development: How the World Bank, Food and Agricultural Organization, and the World Health Organization Changed the World, 1945–1965* (Kent, OH: Kent State University, 2006).

4. Atlantic Charter, August 14, 1941, available at The Avalon Project: Documents in Law, History, and Diplomacy, http://avalon.law.yale.edu/wwii/atlantic.asp.

5. Staples, *Birth of Development*, 8–18.

6. Philip Musgrove, "Health Insurance: The Influence of the Beveridge Report," *Bulletin of the World Health Organization* 78, 6 (2000): 845–846.

7. Nicola Swainson, *The Development of Corporate Capitalism in Kenya, 1918–1977* (Berkeley: University of California Press, 1980), 106.

8. Frederick Cooper, *Decolonization and African Society: The Labor Question in French and British Africa* (Cambridge: Cambridge University Press, 1996); Jonathan H. Frimpong-Ansah, *The Vampire State in Africa: The Political Economy of Decline in Ghana* (Trenton, NJ: Africa World Press, 1992), 24.

9. Sunil S. Amrith, *Decolonizing International Health: India and Southeast Asia, 1930–65* (New York: Palgrave Macmillan, 2006).

10. Sunil S. Amrith, "Political Culture of Health in India: A Historical Perspective," *Economic and Political Weekly* 42, 2 (2007): 114–121.

11. Government of India, Ministry of Health, *Report of the Health Survey and Development Committee*, vol. 2 (Delhi: Manager of Publications, 1946), 2.

12. Amrith, *Decolonizing International Health*, 61–63.

13. Michael R. Grey, *New Deal Medicine: The Rural Health Programs of the Farm Security Administration* (Baltimore: Johns Hopkins University Press, 1999).

14. George Marshall, opening address, Fourth International Congress on Tropical Diseases and Malaria, Washington, DC, 1948.

15. Charles-Edward A. Winslow, *The Cost of Sickness and the Price of Health*, WHO Monograph Series 7 (Geneva: World Health Organization, 1951), 81.

16. Paul Russell, *Man's Mastery of Malaria* (New York: Oxford University Press, 1955), 257.

17. The International Labour Organization (ILO), which had been a major partner in the LNHO's efforts to improve health and nutrition before World War II, remerged after the war as an advocate for social insurance aimed at securing the health and well-being of working families. The chairman of the ILO governing body was US economist Carter Goodrich. Goodrich was well connected with the radical side of the New Deal and pushed for the ILO to become more active in planning for a postwar world. Edward Phelan, the acting director, argued that social insurance must be at the heart of postwar reconstruction, not merely to protect living standards, but to build social solidarity and world peace. Accordingly, the 1941 ILO conference in New York City called for a "social mandate" to protect the health and welfare of workers by ensuring full employment, extending training, establishing a minimum wage, protecting nutrition, and extending social insurance. Frances Perkins, Roosevelt's labor secretary, opened this conference with a call for public responsibility after the war for health, nutrition, and housing. The ILO went further at its 1944 conference in Philadelphia, calling for the creation of universal medical services, funded by compulsory insurance without income limits. The systems would be run by central governmental agencies and only allowed a limited role for private, fees-for-service medical services. The 1944 conference also proposed that these rights be extended to workers in non-self-governing territories.

18. James A. Gillespie, "International Organizations and the Problem of Child Health, 1945–1960," *Dynamis* (Granada, Spain) 23 (2003): 115–142.

19. See Jessica Reinisch, "Internationalism in Relief: The Birth (and Death) of UNRRA" *Past & Present* 31, Suppl. 6 (2011): 258–289.

20. "Establishment of the Interim Commission of Food and Agriculture," sent by the acting secretary of state to the chargé in the United Kingdom, telegram 550.AD1/b, March 8, 1943, UW–Madison Libraries Digital Collections, http://digicoll.library.wisc.edu/cgi-bin/FRUS/FRUS-idx?type=div&did=FRUS.FRUS1943v01.i0020&isize=L/.

21. "Final Act of the United Nations Conference on Food and Agriculture, Hot Springs, June 3rd, 1943," RG 3, folio 233, FAO Archives, Rome, Italy. Cited in Ruth Jachertz and Alexander Nützenadel, "Coping with Hunger? Visions of a Global Food System, 1930–1960," *Journal of Global History* 6, 1 (2011): 99–119.

22. Jachertz and Nützenadel, "Coping with Hunger."

23. Staples, *Birth of Development*, 83–104.

24. United Nations, "The Universal Declaration of Human Rights," www.un.org/en/documents/udhr/.

25. Javed Siddiqi, *World Health and World Politics: The World Health Organization and the World System* (Columbia: University of South Carolina Press, 1995), 1956–1959.

26. It should be noted that Szeming Sze, China's representative to the TPC, claims to have written the phrase "Health is a state of physical fitness and of mental and social well-being, and not only the absence of infirmity and disease." See Szeming Sze, *The Origins of the World Health Organization: A Personal Memoir, 1945–1948* (Boca Raton, FL: L.I.S.Z., 1982), 15.

27. James A. Gillespie, "Social Medicine, Social Security, and International Health," in *The Politics of the Health Life: An International Perspective*, ed. Esteban Roderiquez Ocaña (Sheffield, UK: European Association for the History of Medicine and Health Publications, 2002), 221–228.

28. World Health Organization, "First World Health Assembly," Global Health Histories, www.who.int/global_health_histories/first_world_health_assembly/en/index.html.

29. Frederick Dodge Mott and Milton Irwin Roemer, *Rural Health and Medical Care* (New York: McGraw-Hill, 1948).

30. John Farley, *Brock Chisholm, the World Health Organization, and the Cold War* (Vancouver: University of British Columbia Press, 2008). Roemer's views on national health insurance made him a target of conservative politicians and physician groups in the United States. In 1948, the Board of Inquiry on Employee Loyalty, part of the Federal Security Agency, notified him that "reasonable grounds may exist for belief that you are disloyal to the Government of the United States." The six charges against him included membership in the American-Soviet Medical Society. Like his mentor Sigerist, Roemer was an

admirer of the Soviet Union's state-run health services. Roemer succeeded in absolving himself of the charges, but his troubles with the political Right in the United States were only just beginning. Soon after clearing his name, Roemer received his invitation to join WHO, and he welcomed the opportunity to leave the country. While Roemer's views made him a target of right-wing politicians in the United States, they made him an attractive addition to Chisholm's directorate in Geneva.

31. Farley, *Brock Chisholm*, 114–116.

32. Expert Committee on Professional and Technical Education of Medical and Auxiliary Personnel, *Report on the First Session, Geneva, 6–10 February 1950*, Technical Report Series 22 (Geneva: World Health Organization, 1950), 5, 9.

33. Expert Committee on Professional and Technical Education of Medical and Auxiliary Personnel, *Second Report [Second Session, Nancy, 3–9 December 1952]*, Technical Report Series 69 (Geneva: World Health Organization, 1953), 4, 5.

34. Expert Committee on Public-Health Administration, *First Report [First Session, Geneva, 3–7 December 1951]*, Technical Report Series 55 (Geneva: World Health Organization), 4.

35. Charles-Edward A. Winslow, *Evolution and Significance of the Modern Public Health Campaign* (New Haven, CT: Yale University Press, 1923), 1.

36. Expert Committee on Public-Health Administration, *First Report*, 5–6, 8.

Chapter Six: A Narrowing Vision

1. John Farley, *Brock Chisholm, the World Health Organization, and the Cold War* (Vancouver: University of British Columbia Press, 2008), 50–51.

2. League of Nations, *Food, Famine, and Relief, 1940–1946* (Geneva: League of Nations, 1946), 6.

3. Frank Snowden, *The Conquest of Malaria: Italy, 1900 to 1962* (New Haven, CT: Yale University Press, 2006).

4. "Meeting of the Board of Scientific Directors of the IHD at the Rockefeller Institute in New York," September 19, 1947, Paul F. Russell Diaries, Rockefeller Archives Center, Sleepy Hollow, New York.

5. Snowden, *Conquest of Malaria*, 206–207.

6. Randall M. Packard, *The Making of a Tropical Disease: A Short History of Malaria* (Baltimore: Johns Hopkins University Press, 2007), 143.

7. Leo Slater, "A Malaria Chemotherapy and the 'Kaleidoscopic' Organization of Biomedical Research During World War II," *Ambix: The Journal of the Society for the Study of Alchemy and Early Chemistry* 51, 2 (2004): 107–134.

8. US Department of State, *Point Four: Cooperative Program for AID in the Development of Underdeveloped Areas*, Economic Cooperation Series, Department of State Publications 3719 (Washington, DC: Government Printing Office, 1949), 24.

9. Arturo Escobar, *Encountering Development: The Making and Unmaking of the Third World* (Princeton, NJ: Princeton University Press, 2011), 3–4.

10. Frederick Cooper and Randall M. Packard (eds.), *International Development and the Social Sciences: Essays on the History and Politics of Knowledge* (Berkeley: University of California Press, 1997), 7–8.

11. Escobar, *Encountering Development*, chapter 3.

12. Simon Szreter, "The Idea of Demographic Transition and the Study of Fertility Change," *Population and Development Review* 19, 4 (1993): 659–701.

13. André Luiz Virira de Campos, "The Institute of Inter-American Affairs and Its Health Policies in Brazil During World War II," *Presidential Studies Quarterly* 28, 3 (1998): 523–534. US officials were concerned about the growing German influence in the region. During the 1930s, Latin American countries had been the leading providers of raw materials for rebuilding the German war machine. The United States also feared that German military activities in North Africa could be extended into the Americas. Of particular concern was northeastern Brazil, which could easily be reached from West Africa by the power of German aircraft. If the Germans conquered the Brazilian "hump," they could reach the Caribbean and the Panama Canal. To defend against this possibility, the United States entered into agreements with several Latin American countries to construct US military installations within their borders. At the same time, US business leaders, faced with a loss of markets in Europe and eastern Asia, saw an opportunity to expand US trade and investment in Latin America. Latin American countries possessed certain raw materials that were viewed as critical to the war effort. The most important of these was rubber, which was widely grown in the Amazon region of Brazil and was in short supply once the Japanese blocked access to sources in Southeast Asia. Other strategic materials produced in the region were quartz, tantalite, chromium ore, mica, and iron ore. In order to protect the health of US troops stationed at Latin American bases and of local workers producing rubber and strategic minerals, the United States provided technical assistance to set up sanitary and disease-control programs, many of which were aimed at controlling malaria. These activities were overseen by the Office for Inter-American Affairs, established in 1941; Nelson Rockefeller, a strong advocate of strengthening economic and cultural ties in the region, was its first director. By 1951, when the IIAA completed its operations, over US$30 million had been spent on health programs in Latin America.

14. National Academy of Sciences, "George McClelland Foster Jr.," *Biographical Memoirs*, vol. 90 (Washington, DC: National Academy of Sciences, 2009).

15. George M. Foster, *A Cross-Cultural Anthropological Analysis of a Technical Aid Program* (Washington, DC: Smithsonian Institution, 1951), 3.

16. George M. Foster, "An Anthropologist's Life in the Twentieth Century: Theory and Practice at UC Berkeley, the Smithsonian, in Mexico, and with the World Health Organization," transcript of interviews conducted by Suzanne Riess in 1988 and 1999, Regional Oral History Office, Bancroft Library, University of California–Berkeley.

17. Socrates Litsios, "Malaria Control, the Cold War, and the Postwar Reorganization of International Assistance," *Medical Anthropology* 17, 3 (1997): 255–278.

18. Randall M. Packard, "Malaria Dreams: Postwar Visions of Health and Development in the Third World," *Medical Anthropology* 17, 3 (1997): 279–296.

19. International Development Advisory Board, *Malaria Eradication: Report and Recommendations of the International Development Advisory Board* (Washington, DC: International Cooperation Agency, 1956), 14.

20. Ibid., 11.

21. Ibid., 8.

22. Waldemar A. Nielsen and Zoran S. Hodjera, "Sino-Soviet Bloc Technical Assistance—Another Bilateral Approach," *Annals of the American Academy of Political and Social Science* 323, 1 (1959): 40–49.

23. Wilbur Sawyer, former director of the Rockefeller Foundation's International Health Division, became UNRRA's first director. Sawyer had overseen and supported eradication work on Soper's disease in Brazil and Egypt and was wedded to the IHD's disease-centered vision of international health. See John Farley, *To Cast Out Disease: A History of the International Health Division of the Rockefeller Foundation (1913–1951)* (New York: Oxford University Press, 2004), 144.

24. James A. Gillespie, "International Organizations and the Problem of Child Health, 1945–1960," *Dynamis* (Granada, Spain) 23 (2003): 115–142.

25. Jessica Reinisch, "'Auntie UNRRA' at the Crossroads," *Past & Present* 218, Suppl. 8 (2013): 70–97.

26. James A. Gillespie, "Europe, America, and the Space of International Health," in *Shifting Boundaries of Public Health*, ed. Susan Gross Solomon, Lion Murard, and Patrick Zylberman (Rochester, NY: University of Rochester, 2008), 122–123; Katerina Gardikas, "Relief Work and Malaria in Greece, 1943–1947," *Journal of Contemporary History* 43, 3 (2008): 493–508.

27. Gillespie, "Europe, America," 124.

28. Reinisch, "'Auntie UNRRA.'"

29. Silvia Salvatici, "'Help the People to Help Themselves': UNRRA Relief Workers and European Displaced Persons," *Journal of Refugee Studies* 25, 3 (2112): 428–451.

30. Reinisch, "'Auntie UNRRA.'"

31. Ruth Jachertz and Alexander Nützenadel, "Coping with Hunger? Visions of a Global Food System, 1930–1960," *Journal of Global History* 6, 1 (2011): 99–119.

32. While the United States had supported the creation of the World Bank and the IMF, both of which could affect its international financial relations, the voting arrangements for these organizations were based on a member nation's contributions to the organization's budget. The United States was by far the greatest contributor to the budgets of the World Bank and IMF and, as a result, was assured virtual veto power over their activities. The FAO's voting arrange-

ments gave the United States less control, though it was still an influential partner and was able to sidetrack Orr's proposals.

33. Amy Staples, *The Birth of Development: How the World Bank, Food and Agricultural Organization, and the World Health Organization Changed the World, 1945–1965* (Kent, OH: Kent State University, 2006), 91–92.

34. Ibid., 96–98.

35. Ibid., 99. A Technical Assistance Board, made up of representatives of member nations (whose voting power reflected their contributions to the EPTA budget), controlled the allocation of EPTA funds. This meant that the United States, as the largest donor, could dictate how the funding was used.

36. Ibid., 98–99.

37. Jachertz and Nützenadel, "Coping With Hunger," 117.

38. Anne-Emanuelle Birn, "'No More Surprising Than a Broken Pitcher'? Maternal and Child Health in the Early Years of the Pan American Sanitary Bureau," *Canadian Bulletin of the History of Medicine* 19, 1 (2002): 17–46.

39. Before the San Francisco Conference on International Health opened on April 25, 1945, the US and UK delegates had consulted each other and agreed that no questions in the field of health would be included on the conference agenda. Unaware of the US/UK consultations, Szeming Sze, Geraldo de Paula Souza, and Karl Evang—from the Chinese, Brazilian, and Norwegian delegations, respectively—agreed that a proposal for the creation of a new international-health organization should be put on the conference agenda. When their proposal became bogged down in committee discussions, Sze and Souza entered it into the conference discussions as a declaration, which stated that an international conference should be held with the aim of establishing an international-health organization. The declaration was adopted unanimously, as the United States and Britain chose to avoid the embarrassment of opposing it. See Staples, *Birth of Development*, 132.

40. Ibid., 135.

41. Farley, *Brock Chisholm*, 49.

42. Ibid., 63.

43. James A. Gillespie, "Social Medicine, Social Security, and International Health," in *The Politics of the Health Life: An International Perspective*, ed. Esteban Roderiquez Ocaña (Sheffield, UK: European Association for the History of Medicine and Health Publications, 2002), 219–240.

44. Farley, *Brock Chisholm*, 97–100.

45. Litsios, "Malaria Control."

46. The topics of discussion and the identity of the committee members during this period is based on a review of the joint FAO/WHO committee reports, 1949–1952.

47. See Jennifer Tappan's study, *Medicalizing Malnutrition: Nearly a Century of Kwashiorkor Work in a Small Part of Africa* (Athens: Ohio State University Press, forthcoming).

48. Litsios, "Malaria Control," 259.

49. Emilio J. Pampana and Paul F. Russell, "Malaria—A World Problem," *Chronicle of the World Health Organization* 9, Special Issue (1955): 31–100.

50. Niels Brimnes, "Vikings Against Tuberculosis: The International Tuberculosis Campaign in India, 1948–1951," *Bulletin of the History of Medicine* 81, 2 (2007): 407–430.

51. D. Behera, "Tuberculosis Control in India—a View Point," *Indian Journal of Tuberculosis* 54, 2 (2007): 63–65.

52. Jessica Reinisch, "Internationalism in Relief: The Birth (and Death) of UNRRA," *Past & Present* 31, Suppl. 6 (2011): 258–289.

53. Marta Balinska, *For the Good of Humanity: Ludwik Rajchman, Medical Statesman* (Budapest: Central European University Press, 1995), 201–233.

54. Not all of this amount was available for use, because of a US governmental stipulation that the amount provided by the United States had to be matched by other governments; such support was slow to develop. See Maggie Black, *The Children and the Nations: The Story of UNICEF* (New York: UNICEF, 1989), 67.

55. Ibid., 66–67.

56. Balinska, *For the Good of Humanity*, 210.

57. Ibid., 210–214.

58. Christian McMillan, *Discovering Tuberculosis: A Global History, 1900 to the Present* (New Haven, CT: Yale University Press, 2015), 84.

59. Farley, *Brock Chisholm*, 128–130.

60. WHO also faced competition from within. The decentralized structure of the organization—with the world being divided into six regions, each with its own semiautonomous health organization—would prove unwieldy, for it allowed regions to develop their own health agendas. This arrangement was, in large part, the result of US insistence on the Pan American Sanitary Bureau retaining its autonomy. While the PASB became part of WHO, changing its name to the Pan American Sanitary Organization, and later to the Pan American Health Organization, it charted its own course on a number of occasions. The Pan American Sanitary Organization's decisions regarding malaria and smallpox eradication played an important role in forcing WHO to initiate these campaigns (part IV). More generally, competition for leadership in international health played a significant role in causing WHO to move decidedly in the direction of disease control and eradication over the next three decades. See Javed Siddiqi, *World Health and World Politics: The World Health Organization and the World System* (Columbia: University of South Carolina Press, 1995), 60–72; Farley, *Brock Chisholm*, 20–21, 101–103.

61. McMillan, *Discovering Tuberculosis*, 93.

62. Ibid., 104.

Part IV: The Era of Eradication

1. These were not insignificant, and they touched on people's lives. For example, maternal and child health initiatives extended missionary and colonial

efforts to promote breastfeeding and Western birth practices, contributing to the global medicalization of childbirth. See Carla AbouZahr, "Safe Motherhood: A Brief History of the Global Movement, 1947–2002," *British Medical Journal* 67, 1 (2003): 13–25; Kalpana Ram and Margaret Jolly (eds.), *Maternities and Modernities: Colonial and Postcolonial Experiences in Asia and the Pacific* (Cambridge: Cambridge University Press, 1998); Nancy Rose Hunt, *A Colonial Lexicon: Of Birth, Ritual, Medicalization, and Mobility in the Congo* (Durham, NC: Duke University Press, 1999); Dilek Cindoglu and Feyda Sayan-Gengiz, "Medicalization Discourse and Modernity: Contested Meanings over Childbirth in Contemporary Turkey," *Health Care for Women International* 31, 3 (2010): 221–243; Celia Van Hollen, *Birth on the Threshold: Childbirth and Modernity in South India* (Berkeley: University of California Press, 2003).

2. See, for example, Donald A. Henderson, *Smallpox: The Death of a Disease; The Inside Story of Eradicating a Worldwide Killer* (New York: Prometheus Books, 2009); William Foege, *House on Fire: The Fight to Eradicate Smallpox* (Berkeley: University of California Press, 2012); Sanjoy Bhattacharya, *Expunging* Variola*: The Control and Eradication of Smallpox in India, 1947–1977*, New Perspectives in South Asian History 14 (New Delhi: Orient Longman, 2006); Paul Greenough, "Intimidation, Coercion, and Resistance in the Final Stages of the South Asian Smallpox Eradication Campaign, 1973–1975," *Social Science & Medicine*, 41, 5 (1995): 633–645; Sanjoy Bhattacharya, "International Health and the Limits of Its Global Influence: Bhutan and the Worldwide Smallpox Eradication Programme," *Medical History* 57, 4 (2013): 461–486; Anne-Emanuelle Birn, "Small(pox) Success?" *Ciência & Saude Coletiva* 16 (2011): 591–597.

Chapter Seven: Uncertain Beginnings

1. World Health Organization, "Malaria Eradication: Proposal by the Director-General," A8/P&B/10, May 3, 1955.

2. Randall M. Packard, "'No Other Logical Choice': Global Malaria Eradication and the Politics of International Health in the Post-War Era," *Parassitologia* 40, 1–2 (1998): 222.

3. J. N. Togba, from Liberia, claimed that the amount allocated for helping Liberia would not eradicate malaria in a single village. As it turned out, initial calculations of the proposed cost of eradication wildly underestimated the actual expense.

4. Expert Committees were composed of specialists in a particular field. They would meet on a regular basis to review developments in their respective fields and recommend policies. Their recommendations did not express official WHO positions.

5. Packard, "'No Other Logical Choice,'" 224.

6. "The United States Delegation is deeply impressed by the challenge which the present world malaria situation presents to the World Health Organization. At the same time it feels that the WHO has an opportunity to

make a great contribution toward world health, economic stability, and peace in an attack upon this disease. . . . Malaria is a direct and important contributing cause of the current world food shortage. This situation is being allowed to continue in the face of new discoveries of major importance in the field of malaria control. Dramatic control even to the point of eradication of malaria has been accomplished in small scattered areas. The United States Delegation believes that the WHO should direct a major share of its energy and resources during its first years to the application of such measures to larger areas with particular attention to major food producing areas afflicted with malaria." See "Proposal from the Representative of the United States of America," Fifth Session, Malaria Programme, WHO.IC.152, WHO.IC/Mal/11, January 23, 1948.

7. "An international programme for malaria control on a world-wide basis is urgently required in view of . . . the serious loss of working efficiency among malaria ridden populations, which contributes directly to the critical shortage of food and causes grave interference with industrial and agricultural activities and development." See Expert Committee on Malaria, "Report on the Second Session, May 19–25, 1948," WHO.I/205, June 8, 1948, 44.

8. John H. Perkins, "Reshaping Technology in Wartime: The Effect of Military Goals on Entomological Research and Insect-Control Practices," *Technology and Culture* 19 (1978): 182.

9. Robert Van den Bosch, *The Pesticide Conspiracy* (Garden City, NY: Doubleday, 1978), 21.

10. Packard, "'No Other Logical Choice,'" 222; Marcos Cueto, *Cold War, Deadly Fevers: Malaria Eradication in Mexico, 1955–1975* (Baltimore: Johns Hopkins University Press, 2007), 62–63.

11. Harold K. Jacobson, "WHO: Medicine, Regionalism, and Managed Politics," in *The Anatomy of Influence: Decision Making in International Organization*, ed. Robert W. Cox and Harold K. Jacobsen (New Haven, CT: Yale University Press, 1973).

12. World Health Organization, "Minutes of Discussion on Committee on Programme and Budget," Eighth World Health Assembly, 1955, 205.

13. "NIH Malaria Conference, Washington, D.C.," January 8, 1948, Paul F. Russell Diaries, Rockefeller Archives Center, Sleepy Hollow, New York.

14. "NIH Malaria Study Section Meetings, Washington, D.C.," September 27, 1948, Paul F. Russell Diaries, Rockefeller Archives Center.

15. Frederick L. Soper, "Hemisphere-Wide Malaria Eradication," statement to the Pan American Sanitary Bureau, Washington, DC, February 14, 1955, 28.

16. World Health Organization, "Report of the Meeting of the Eighth World Health Assembly, Geneva, 1955," 232.

17. "Dr. M. G. Candau and W.H.O.," *British Medical Journal* 2, 5864 (1973): 433–434.

18. Donald A. Henderson, *Smallpox: The Death of a Disease; The Inside Story of Eradicating a Worldwide Killer* (New York: Prometheus Books, 2009).

19. Frank Fenner, Donald A. Henderson, Isao Arita, Zdenek Jezek, and Ivan Danilovich Ladnyi, *Smallpox and Its Eradication* (Geneva: World Health Organization, 1988), 395.

20. World Health Organization, *Eighteenth World Health Assembly, Part II* (Geneva: World Health Organization, 1965), 313.

21. Fenner et al., *Smallpox and Its Eradication*, 417.

22. Ibid., 418.

23. Erez Manela, "A Pox on Your Narrative: Writing Disease Control into Cold War History," *Diplomatic History* 34, 2 (2010): 299–324.

24. Fenner et al., *Smallpox and Its Eradication*, 414.

25. Nancy Stepan, *Eradication: Ridding the World of Diseases Forever?* (London: Reaktion Books, 2011), 198–199. Stepan provides a detailed account of the smallpox-eradication campaign in the Americas before the launch of the WHO program.

26. Jack Hopkins gives this figure for what the United States spent in smallpox control in 1968. Nancy Stepan gives a figure of US$30 million a year for the 1950s. See Jack W. Hopkins, "The Eradication of Smallpox: Organizational Learning and Innovation in International Health Administration," *Journal of Developing Areas* 22, 3 (1988): 321–332; Stepan, *Eradication*, 193.

27. Manela, "Pox on Your Narrative," 309–310.

Chapter Eight: The Good and the Bad Campaigns

1. An earlier version of this chapter was presented in 2014, at the annual meeting of the American Association of the History of Medicine in Chicago.

2. WHO also created organizational structures and boundaries that did not necessarily correspond with epidemiological realities.

3. Dharmavadani K. Viswanathan, *The Conquest of Malaria in India: An Indo-American Co-Operative Effort* (Madras: Company Law Institute Press, 1958), 69–70.

4. Randall M. Packard, *The Making of a Tropical Disease: A Short History of Malaria* (Baltimore, Johns Hopkins University Press, 2007), 170–171.

5. "Moreover, the global eradication campaign was based on the assumption that all the necessary knowledge for eradication was available, that further research was superfluous, and that eradication required a rigid discipline in which local deviations from a centrally defined plan must be prevented." See José A. Nájera, Matiana González-Silva, and Pedro L Alonso, "Some Lessons for the Future from the Global Malaria Eradication Programme (1955–1969)," *PLoS Medicine* 8, 1 (2011): e1000412.

6. Randall M. Packard, "'No Other Logical Choice': Global Malaria Eradication and the Politics of International Health in the Post-War Era," *Parassitologia* 40, 1–2 (1998): 217–229.

7. The phrase "stabilize uncertainty" was used by economist Michael Carter at a Social Science Research Council conference on International Development

and the Social Sciences (held at the University of California–Berkeley in October 1994) to describe how economic formulas reduced complex sets of social, economic, and political variables that might effect an outcome to single variables.

8. Marcos Cueto, *Cold War, Deadly Fevers: Malaria Eradication in Mexico, 1955–1975* (Baltimore: Johns Hopkins University Press, 2007), 91.

9. Nancy Stepan, *Eradication: Ridding the World of Diseases Forever?* (London: Reaktion Books, 2011), 158.

10. There are many reports of local reactions against spraying and the negative iatrogenic, or treatment-caused, effects that spraying had on local ecologies. For an example from Mexico, see Cueto, *Cold War, Deadly Fevers*, 136–139.

11. Ibid., 117–119.

12. Melvin Griffith to A. D. Hass, November 25, 1954, Mission to Thailand Executive Office Subject Files 1950–54, Box 54, RG 469, US National Archives, Washington, DC.

13. Perez Yekutiel, "Problems of Epidemiology in Malaria Eradication," *Bulletin of the World Health Organization* 2, 4 (1960): 669–683.

14. Ministère de la Santé et de la Prévoyance Sociale, République Socialiste de Roumanie, *L'éradication du paludisme en Roumanie* (Bucharest: Éditions Médicales, 1966). Some of the problems programs encountered with resistance turned out to be related to behavioral tendencies of the primary *Anopheles* transmitters of malaria and not to the use of pesticides. Thus some mosquitoes preferred to feed outdoors or did not rest on walls. Inadequate entomological research before the spraying program was initiated prevented programs from planning for this problem. A lack of research had been encouraged by the sixth Expert Committee, which indicated that elaborate entomological resources were no longer needed in the face of new control methods.

15. World Health Organization, *WHO Expert Committee on Malaria: Eighth Report*, WHO Technical Report Series 205 (Geneva, World Health Organization, 1961).

16. Ka-che Yip, "Malaria Eradication: The Taiwan Experience," *Parassitologia* 42, 1–2 (2000): 117–126.

17. William Foege, *House on Fire: The Fight to Eradicate Smallpox* (Berkeley: University of California Press, 2012), 53–79.

18. World Health Organization, *The Third Ten Years of the World Health Organization, 1968–1977* (Geneva: World Health Organization, 2008), 180.

19. Foege, *House on Fire*, 57.

20. Alan Schnur, "Innovation as an Integral Part of Smallpox Eradication: A Fieldworker's Perspective," in *The Global Eradication of Smallpox*, ed. Sanjoy Bhattacharya and Sharon Messenger (New Delhi: Orient Black Swan, 2010), 106–150.

21. Donald A. Henderson. *Smallpox: The Death of a Disease; The Inside Story of Eradicating a Worldwide Killer* (New York: Prometheus Books, 2009), 192–193.

22. Jack W. Hopkins, "The Eradication of Smallpox: Organizational Learning and Innovation in International Health Administration," *Journal of Developing Areas* 22, 3 (1988): 321–332.

23. Schnur, "Innovation," 141.

24. Paul Greenough, "Intimidation, Coercion, and Resistance in the Final Stages of the South Asian Smallpox Eradication Campaign, 1973–1975," *Social Science & Medicine* 41, 5 (1995): 633–645.

25. Ibid.

26. Foege, *House on Fire*, 63–66.

27. For an extended biosocial analysis of global health, see Paul Farmer, Jin Yong Kim, Arthur Kleinman, and Matthew Basilico. *Reimagining Global Health: An Introduction* (Berkeley: University of California Press, 2013).

28. Asymptomatic cases were a problem in the Americas in areas where *Variola minor* dominated. This meant that surveillance had to be more intensive and, in some cases, mass vaccinations had to be practiced.

29. Maggie Black, *The Children and the Nations: The Story of UNICEF* (New York: UNICEF, 1989), 121.

30. Georganne Chapin and Robert Wasserstrom, "Agricultural Production and Malaria Resurgence in Central America and India," *Nature* 263, 5829 (1981): 181–185.

31. Randall M. Packard, "Maize, Cattle, and Mosquitoes: The Political Economy of Malaria Epidemics in Colonial Swaziland," *Journal of African History* 25, 2 (1984): 189–212.

32. Black, *Children and the Nations*, 129–130.

33. Perez Yekutiel, "Lessons from the Big Eradication Campaigns," *World Health Forum* 2, 4 (1981): 465–490.

34. Sunil S. Amrith, "In Search of a 'Magic Bullet' for Tuberculosis: South India and Beyond, 1955–1965," *Social History of Medicine* 17, 1 (2004): 113–130.

Part V: Controlling the World's Populations

1. Alison Bashford, *Global Population, History, Geopolitics, and Life on Earth* (New York: Columbia University Press, 2014).

2. Walter W. Rostow, *The Stages of Economic Growth: A Non-Communist Manifesto* (Cambridge: University Press, 1960).

3. Frank W. Notestein, "Population—The Long View," in *Food for the World*, ed. Theodore Schultz (Chicago: University of Chicago Press, 1945), 36–57.

4. Ibid., 51–52.

5. Kingsley Davis, "The World Demographic Transition," *Annals of the American Academy of Political and Social Sciences* 237, 1 (1945): 1–11.

6. I am indebted to Heidi Morefield, whose groundbreaking research on the development of US foreign assistance related to basic drug policies has revealed the growing importance of NGOs and the private sector in USAID programs under the Carter administration.

Chapter Nine: The Birth of the Population Crisis

1. Edward Alsworth Ross, *Standing Room Only?* (New York: century, 1927); Alison Bashford, *Global Population, History, Geopolitics, and Life on Earth* (New York: Columbia University Press, 2014), 118–124.

2. Radhakamal Mukerjee, *Migrant Asia* (Rome: I. Failli, 1936); Bashford, *Global Population*, 149–153.

3. For a history of eugenics in the United States, see Alexandra Stern, *Eugenics Nation: Faults and Frontiers of Better Breeding in Modern America* (Berkeley: University of California Press, 2012); Nathaniel Comfort, *The Science of Human Perfection: How Genes Became the Heart of American Medicine* (New Haven, CT: Yale University Press, 2012).

4. Bashford, *Global Population*, 239–264.

5. Laura Briggs, *Reproducing Empire: Race, Sex, Science, and U.S. Imperialism in Puerto Rico* (Berkeley: University of California Press, 2002), 102.

6. Ibid., 107.

7. Any attempt to describe the history of population programs is indebted to Matthew James Connelly's monumental history of population planning, *Fatal Misconception: The Struggle to Control World Population* (Cambridge, MA: Belknap Press of Harvard University Press, 2008). Many parts of the story I tell in this chapter are informed by Connelly's path-breaking work.

8. John Sharpless, "World Population Growth, Family Planning, and American Foreign Policy," *Journal of Policy History* 7, 1 (1995): 72–102.

9. Nilanjana Chatterjee and Nancy E. Riley, "Planning an Indian Modernity: The Gendered Politics of Fertility Control," *Signs* 26, 3 (2001): 811–845.

10. Connelly, *Fatal Misconception*, 71.

11. Randall M. Packard, "'Roll Back Malaria, Roll in Development?' Reassessing the Economic Burden of Malaria," *Population and Development Review* 35, 1 (2009): 53–87.

12. Connelly, *Fatal Misconception*, 115.

13. William Vogt, *Road to Survival* (New York: W. Sloane Associates, 1948), 226–227. Vogt added: "The dilemma is neatly stated by Dr. Chandrasekhar, who says: 'India's population today exceeds 400 million and at the lowest minimum of 1,400 calories she can only feed less than 300 million people! . . . More than a hundred million people . . . are either starving or on the brink of starvation.' "

14. Bashford, *Global Population*, 281.

15. Fairfield Osborne, *Our Plundered Planet* (Boston: Little, Brown, 1948), 170.

16. Connelly, *Fatal Misconception*, 187–188.

17. Paul Ehrlich, *The Population Bomb* (New York, Ballantine Books, 1968).

18. Simon Szreter has suggested that before Notestein's trip to Japan in 1948, he had begun to consider the need for finding ways to stimulate changes in reproductive behavior in agrarian societies before social and economic forces forced them to occur. Szreter cites an article Notestein published in 1947 that

seems to raise this possibility. Szreter may be correct. It is clear, however, that Notestein's Far East trip did much to solidify his views regarding the desirability of birth control policies. See Simon Szreter, "The Idea of Demographic Transition and the Study of Fertility Change: A Critical Intellectual History," *Population* (English edition) 19, 4 (2012): 671–672.

19. Connelly, *Fatal Misconception*, 134–135; John Sharpless, "Population Science, Private Foundations, and Development Aid: The Transformation of Demographic Knowledge in the United States, 1945–1965," in *International Development and the Social Sciences: Essays on the History and Politics of Knowledge*, ed. Frederick Cooper and Randall M. Packard (Berkeley: University of California Press, 1997), 176–202.

20. Similar conclusions were expressed by Koya Yosha, director of the Japanese National Institute for Public Health. He found that 92% of married couples with fewer than three children enthusiastically wanted to practice contraception. See Bashford, *Global Population*, 313.

21. Marshall C. Balfour, Roger F. Evans, Frank W. Notestein, and Irene B. Tauber, *Public Health and Demography in the Far East* (New York: Rockefeller Foundation, 1950); Sharpless, "Population Science," 180–181.

22. Kingsley Davis, *The Population of India and Pakistan* (Princeton, NJ: Princeton University Press, 1951); Szreter, "Idea of Demographic Transition," 659–701.

23. Davis argued that "the demographic problems of the underdeveloped countries, especially in areas of non-Western culture, make these nations more vulnerable to Communism." See Kingsley Davis, "Population and Power in the Free World," in *Population Theory and Policy*, ed. Joseph J. Spengler and Otis D. Duncan (Chicago: Free Press, 1956), 342–356.

24. Other economists and demographers argued that population growth would offset the economic benefits of disease-control programs and undermine investments in economic development during the 1950s and 1960s. See Randall M. Packard, "'Roll Back Malaria,'" 53–87.

25. Chatterjee and Riley, "Planning and Indian Modernity," 811–845.

26. At that time, India's population stood at 362 million, having increased by 43 million in ten years, despite World War II, the 1943 Bengal famine, and the population massacres that accompanied partition in 1947. See Connelly, *Fatal Misconception*, 145–146.

27. Laura Bier, "'The Family Is a Factory': Gender, Citizenship, and the Regulation of Reproduction in Postwar Egypt," *Feminist Studies* 2, 2 (2010): 404–432.

28. Connelly, *Fatal Misconception*, 126.

29. John Farley, *Brock Chisholm, the World Health Organization, and the Cold War* (Vancouver: University of British Columbia Press, 2008), 180–181.

30. Maggie Black, *The Children and the Nations: The Story of UNICEF* (New York: UNICEF, 1989), 246–247.

31. Marc Frey, "Neo-Malthusianism and Development: Shifting Interpretations of a Contested Paradigm," *Journal of Global History* 6, 1 (2011): 75–97.

32. Connelly, *Fatal Misconception*, 198.

33. Ibid., 181.

34. Helen Desfosses, "Population as a Global Issue: The Soviet Prism," in *Soviet Population Policy: Conflicts and Constraints*, ed. Helen Desfosses (New York: Pergamon Press, 1981), 179–202.

35. Connelly, *Fatal Misconception*, 157–161.

36. Ibid., 170.

37. Sharpless, "Population Science," 182–183.

38. Johns Hopkins University researchers working in the Eastern Health District of Baltimore in the late 1940s had pioneered this method of data collection. I am indebted here to the work of Karen Thomas and her forthcoming history of the Johns Hopkins School of Public Health.

39. Mahmood Mamdani, *The Myth of Population Control: Family, Caste, and Class in an Indian Village* (New York: Monthly Review Press, 1973).

40. Raúl Necochea, "Gambling on the Protestants: The Pathfinder Fund and Birth Control in Peru, 1958–1965," *Bulletin of the History of Medicine* 88, 2 (2014): 344–372.

Chapter Ten: Accelerating International Family-Planning Programs

1. Elizabeth Watkins, *On the Pill: A Social History of Oral Contraceptives, 1950 to 1970* (Baltimore: Johns Hopkins University Press, 2001); Laura Briggs, *Reproducing Empire: Race, Sex, Science, and U.S. Imperialism in Puerto Rico* (Berkeley: University of California Press, 2002).

2. Matthew James Connelly, *Fatal Misconception: The Struggle to Control World Population* (Cambridge, MA: Belknap Press of Harvard University Press, 2008), 174.

3. Briggs, *Reproducing Empire*, 134–137.

4. At the same time, Pincus pointed to the finding that the pregnancy rate of those who stopped taking the pill was 79% within four months and argued that this was evidence of the pill's efficacy. In reality, however, this high pregnancy rate was partially due to fertility increases after coming off the pill, due to hormonal changes.

5. Peter J. Donaldson, *Nature Against Us: The United States and the World Population Crisis, 1965–1980* (Chapel Hill: University of North Carolina Press, 1990), 37–38.

6. Connelly, *Fatal Misconception*, 201–213.

7. In their 1958 book, economist Edgar Hoover and demographer Ansley Coale argued that population growth could undermine the positive effects of saving lives through health interventions. Their study questioned the claim that decreases in childhood mortality brought on by disease-control programs reduced nonproductive expenditures on persons who would die before they contributed to

the economy, lessening the "burden of dependency." This had been a popular argument among those advocating disease-control programs, described previously by Charles-Edward A. Winslow in *The Cost of Sickness and the Price of Health*, WHO Monograph Series 7 (Geneva: World Health Organization, 1951). Coale and Hoover posited instead "that if more children are enabled to survive to their adult years, there will be more workers but also more parents; the larger number of parents, if fertility rates remains unchanged, will produce more children." They argued that demographic analyses showed that the rise in the number of children would be greater than the rise in the number of workers. "So while it is true that a decrease in childhood mortality will yield a larger population at the working ages than would otherwise have resulted, it produces an even greater rise in the number of children whom the people of working ages must support. . . . In short, the 'waste' of always supporting a much larger next generation . . . replaces the waste of spending on persons who later die." See Ansley J. Coale and Edgar M. Hoover, *Population Growth and Economic Development in Low-Income Countries: A Case Study of India's Prospects* (Princeton, NJ: Princeton University Press, 1958), 23.

8. Connelly, *Fatal Misconception*, 211–213.

9. Donaldson, *Nature Against Us*, 40.

10. Ibid., 39.

11. Connelly, *Fatal Misconception*, 221.

12. Donaldson, *Nature Against Us*, 44.

13. Ibid., 79.

14. Ibid., 97–112.

15. Connelly, *Fatal Misconception*, 290.

16. Etienne Gosselin, "Constructing International Health: The Communicable Disease Center, Field Epidemiologists, and the Politics of Foreign Assistance (1948–1972)," PhD dissertation, McGill University, 2011, 342.

17. Amy Kaler describes how white Rhodesian authorities very consciously used birth control programs as a way of restricting the birth rates in that country's black populations. See Amy Kaler, *Running After Pills: Politics, Gender, and Contraception in Colonial Zimbabwe* (Portsmouth, NH: Heinemann, 2003).

18. Maggie Black, *The Children and the Nations: The Story of UNICEF* (New York: UNICEF, 1989), 249–251.

19. Laura Bier, "'The Family Is a Factory': Gender, Citizenship, and the Regulation of Reproduction in Postwar Egypt," *Feminist Studies* 2, 2 (2010): 417.

20. Cheryl Benoit Mwaria, "Rural Urban Migrations and Family Arrangements among the Kamba," in *Transformation and Resiliency in Africa: As Seen by Afro-American Scholars*, ed. Pearl T. Robinson and Elliott P. Skinner (Washington, DC: Howard University Press, 1983), 29–45.

21. Daniel Smith, "Contradictions in Nigeria's Fertility Transition: The Burdens and Benefits of Having People," *Population and Development Review* 30, 2 (2004), 221–238.

22. Kenneth Newell, *Health By the People* (Geneva: World Health Organization, 1975).

Chapter Eleven: Rethinking Family Planning

1. Matthew James Connelly, *Fatal Misconception: The Struggle to Control World Population* (Cambridge, MA: Belknap Press of Harvard University Press, 2008), 305.

2. Ibid., 321.

3. Harriet B. Presser, "The Role of Sterilization in Controlling Puerto Rican Fertility," *Population Studies* 23, 3 (1969): 343–361.

4. Laura Briggs, *Reproducing Empire: Race, Sex, Science, and U.S. Imperialism in Puerto Rico* (Berkeley: University of California Press, 2002), 142–161.

5. Iris Lopez, "Agency and Constraint: Sterilization and Reproductive Freedom among Puerto Rican Women in New York City," *Urban Anthropology and Studies of Cultural Systems and World Economic Development* 22, 3 (1993): 299–323.

6. Marc Frey, "Neo-Malthusianism and Development: Shifting Interpretations of a Contested Paradigm," *Journal of Global History* 6, 1 (2011): 75–97; Susan Greenhalgh, "Missile Science, Population Science: The Origins of China's One-Child Policy," *China Quarterly* 182 (2005): 253–276.

7. "40 Greatest Hits: Laparoscopy Paves the Way for Jhpiego's Work in Reproductive Health," Jhpiego, www.jhpiego.org/content/40-greatest-hits-laparoscopy-paves-way-jhpiegos-work-in-reproductive-health/. Jhpiego is an affiliate of Johns Hopkins University, but it has never been administered under either the School of Medicine or the School of Public Health. Instead, its head reports directly to the president of the university.

8. Connelly, *Fatal Misconception*, 306.

9. Kingsley Davis, "Population Policy: Will Current Programs Succeed?" *Science* 158, 3802 (1967): 730–739.

10. Connelly, *Fatal Misconception*, 308.

11. Ibid., 314–315.

12. Susan George, *How the Other Half Dies: The Real Reason for World Hunger* (Montclair, NJ: Allanheld, Osmun, 1979).

13. Richard Franke and Barbara Chasin, *Seeds of Famine* (Montclair, NJ: Allanheld, Osmun, 1986); Michael Watts, *Silent Violence* (Berkeley: University of California Press, 1983).

14. Timothy Mitchell, "The Object of Development: America's Egypt," in *Power of Development*, ed. Jonathan Crush (New York: Routledge, 1995), 129–157.

15. Peter J. Donaldson, *Nature Against Us: The United States and the World Population Crisis, 1965–1980* (Chapel Hill: University of North Carolina Press, 1990), 57.

16. Connelly, *Fatal Misconception*, 354

17. For the complete text of the speech, see Michael S. Teitelbaum and Jay M. Winter, *The Fear of Population Decline* (London: Academic Press, 1984), appendix C.

18. See, for example, Caroline Bledsoe's insightful study of women's reproductive decision making in the Gambia, *Contingent Lives: Fertility, Time, and Aging in West Africa* (Chicago: University of Chicago Press, 2002).

19. Emi Suzuki, "Between 1960 and 2012, the World Average Fertility Rate Halved to 2.5 Births per Woman," World Bank Open Data: The World Bank Data Blog, http://blogs.worldbank.org/opendata/between-1960-and-2012-world-average-fertility-rate-halved-25-births-woman/.

20. Connelly, *Fatal Misconception*, 374.

21. While there is strong evidence from a number of countries that the education of women correlates with declining fertility, the relationship between education and fertility remains unclear and is debated. See Susan Cochrane, *Fertility and Education: What Do We Really Know?* (Baltimore: Johns Hopkins University Press, 1979); Carolyn Bledsoe, John B. Casterline, Jennifer A. Johnson-Kuhn, and John G. Haaga, *Critical Perspectives on Schooling and Fertility in the Developing World* (Washington, DC: National Academy Press, 1999).

Part VI: The Rise and Fall of Primary Health Care

1. World Health Organization, Constitution of the World Health Organization, www.who.int/governance/eb/who_constitution_en.pdf.

Chapter Twelve: Rethinking Health 2.0

1. The makeup of the WHA had changed since the Malaria Eradication Programme had been launched (part V). By the late 1970s, decolonization had occurred nearly everywhere in the world. Many of the newly independent countries that joined WHO in the 1960s and 1970s possessed limited health-care resources, with a high concentration of services located in urban centers and few within the rural areas where the majority of their populations lived. Developing better access to health care in these countries was both a popular demand and a stated goal of their newly established governments.

2. World Health Organization, *The Third Ten Years of the World Health Organization: 1968–1977* (Geneva: Apia Way Press, 2008), 117.

3. Socrates Litsios, "The Long and Difficult Road to Alma-Ata: A Personal Reflection," *International Journal of Health Services: Planning, Administration, Evaluation* 32, 4 (2002): 717.

4. Hafdan Mahler, "An International Health Conscience," *WHO Chronicle* 28, 5 (1974): 207–211.

5. Walter W. Rostow, *The Stages of Economic Growth: A Non-Communist Manifesto* (Cambridge: University Press, 1960).

6. International Labour Office, *Poverty and Minimum Living Standards: The Role of the ILO*, Report of the Director-General (Geneva: International Labour

Office, 1970), cited in "Briefing Note Number 7," *UN Intellectual History Project*, May 2010, 4.

7. International Labour Office, *Employment Growth and Basic Needs: A One-World Problem* (Geneva: International Labour Office, 1976), cited in "Briefing Note Number 7," *UN Intellectual History Project*, May 2010, 4.

8. Cited in the chapter by Martha Finnemore, "Redefining Development at the World Bank," in *International Development and the Social Sciences: Essays on the History and Politics of Knowledge*, ed. Frederick Cooper and Randall M. Packard (Berkeley: University of California Press, 1997), 211.

9. Great Britain, Ministry of Overseas Development, *The Changing Emphasis of British Aid Policies: More Help for the Poorest* (London: Her Majesty's Stationery Office, 1975), cited in the chapter by Owen Barder, "Reforming Development Assistance: Lessons from the U.K. Experience," in *Security by Other Means: Foreign Assistance, Global Poverty, and American Leadership*, ed. Lael Brainard (Washington, DC: Brookings Institute, 2007).

10. Kathryn Sikkink, "Development Ideas in Latin America: Paradigm Shift and the Economic Commission of Latin America," in *International Development*, ed. Cooper and Packard, 228–256.

11. Countries that attempted to implement the ISI strategy, however, often read too much into Prebisch's arguments and created policies that led to a neglect of the production and export of primary goods and to declining participation in world trade.

12. Walter Rodney, *How Europe Underdeveloped Africa* (London: Bogle-L'Ouverture, 1972).

13. Basil Davidson, *Black Star: A View of the Life and Times of Kwame Nkrumah* (Boulder, CO: Westview Press, 1989).

14. Nitsan Chorev, *The World Health Organization between North and South* (Ithaca, NY: Cornell University Press, 2002), 47–48.

15. David Werner, *Where There Is No Doctor* (Berkeley, CA: Hesperian Books, 1973).

16. Ivan Illich, *Medical Nemesis: The Expropriation of Health* (New York: Pantheon Books, 1976), 4.

17. Thomas McKeown, *The Modern Rise of Population* (New York: Academic Press, 1976).

18. Victor W. Sidel, "The Barefoot Doctors of the People's Republic of China," *New England Journal of Medicine* 286, 24 (1972): 1292–1300.

19. Litsios, "Long and Difficult Road," 714. Litsios provides an insightful insider's view of the events that led up to the Alma-Ata conference. See also Marcos Cueto, "The Origins of Primary Health Care and Selective Primary Health Care," *American Journal of Public Health* 94, 11 (2004): 1864–1874.

20. V. Djukanović and E. P. Mach (eds.), *Alternative Approaches to Meeting Basic Health Needs in Developing Countries* (Geneva: World Health Organization, 1975).

21. Kenneth Newell, *Health by the People* (Geneva: World Health Organization, 1975).

22. Litsios, "Long and Difficult Road," 714.

23. Ibid., 716–718.

24. Vera Mann, Alex Eble, Chris Frost, Ramaswamy Premkumar, and Peter Boone, "Retrospective Comparative Evaluation of the Lasting Impact of a Community-Based Primary Health Care Programme on Under-5 Mortality in Villages around Jamkhed, India," *Bulletin of the World Health Organization* 88, 10 (2010): 727–736; U. N. Chakravorty, "A Health Project That Works—Progress in Jamkhed," *World Health Forum* 4, 1 (1983); Raj and Mabelle Arole, *Jamkhed: A Comprehensive Rural Health Care Project* (New York, McMillan, 1994), 38–40.

25. Socrates Litsios, "The Christian Medical Commission and the Development of the World Health Organization's Primary Health Care Approach," *American Journal of Public Health* 94, 11 (2004): 1884–1893.

26. Comprehensive Rural Health Project, Jamkhed, www.jamkhed.org.

Chapter Thirteen: Challenges to Primary Health Care

1. Gillian Walt and Angela Melamed, *Mozambique: Toward a People's Health Service* (London: ZED Books, 1983).

2. As Malcolm Segal noted in a 1983 review of the progress of PHC efforts: "Tenant farmers and landless labourers on the one hand, and landlords on the other, do not have a community of interest such that they can be characterized meaningfully as members of the same community, who happen to perform different roles. Poor homesteaders may well have different community interests from those who hold some capital, and both are very likely to have different interests from the owners or managers of large scale commercial farms or plantations. Differential ownership of productive resources creates different class interests, and it confers great influence in community decision making on those holding economic power." See Malcolm Segall, "The Politics of Primary Health Care," *IDS Bulletin* (Brighton, England) 14, 4 (1983): 27–37.

3. Benjamin D. Paul and William J. Demarest, "Citizen Participation Overplanned: The Case of a Health Project in the Guatemalan Community of San Pedro La Laguna," *Social Science & Medicine* 19, 3 (1984): 185–192.

4. Judith Justice, *Policies, Plans, and People: Foreign Aid and Health Development* (Berkeley: University of California Press, 1989).

5. Given the movement of international-health consultants from country to country, it may be impossible to provide them with the local knowledge they will need to do their jobs effectively in particular places. But it would be possible to train them in how to better acquire this knowledge and in how to develop cultural literacy.

6. Stacey Pigg, "'Found in Most Traditional Societies': Traditional Medical Practitioners between Culture and Development," in *International Development*

and the Social Sciences, ed. Frederick Cooper and Randall Packard (Berkeley: University of California Press, 1997), 259–290.

7. Linda Stone, "Primary Health Care for Whom? Village Perspectives from Nepal," *Social Science & Medicine* 22, 3 (1986): 293–302.

8. A number of the basic arguments in this chapter have been made by others. See, for example, Marcus Cueto, "The Origins of Primary Health Care and Selective Primary Health Care," *American Journal of Public Health* 94, 11 (2004): 1864–1874.

9. Ali Maow Maslin died of malaria in 2013.

10. Julia A. Walsh and Kenneth S. Warren, "Selective Primary Health Care: An Interim Strategy for Disease Control in Developing Countries," *Social Science & Medicine, Part C: Medical Economics* 14, 2 (1980): 145–163. Total primary health care was equated with the Alma-Ata model and included the "development of all segments of the economy, ready and universal access to comprehensive curative care, prevention of endemic disease, proper sanitation and safe water supplies, immunization, nutrition promotion, health education, maternal and child care, and family planning." This approach, the authors asserted, exceeded the resources available for most health programs then or in the near future. Basic primary health care was more limited, focusing on providing communities with health workers and establishing clinics for treating all illnesses within a population. This, too, was said to be very costly. Walsh and Warren cited a World Bank estimate that "supplying one health post with 1 vehicle per 10,000 people and training 125 auxiliary nurse midwives and 250 health workers would cost $2,500,000," but then added a laundry list of additional expenses that this model would entail. Vector disease-control and nutrition programs were also rejected. Walsh and Warren acknowledged that vector-control programs had had a significant impact on mortality but argued that such programs required perpetual maintenance; they did not eliminate diseases. Nutritional-supplement programs had a limited impact on child and infant mortality (though this was before studies demonstrated the impact of vitamin A supplementation on child mortality). Those nutrition programs that had a positive effect, such as Carl Taylor's Narangwal project in India, were said to be too expensive in terms of cost per death averted. In children one to three years old, the cost was $13,000, 1.5-3 times higher than the cost with medical care alone.

11. In another paper presented at the Bellagio conference, Sharon Russell and Stephen Joseph estimated that the cost of achieving health for all by 2000 using a PHC model would be between US$40 and US$85 billion, based on 1978 dollars. These figures would increase to $49–$103 billion when a population growth rate of 2% per year is factored in. See Sharon Stanton Russell and Stephen C. Joseph, "Is Primary Care the Wave of the Future?" *Social Science & Medicine, Part C: Medical Economics* 14, 2 (1980): 137–144; Stephen C. Joseph, "Outline of National Primary Health Care System Develop-

ment: A Framework for Donor Involvement," *Social Science & Medicine, Part C: Medical Economics* 14, 2 (1980): 177–180.

12. Ben Wisner, "GOBI versus PHC? Some Dangers of Selective Primary Health Care," *Social Science & Medicine* 26, 9 (1988): 963–969.

13. Jean-Pierre Habicht and Peter A. Berman, "Planning Primary Health Services from a Body Count," *Social Science & Medicine, Part C: Medical Economics* 14, 2 (1980): 129–136; Russell and Joseph, "Primary Care."

14. June Goodfield, *A Chance to Live: The Global Campaign to Immunize the World's Children* (New York: McMillan, 1991), 35–36.

15. Jon E. Rhode, "Early Influences in the Life of James Grant," in *Jim Grant, UNICEF Visionary*, ed. Richard Jolly (New York: UNICEF, 2002), 39–45.

16. Peter Adamson, "The Mad American," in *Jim Grant, UNICEF Visionary*, ed. Richard Jolly (New York: UNICEF, 2002), 19–28.

17. Ibid., 22.

18. Goodfield, *Chance to Live*, 36

19. Ibid., 37.

20. Ibid., 40.

21. Inflation meant that the interest paid by borrowing countries increased over the 1970s. Debt service (interest payments and the repayment of principal) in Latin America grew even faster, reaching US$66 billion in 1982, up from US$12 billion in 1975. African countries also took on debts during the 1970s, much of it owed to foreign governments and development banks.

22. Caroline Swinburne, "Zambia's Nursing Brain Drain," *BBC News*, September 23, 2006, http://news.bbc.co.uk/1/hi/world/africa/5362958.stm.

23. Randall M. Packard, *The Making of a Tropical Disease: A Short History of Malaria* (Baltimore, Johns Hopkins University Press, 2007).

24. World Bank, "Zambia—Health Sector Support Project," Staff Appraisal Report, October 14, 1994. It is unclear whether patients stopped attending clinics because of the cost, or because the clinics did not provide quality service and the patients refused to pay for poor service.

25. Yann Derriennic, Katherine Wolf, Maureen Daura, Mwangala Namushi Kalila, Ruth Hankede, and C. Natasha Hsi, "Impact of a User Fee Waiver Pilot on Health Seeking Behavior of Vulnerable Populations in Kafue District, Zambia," paper presented at the American Public Health Association annual meeting, Philadelphia, Pennsylvania, December 10–14, 2005.

26. National Malaria Control Centre, Central Board of Health, *Malaria in Zambia, Situation Analysis* (Lusaka: Central Board of Health, 2000), 6–7.

27. For a detailed description of the child-survival initiative, see Goodfield, *Chance to Live*.

28. World Bank, *World Development Report 1993: Investing in Health* (New York: Oxford University Press, 1993).

Part VII: Back to the Future

1. Bill & Melinda Gates Foundation, *Malaria Forum, Final Report*, October 16–18, 2007, https://docs.gatesfoundation.org/Documents/MalariaForumReport.pdf.

2. The role of gender in international health has received very little attention. For example, the first woman to become director-general of WHO, Gro Harlem Brundtland, oversaw the launching of the Roll Back Malaria Partnership; took on the tobacco companies by initiating an attack on tobacco use; and created the Commission of Macroeconomics and Health, which laid out economic arguments for investments in health. These were important initiatives. But it is difficult to know how, if at all, they reflected the fact that the director-general was a woman. Margaret Chan, on the other hand, placed "accelerating existing initiatives to make pregnancy safer, and integrating the management of childhood illness, and immunizing children against vaccine-preventable diseases" at the top of her list of commitments as director-general. She also oversaw the creation of a Commission on Information and Accountability for Women's and Children's Health. Concerns about the health of women and children have a long history, predating the rise of women into positions of leadership. But Chan clearly planted a flag in this arena.

3. Randall M. Packard, *The Making of a Tropical Disease: A Short History of Malaria* (Baltimore: Johns Hopkins University Press, 2007), 217.

Chapter Fourteen: AIDS and the Birth of Global Health

1. Michael D. Lemonick and Alice Park, "The Truth About SARS," *Time*, April 27, 2003.

2. The breakup of the Soviet Union was followed by a series of health crises, in the form of injuries caused by ethnic violence in countries that had been held together by Soviet rule, as well as by declining health conditions fostered by the economic crisis that followed the breakup, the dissipation of state-run health systems, and deteriorating environments. Many populations joined the ranks of impoverished people who needed medical aid.

3. Thus elements of neoliberalism played an important role in policy discussions in the United States during the Bill Clinton years; within the Labour government under Tony Blair in the UK; as well as in *liberalisme* in France under Jacques Chirac and, even more so subsequently, under Nicolas Sarkozy. See Randall Packard, Will Dychman, Leo Ryan, Eric Sarriott, and Peter Winch, "The Global Fund: Historical Antecedents and First Five Years of Operation," in *Evaluation of the Organizational Effectiveness and Efficiency of the Global Fund to Fight AIDS, Tuberculosis and Malaria: Results from Study Area 1 of the Five-Year Evaluation*, Annex 2, Technical Evaluation Reference Group, Study Area 1 Methodology, 2009, 265–274.

4. UNESCO, "Non-Governmental Organizations Accredited to Provide Advisory Services to the Committee," http://www.unesco.org/culture/ich/en/accredited-ngos-00331.

5. Jim Igoe and Tim Kelsall (eds.), *Between a Rock and a Hard Place: African NGOs, Donors, and the State* (Durham, NC: Carolina Academic Press, 2005), 7.

6. Nitsan Chorev, *The World Health Organization between North and South* (Ithaca, NY: Cornell University Press, 2012), 8–9.

7. Alan Brandt, "How AIDS Invented Global Health," *New England Journal of Medicine* 368, 234 (2013): 2149–2152.

8. John Illife, *The African AIDS Epidemic* (London: Oxford University Press, 2008).

9. Lindsey Knight, *UNAIDS: The First Ten Years, 1996–2007* (Geneva: Joint United Nations Programme on HIV/AIDS, 2008), 13.

10. Fiona Godlee, "WHO's Special Programmes: Undermining from Above." *British Medical Journal (Clinical Research Edition)* 310, 6973 (1995): 178–182.

11. Theodore M Brown, Marcus Cueto, and Elizabeth Fee, "The World Health Organization and the Transition from 'International' to 'Global' Health," in *Medicine at the Border: Disease, Globalization, and Security, 1850 to the Present*, ed. Alison Bashford (New York: Palgrave Macmillan, 2006), 79–94.

12. Knight, *UNAIDS*, 17.

13. This strategy had been successful in slowing the epidemic in Southeast Asia. But these methods were less successful once AIDS spread into the general population, as it had in many African countries by the 1990s. See Randall M. Packard and Paul Epstein, "Epidemiologists, Social Scientists, and the Structure of Medical Research on AIDS in Africa," *Social Science & Medicine* 33, 7 (1991): 771–783.

14. Those organizing the early response to AIDS in Africa also failed to acknowledge the role that the medical system itself was playing in the spread of the disease through the reuse of needles and transfusions, since weakened health systems lacked medical supplies. Given the limited medical budgets (averaging US$5 per capita) and shortages in foreign exchange in most African countries (a product of the declining value of African exports on world commodity markets and IMF-instigated currency devaluations in the 1980s and early 1990s), disposable needles, sterilizers, and even the chemicals needed for sterilization were often in short supply. The energy costs involved in running sterilizers, even where they existed, sometimes limited the sterilization of needles. See Packard and Epstein, "Epidemiologists," 779.

15. Richard Parker, Delia Easton, and Charles H. Klein, "Structural Barriers and Facilitators in HIV Prevent a Review of International Research," *AIDS* 14, Suppl. 1 (2000): 522–532; Merrill Singer (ed.), *The Political Economy of AIDS* (New York: Baywood, 1998); Paul Farmer, *AIDS and Accusation* (Berkeley: University of California Press, 1993); Richard Parker, *The Global Political Economy of AIDS* (London: Routledge, 2000).

16. Brook G. Schoepf, "Women, AIDS, and Economic Crisis in Central Africa," *Canadian Journal of African Studies* 22, 3 (1988), 625–644.

17. Knight, *UNAIDS*, 1996–2007.

18. Jesse Bump, "The Lion's Gaze: African River Blindness from Tropical Curiosity to International Development," PhD dissertation, Johns Hopkins University, 2004, 306–372.

19. Robin Barlow and Lisa M. Grobar, "Costs and Benefits of Controlling Parasitic Diseases," PHN Technical Note 85–17 (Washington, DC: Population, Health, and Nutrition Department, World Bank, 1986).

20. World Bank, *World Development Report 1993: Investing in Health* (New York: Oxford University Press, 1993), 17.

21. Sudhir Anand and Kara Hanson, "Disability-Adjusted Life Years: A Critical Review," *Journal of Health Economics* 16, 6 (1997): 685–702.

22. World Bank, *Investing in Health*, 117.

23. Jennifer Prah Ruger, "The Changing Role of the World Bank in Global Health," *American Journal of Public Health* 95, 1 (2005): 60–70.

24. William A. Muraskin, *The Politics of International Health: The Children's Vaccine Initiative and the Struggle to Develop Vaccines for the Third World* (Albany, NY: State University of New York Press, 1998).

25. Randall M. Packard, *The Making of a Tropical Disease: A Short History of Malaria* (Baltimore: Johns Hopkins University Press, 2007), 231.

26. Jennifer Chan, *Politics in the Corridor of Dying: AIDS Activism and Global Health Governance* (Baltimore: Johns Hopkins University Press, 2015), 135.

27. Charles Wendo, "Uganda and the Global Fund Sign Grant Agreement," *Lancet* 361, 9361 (2003): 942.

28. The objective of the first phase was to increase access to HIV/AIDS prevention, care, and treatment programs, with an emphasis on vulnerable groups, such as youth, women of childbearing age, and others at high risk. The second phase, MAP2, approved in February 2002, added pilot testing of antiretroviral therapies and support for cross-border initiatives. MAP2 received an additional US$500 million. See Ruger, "Changing Role," 60–70.

29. World Bank, "Meeting the Challenge: The World Bank and HIV/AIDS," April 3, 2013, www.worldbank.org/en/results/2013/04/03/hivaids-sector-results-profile.

Chapter Fifteen: The Global Fund, PEPFAR, and the Transformation of Global Health

1. Luke Messac and Krishna Prabhu, "Redefining the Possible: The Global AIDS Response," in *Reimagining Global Health: An Introduction*, ed. Paul Farmer, Arthur Kleinman, Jim Yong Kim, and Matthew Basilico (Berkeley: University of California Press, 2013), 117–119.

2. Ibid.

3. Paul Farmer, Arthur Kleinman, Jim Yong Kim, and Matthew Basilico (eds.), *Reimagining Global Health: An Introduction* (Berkeley: University of California Press, 2013).

4. Messac and Prabhu, "Redefining the Possible," 119.

5. Jennifer Chan, *Politics in the Corridor of Dying: AIDS Activism and Global Health Governance* (Baltimore: Johns Hopkins University Press, 2015), 75–77.

6. Messac and Prabhu, "Redefining the Possible," 126.

7. Amir Attran and Jeffrey Sachs, "Defining and Refining International Donor Support for Combating the AIDS Pandemic," *Lancet* 357, 2924 (2001): 57–61.

8. The following section on the history of the Global Fund is based on a study I coauthored for the five-year evaluation of the fund. See Randall Packard, Will Dychman, Leo Ryan, Eric Sarriott, and Peter Winch, "The Global Fund: Historical Antecedents and First Five Years of Operation," *Evaluation of the Organizational Effectiveness and Efficiency of the Global Fund to Fight AIDS, Tuberculosis and Malaria: Results from Study Area 1 of the Five-Year Evaluation*, Annex 2, Technical Evaluation Reference Group, Study Area 1 Methodology, 2009, 265–274.

9. These concerns were openly expressed by a number of the major institutional actors involved in the fund's creation, including representatives of the G8 countries; the US government, which became the primary source of public financing for the Global Fund; the Bill & Melinda Gates Foundation, which was the largest private donor and was represented on the fund's board; and the World Bank, which became the trustee for the fund. The concerns of these groups played an important role in the Global Fund's design and, in turn, in reshaping the terrain of global-health funding.

10. The very need for a permanent Secretariat to run the Global Fund was a subject of debate by its board, and the board expressed a desire from the outset to limit the size and budget of the Secretariat (e.g., by hiring temporary rather than permanent staff). About 50 staff members were seconded from WHO and UNAIDS, and additional staff were quickly hired to serve as fund managers. Many of these managers were young, with relatively little work experience.

11. In rounds 1 through 4, applicants had the option of applying for "integrated" HSS projects. Yet the definition of what constituted fundable HSS proposals was unclear, and only a single HSS proposal out of the 10 submitted received funding. In round 5, the Global Fund attempted to clarify its position on HSS by instituting a separate, HSS-specific funding option. But this time, only 10%, or 3 out of 30, of the HSS proposals submitted were approved. This funding option was subsequently removed in rounds 6 and 7, during which the Global Fund attempted to further clarify its requirements regarding the specific linking of HSS activities to the three target diseases.

12. "Cancer," World Health Organization Media Centre, updated February 2015, www.who.int/mediacentre/factsheets/fs297/en/.

13. Julie Livingston, *Improvising Medicine: An Oncology Ward in an Emerging Cancer Epidemic* (Durham, NC: Duke University Press, 2012).

14. World Economic Forum and Harvard School of Public Health, *The Global Economic Burden of Non-Communicable Diseases: A Report* (Geneva: World Economic Forum, 2011).

15. Thomas Insel, "Director's Blog: The Global Cost of Mental Health," National Institute of Mental Health, September 28, 2011, www.nimh.nih.gov/about/director/2011/the-global-cost-of-mental-illness.shtml.

16. Sam Lowenberg, "The World Bank under Jim Kim," Special Report, *Lancet* 386, 9991 (2015): 315.

17. President George W. Bush and his advisors saw AIDS as both a humanitarian crisis and a threat to global economic development and political security. Their views were influenced by an unclassified report by the US National Security Council, which projected massive increases in AIDS in five countries over the next decade, exceeding 50 million new cases. This would eclipse the projected 30 million cases in southern Africa. Describing the impact of the epidemic on two of the United States' traditional regional partners, Nigeria and Ethiopia, the report pointed to the high toll the disease was taking on elites in key positions in the government and on the economy. In particular, it expressed concerns about the impact of the epidemic on military leadership and overall manpower and, ultimately, about the inability of the two nations to continue to play a regional leadership role, raising fears of regional political instability. See David F. Gordon, "The Next Wave of HIV/AIDS: Nigeria, Ethiopia, Russia, India, and China," Intelligence Community Assessment 2002–04 D ([Washington, DC?]: National Intelligence Council, 2002).

18. John W. Dietrich, "The Politics of PEPFAR: The President's Emergency Plan for AIDS Relief," *Ethics & International Affairs* 21, 3 (2007): 277–292.

19. Ibid., 236.

20. Donald G. McNiel Jr., "U.S. Push for Abstinence in Africa Is Seen as Failure against H.I.V.," *New York Times*, February 26, 2015. The article described a study by a Stanford University medical student, Nathan Lo.

21. Patricia Rodney, Yassa D. Ndjakani, Fatou K. Ceesay, and Nana O. Wilson, "Addressing the Impact of HIV/AIDS on Women and Children in Sub-Saharan Africa: PEPFAR, the U.S. Strategy," *Africa Today* 57, 1 (2010): 65–76.

22. Eran Bendavid and Jayanta Bhattacharya, "The President's Emergency Plan for AIDS Relief in Africa: An Evaluation of Outcomes," *Annals of Internal Medicine* 150, 10 (2009): 670.

23. James Pfeiffer, "The Struggle for a Public Sector: PEPFAR in Mozambique," in *When People Come First*, ed. João Biehl and Adrianna Petryna (Princeton, NJ: Princeton University Press, 2013), 166–181.

24. Susan Whyte, Michael Whyte, Lotte Meinert, and Betty Kyaddondo, "Treating AIDS: Dilemmas of Unequal Access in Uganda," in *Global Pharma-*

ceuticals: Ethics, Markets, Practices, ed. Adriana Petryna, Andrew Lakoff, and Arthur Kleinman (Durham, NC: Duke University Press, 2007), 240–262.

25. Matthew M. Kavanagh, "The Politics and Epidemiology of Transition: PEPFAR and AIDS in South Africa," *Journal of Acquired Immune Deficiency Syndromes* 65, 3 (2014): 247–250; Ingrid T. Katz, Ingrid V. Basset, and Alexi A. Wright, "PEPFAR in Transition—Implications for HIV Care in South Africa," *New England Journal of Medicine* 369, 15 (2014): 1385–1387.

26. The United States' President's Emergency Plan fort AIDS Relief, "Health Systems Strengthening (HSS)," www.pepfar.gov/about/strategy/ghi/134854.htm.

27. Ibid.

28. Whyte et al., "Treating AIDS."

29. Laurie Garrett, "The Challenge of Global Health," *Foreign Affairs* 86, 1 (2007): 1–16.

Chapter Sixteen: Medicalizing Global Health

1. João Biehl, *Vita: Life in a Zone of Social Abandonment* (Berkeley: University of California Press, 2005).

2. Ronald Bayer, "The Medicalization of HIV Prevention: New Opportunities Beset by Old Challenges," *Milbank Quarterly* 92, 3 (2014): 434–437.

3. Bertran Auvert, Dirk Taljaard, Emmanuel Lagarde, Joëlle Sobngwi-Tambekou, Rémi Sitta, and Adrian Puren, "Randomized, Controlled Intervention Trial of Male Circumcision for Reduction of HIV Infection Risk: The ANRS 1265 Trial," *PLoS Medicine* 2, 11 (2005): e298; "WHO/UNAIDS Technical Consultation on Male Circumcision and HIV Prevention, *New Data on Male Circumcision and HIV Prevention: Policy and Programme Implications* (Geneva: World Health Organization, 2007), 4.

4. HIV/UNAIDS/WHO/SACEMA Expert Group on Modelling the Impact and Cost of Male Circumcision, "Male Circumcision for HIV Prevention in High HIV Prevalence Settings: What Can Mathematical Modelling Contribute to Informed Decision Making?" *PLoS Medicine* 6, 9 (2009): e1000109. Researchers from Johns Hopkins working on the circumcision issue in Uganda's Rakai Health Project claim that risk compensation did not undermine the effectiveness of circumcision in their study groups. Yet the Rakai project has been working in this part of Uganda for so long, and the population been exposed to so much information about AIDS, that one has to question whether the absence of risk compensation can be viewed as representative of populations that lack similar influences.

5. U.S. President's Emergency Plan fort AIDS Relief, "Male Circumcision," www.pepfar.gov/documents/organization/107820.pdf.

6. Innovations for Poverty Action, "About," www.poverty-action.org/about/.

7. Cecil George Sheps, *Higher Education for Public Health: A Report of the Milbank Fund Commission* (New York: PRODIST, 1976).

8. In a recent graduate seminar on the Ebola epidemic, given in the International Health Department at the Johns Hopkins Bloomberg School of Public Health, several participants stated that they were working on vaccine initiatives. I asked if any of them felt they could go to an African country and set up a vaccine program. They all said that they did not know how they would do that. One student, however, added that they would be able to set up a trial to test the program's impact!

9. Peter Lurie and Sidney M. Wolfe, "Unethical Trials of Interventions to Reduce Perinatal Transmission of the Human Immunodeficiency Virus in Developing Countries," *New England Journal of Medicine* 337, 12 (1997): 853–856; Marcia Angel, "The Ethics of Clinical Research in the Third World," *New England Journal of Medicine* 337, 12 (1997): 847–849; Gambia Government/Medical Research Council Joint Ethical Committee, "Ethical Issues Facing Medical Research in Developing Countries," *Lancet* 351, 9098 (1998): 286–287.

10. Johanna Tayloe Crane, *Scrambling for Africa: AIDS, Expertise, and the Rise of American Global Health Science* (Ithaca, NY: Cornell University Press, 2013), 54–80.

11. There is a growing body of scholarly research on these collaborations. See, for example, Adrianna Petryna, "Clinical Trials Offshored: On Private Sector Science and Public Health," *BioSocieties* 2, 1 (2007): 21–40; Crane, *Scrambling for Africa.*

12. Marissa Mika, a graduate student in the History and Sociology of Science Department at the University of Pennsylvania, is currently completing her dissertation on the history of Uganda's National Cancer Institute. I wish to thank Marissa for sharing this information.

13. Nguyen Vinh-Kim, *The Republic of Therapy: Triage and Sovereignty in West Africa's Time of AIDS* (Durham, NC: Duke University Press, 2010).

14. Petryna, "Clinical Trials Offshored," 21–40; Vinh-Kim, *Republic of Therapy.*

15. Vincanne Adams, "Evidence-Based Global Public Health: Subjects, Profits, Erasures," in *When People Come First*, ed. João Biehl and Adrianna Petryna (Princeton, NJ: Princeton University Press, 2013), 54–90.

16. Dharma S. Manandhar, David Osrin, Bhim Prasad Shrestha, Natasha Mesko, Joanna Morrison, Kirti Man Tumbahangphe, Suresh Tamang et al., "Effect of a Participatory Intervention with Women's Groups on Birth Outcomes in Nepal: Cluster-Randomised Controlled Trial," *Lancet* 364, 9438 (2004): 970–979; Prasanta Tripathy, Nirmala Nair, Sarah Barnett, Rajendra Mahapatra, Josephine Borghi, Shibanand Rath, Suchitra Rath et al., "Effect of a Participatory Intervention with Women's Groups on Birth Outcomes and Maternal Depression in Jharkhand and Orissa, India: A Cluster-Randomised Controlled Trial," *Lancet* 375, 9721 (2010): 1182–1192; Sonia Lewycka, Charles Mwansambo, Mikey Rosato, Peter Kazembe, Tambosi Phiri, Andrew Mganga, Hilda Chapota et al., "Effect of Women's Groups and Volunteer Peer

Counseling on Rates of Mortality, Morbidity, and Health Behaviours in Mothers and Children in Rural Malawi (MaiMwana): A Factorial, Cluster-Randomised Controlled Trial," *Lancet* 381, 9879 (2013): 1721–1735.

17. William Muraskin, "The Strengths and Weaknesses of Philanthropy in International Public Health: Bill Gates and GAVI (Global Alliance for Vaccines and Immunizations) as a Test Case," paper presented at the Critical Global Health Seminar, Johns Hopkins University, March 2, 2010; William Muraskin, "The Global Alliance for Vaccines and Immunizations: Is It a New Model for Effective Public-Private Cooperation in International Public Health?" *American Journal of Public Health* 94, 11 (2004): 1922–1925.

18. Muraskin, "Global Alliance for Vaccines," 4.

19. Donald G. McNeil Jr., "New Meningitis Strain in Africa Brings Call for More Vaccines," *New York Times*, July 31, 2015.

20. James Pfeiffer and Rachel Chapman, "The Anthropology of Aid in Africa," *Lancet* 385, 9983 (2015): 2145.

21. Hannah Brown, "Global Health: Great Expectations," *British Medical Journal (Clinical Research Edition)* 334, 7599 (2007): 874–876; Anne-Emanuelle Birn, "Gates's Grandest Challenge: Transcending Technology as Public Health Ideology," *Lancet* 366, 9484 (2005): 514–519.

22. The Bill & Melinda Gates Foundation has also committed itself to the global eradication of malaria. It initially hoped to achieve this goal through the development of new vaccines and poured hundreds of millions of dollars into this endeavor. More recently, the foundation has concluded that a malaria vaccine is unlikely to be developed in the near- or midterm and has shifted its focus to an eradication strategy based on identifying and eliminating malaria parasites within populations in which malaria transmission has been driven down by other measures, including the use of ITNs. This strategy requires the development of new technologies, in the form of more-sensitive, rapid diagnostic tools that will identify asymptomatic cases of malaria and new combination drugs that will eliminate all the parasites within infected populations. In many respects, the strategy resembles the consolidation stage of the earlier WHO-led Malaria Eradication Programme (part IV). Like the earlier strategy, the Gates Foundation's new initiative is likely to require the development of effective active-surveillance systems, as well a basic health service to treat identified cases. Yet the foundation has been slow to adopt these kinds of system-building activities. Whether such a strategy can be successful in the face of complex social and economic conditions, such as population movements and ongoing development patterns that encourage malaria transmission, is unclear. What is clear is that, for the moment, the Gates Foundation is unable or unwilling to address these complexities.

23. "Rotary and Gates Foundation Extend Fundraising Agreement to End Polio," Rotary news release, June 25, 2013, www.rotary.org/en/rotary-and-gates-foundation-extend-fundraising-agreement-end-polio/.

24. Svea Closser, *Chasing Polio in Pakistan: Why the World's Largest Public Health Initiative May Fail* (Nashville, TN: Vanderbilt University, 2010).

25. Leslie Roberts, "Polio Eradication: Is It Time to Give Up?" *Science* 312, 5775 (2006): 832–835.

26. These costs may increase substantially even after the last cases of polio are eliminated, since there may still be undetected and unvaccinated cases that can spread the virus for years. This means that vaccinations will need to continue beyond the time when the last case of paralytic polio has disappeared. In addition, because OPV can occasionally cause polio, it may be necessary to conduct the post-elimination program with IVP, which will increase the cost of the program. In 2003, a report by the Maryland-based development NGO, Abt Associates, estimated that the switch from OPV to IPV—in its current presentation for all developing countries together—would result in an increase in its total annual cost of US$317 million, averaging US$2.91 per child. See Zingzhu Liu, Ann Levin, Marty Makinen, and Jennifer Day, *OPV vs. IPV: Past and Future Choice of Vaccine in the Global Polio Eradication Program* (Bethesda, MD: Abt Associates, revised June 2003).

27. See, for example, Carl E. Taylor, Felicity E. Cutts, and M. E. Taylor, "Ethical Dilemmas in Current Planning for Polio Eradication," *American Journal of Public Health* 87, 6 (1997): 922–924.

28. Carl E. Taylor, M. E. Taylor, and Felicity E. Cutts, "Ethical Dilemmas in Polio Eradication," *American Journal of Public Health* 88, 7 (1998), 1125.

29. Closser, *Chasing Polio in Pakistan*, 48–52, 55–92.

30. Alfred Sommer, "Vitamin A Deficiency and Clinical Disease: An Historical Overview," *Journal of Nutrition* 138, 10 (2008): 1835–1839.

31. Andrew J. G. Barclay, A. Foster, and Alfred Sommer, "Vitamin A Supplements and Mortality Related to Measles: A Randomised Clinical Trial," *British Medical Journal (Clinical Research Edition)* 294, 6567 (1987): 294–296.

32. Michael Latham, "The Great Vitamin A Fiasco," *World Nutrition* 1, 1 (2010): 12–45.

33. Food and Agriculture Organization, *International Conference on Nutrition: Final Report of the Conference, Rome, December 1992* (Rome: FAO/WHO, 1982), 43.

34. Latham, "Great Vitamin A Fiasco," 16.

35. George H. Beaton, Reynaldo Martorell, Kristan J. Aronson, Barry Edmonston, George McCabe, A. Catherine Ross, and Ben Harvey, *Effectiveness of Vitamin A Supplementation in the Control of Young Child Morbidity and Mortality in Developing Countries*, Nutrition Policy Discussion Paper 13 (Toronto: International Nutrition Program, Department of Nutritional Sciences, Faculty of Medicine, University of Toronto, 1993), 11.

36. Ibid., 12.

37. Ibid., 13.

38. Oscar Neidecker-Gonzales, Penelope Nestel, and Howarth Bouis, "Estimating the Global Costs of Vitamin A Capsule Supplementation: A Review of the Literature," *Food and Nutrition Bulletin* 28, 3 (2007): 307–316.

39. Latham, "Great Vitamin A Fiasco," 12–45.

40. The promotion of biomedical technologies as universal solutions to global-health problems is, in many ways, a product of the use of RCTs to evaluate these interventions. A significant amount of work goes into making sure that the trials will accurately measure the efficacy of the intervention in a particular setting. This may mean developing strategies for distributing and monitoring the use of interventions that take local social and cultural conditions into account. The extent to which researchers need to be aware of these conditions is often greater when the interventions are not vaccines or drugs, but technologies that require those receiving them to adjust their behavior. Thus more work was necessary for those testing insecticide-treated bed nets than for those testing a malaria vaccine. In the case of vitamin A supplementation, a great deal of work went into making sure children took the capsules and in measuring vitamin A levels. Yet this preparation work tends to get "black boxed" (i.e., becomes invisible) once the results of the trials are published. What is touted are the quantitative results, presented in terms of relative risk and confidence intervals. How these results were achieved often fades into the background. In general, once biomedical interventions are scaled up, it is difficult to reproduce the rigorous conditions in which they were tested. This phenomenon is being studied by Kirsten Moore-Sheeley (Department of the History of Medicine, Johns Hopkins School of Medicine) in her dissertation on the history of insecticide-treated bed-net research in Kenya.

41. John Mason, Ted Greiner, Roger Shrimpton, David Sanders, and Joshua Yukich, "Vitamin A Policies Need Rethinking," *International Journal of Epidemiology* 44, 1 (2015): 283–292.

42. Keith West, Alfred Sommer, Amanda Palmer, Werner Schilink, and Jean-Pierre Habicht, "Commentary: Vitamin A Policies Need Rethinking," *International Journal of Epidemiology* 44, 1 (2015): 292–294.

43. Peter Lurie and Sydney M. Wolfe, "Unethical Trials of Interventions to Reduce Perinatal Transmission of the Human Immunodeficiency Virus in Developing Countries," *New England Journal of Medicine* 337, 12 (1997): 853–856.

44. Jon Cohen, "A Hard Look at Global Health Measures," *Science* 345, 6202 (2014): 1264.

45. Alfred Sommer, Keith P. West Jr., and Reynaldo Martorell, "Vitamin A Supplementation in Indian Children: Correspondence," *Lancet* 382, 9892 (2013): 591.

46. Supporters of the DEVTA study's findings have suggested that this delay resulted from the efforts by VAC supporters, who sit on the editorial boards of leading nutrition journals, to block the study's publication.

47. Richard Peto, Shally Awashti, Simon Reed, Sarah Clark, and Donald Bundy, "Vitamin A Supplementation in Indian Children: Authors' Reply," *Lancet* 382, 9892 (2013): 594–596.

48. Sarah K. Wallace, "Global Health in Conflict: Understanding Opposition to Vitamin A Supplementation in India," *American Journal of Public Health* 102, 7 (2012): 1286–1297.

49. In the United States, we only solved our vitamin A deficiency problem after instituting food fortification.

50. I am indebted to Joanne Katz, who has spend much of her career working on vitamin A, for her comments on current discussions surrounding the withdrawal of VACs.

51. Similar questions should be asked of the recent widespread adoption of Nutriset RUTF (Ready-to-Use Therapeutic Food), branded as "Plumpy Nut," to combat acute malnutrition. See Jennifer Tappan, *Medicalizing Malnutrition: Nearly a Century of Kwashiorkor Work in a Small Part of Africa* (Athens: Ohio University Press, forthcoming).

Conclusion: Responding to Ebola

1. Saliou Samb, "WHO Says Guinea Ebola Outbreak Small as MSF Slams International Response," Reuters, April 1, 2014.

2. Ibid.

3. "Ebola Response: Where Are We Now?" Médicins Sans Frontières/Doctors Without Borders, www.doctorswithoutborders.org/document/ebola-response-where-are-we-now/.

4. Norimitsu Onishi, "Empty Ebola Clinics in Liberia Are Seen as Misstep in U.S. Relief Effort," *New York Times*, April 11, 2015.

5. Nelson Dunbar, Director of Research, Ministry of Health and Social Welfare, Republic of Liberia, "Ebola: In the Trenches; Perspectives from Liberia about the Ebola Response," presented at the Johns Hopkins Bloomberg School of Public Health, May 18, 2015.

6. Norimitsu Onishi, "As Ebola Ebbs in Africa, Focus Turns from Death to Life," *New York Times*, January 31, 2015.

7. "The Ebola Hot Zone," *60 Minutes*, November 9, 2014.

8. Peter Redfield, *Life in Crisis: The Ethical Journey of Doctors Without Borders* (Berkeley: University of California Press, 2013).

9. Ibid., 177.

10. MSF's strategy to control Ebola has been organized into six elements: (1) isolation and supportive medical care for cases, including laboratory capacity to confirm an infection; (2) safe-burial activities in case-management facilities and communities; (3) awareness-raising; (4) alerts and surveillance in the community; (5) contact tracing; and (6) access to health care for non-Ebola patients, including the protection of health facilities and health workers. MSF also started distributing antimalarial medicine in Monrovia, where the health

system had collapsed, leaving residents of the city without access to these drugs. All of their work was carried out by their own staff, who had limited engagement with local health authorities.

11. Paul Farmer, *AIDS and Accusation: Haiti and the Geography of Blame* (Berkeley: University of California Press, 1992).

12. People commenting on the PIH website complained about this delay and wondered why they should be donating to PIH.

13. Tomas Kellner, "Coalition Fights Ebola at the Outbreak's Invisible Frontline in Remote African Rainforest," GE Reports, October 14, 2014, www.gereports.com/post/98748927415/coalition-fights-ebola-at-the-outbreaks-invisible/.

14. "Why Ebola Costs Could Top $4bln in 2015," Times Video, December 29, 2014, www.nytimes.com/video/multimedia/100000003421592/why-ebola-costs-could-top-4bln-in-2015.html.

15. "Integrated Delivery: Strategy Overview," Bill & Melinda Gates Foundation, www.gatesfoundation.org/What-We-Do/Global-Development/Integrated-Delivery/.

16. Prabhjot Singh and Jeffrey Sachs, "1 Million Community Health Workers in Sub-Saharan Africa," *Lancet* 383, 9889 (2013): 285.

17. "Economists Declaration," Global Coalition for Universal Health Coverage, http://universalhealthcoverageday.org/economists-declaration/.

18. Matshidiso Moeti, "Op-Ed: Toward Universal Health Coverage in Africa, *Daily Maverick*, December 13, 2015, www.dailymaverick.co.za/article/2015-12-13-op-ed-toward-universal-health-coverage-in-africa/#.Vm7hYuODFBc/.

19. Sheri Fink, "Experimental Ebola Vaccine Tested in Guinea Shows Promise, Report Says," *New York Times*, July 31, 2015.

INDEX